CLUSTERS IN URBAN AND REGIONAL DEVELOPMENT

Clusters have become a key focus of urban and regional policy in advanced economies, as regional specialisation in particular industries has come to be regarded as advantageous in the context of debates about globalization and the knowledge economy. In particular, spatial proximity between associated firms and organisations is claimed to stimulate processes of innovation and learning. Consequently, governments have promoted dynamic clusters as a means of generating competitive advantage in particular cities and regions.

In this collection, these claims are critically assessed by drawing upon the work of leading specialists from Western Europe and North America. Going beyond the celebrated 'hot-spots' of economic development, the book draws upon evidence from a broader range of cities and regions to help fill some important gaps in our knowledge of how clusters operate within the contemporary global economy. Cluster dynamics are situated in time and space; interrogating both how firms, organisations and actors within clusters adapt to changes over time, and how clusters are embedded within broader spatial divisions of labour at regional, national and international scales.

This book was previously published as a special issue of the journal *Urban Studies*.

Andy Cumbers is a Senior Lecturer in Human Geography at the University of Glasgow.

Danny MacKinnon is a Lecturer in Human Geography at the University of Aberdeen.

URBAN STUDIES MONOGRAPHS

Series Editors: Ronan Paddison, Jon Bannister, Andy Cumbers, Ken Gibb, all at the University of Glasgow

In the contemporary world cities have a renewed significance. Trends such as globalisation, neo-liberalism, new technologies, the rise of consumption and consumerism, the re-definition of modes of governance, demographic and social shifts, and the tensions to which such changes give rise, have re-defined the critical role cities play, as well as the problems arising from city life. By focusing on specific issues this series aims to explore the nature of the contemporary urban condition.

Cities of Pleasure
Sex and the Urban Socialscape
Edited by Alan Collins

Clusters in Urban and Regional Development
Edited by Andy Cumbers & Danny MacKinnon

Globalisation and the Politics of Forgetting
Edited by Yong-Sook Lee & Brenda S. A. Yeoh

CLUSTERS IN URBAN AND REGIONAL DEVELOPMENT

Edited by
Andy Cumbers and Danny MacKinnon

LONDON AND NEW YORK

First published 2006
by Routledge
2 Park Square, Milton Park, Abingdon, Oxon OX14 4RN

Simultaneously published in the USA and Canada
by Routledge
270 Madison Avenue, New York, NY 10016

Routledge is an imprint of the Taylor & Francis Group, an informa business

Transferred to Digital Printing 2009

Typeset in Times by Infotype Ltd, Eynsham, Oxfordshire

British Library Cataloguing in Publication Data
A catalogue record for this book is available
from the British Library

Library of Congress Cataloging in Publication Data
A catalog record for this book has been requested

ISBN10: 0-415-36011-0 (hbk)
ISBN10: 0-415-56841-2 (pbk)

ISBN13: 978-0-415-36011-1 (hbk)
ISBN13: 978-0-415-56841-8 (pbk)

CONTENTS

Introduction: Clusters in Urban and Regional Development

Andy Cumbers and Danny MacKinnon

[Paper received in final form, January 2004]

Introduction

Clusters have become a key focus of discussion and analysis in contemporary debates on urban and regional economic development (Feldman, 2000; Porter, 1990; Steiner, 1998). Closely associated with the work of the Harvard business economist, Michael Porter (1990, 2000), the cluster concept has attracted particular interest from academics, consultants and policy-makers concerned with promoting urban and regional growth in an increasingly global economy (Benneworth *et al.*, 2003; Glaeser, 2000; Martin and Sunley, 2003). As Steiner (1998, pp. 1 and 4) observes, clusters have become an object of desire for many cities and regions, resting on the widely accepted assumption that increased specialisation will lead to increased levels of productivity, growth and employment. Cluster-based policies have been adopted by a range of organisations operating at different geographical scales, including regional development agencies within a number of European and North American states, national government units such as the UK government's Department of Trade and Industry (DTI) and supranational bodies such as the OECD and the European Commission (see—for example, DTI, 1998; European Commission, 2002; OECD, 2002). Such policies require the identification of specialist clusters which can then be targeted for support, typically in the form of R&D assistance, bespoke training, venture capital, initiatives which attempt to inculcate a culture of innovation and learning and efforts to build and reinforce a sense of cluster identity amongst constituent firms and organisations (Raines, 2002). As advocates of clusters often argue (Porter 1998, 2000; Botham and Downes, 1999), the approach is not confined to dynamic new clusters of the knowledge economy, but can be applied in a variety of sectoral and spatial contexts, from old industrial regions undergoing structural readjustment to peripheral rural regions seeking new sources of growth and even prospective industrial 'hotspots' in the 'global south' (Altenberg and Maeyer-Stamer, 1999; World Bank, 2000).

Despite a growing literature, reflected in a burgeoning number of articles, journal special issues (*see European Planning Studies, Small Business Economics*)—to which we are, of course, adding another—and official reports, we would argue that there is still a lack of critical intellectual engagement with the cluster concept. Whilst a number of critiques of clusters and concepts associated with what has been termed the 'new regionalism' in economic geography and related fields have appeared in recent years (see, for example, Lovering, 1999; MacLeod, 2001; MacKinnon *et al.*, 2002; Cumbers *et al.*, 2003b), there remains a marked absence of work which critically evaluates the theoretical and policy claims of the cluster concept through detailed empirical research (though see Cumbers *et al.*, 2003a; Keeble *et al.*, 1999; Swann *et al.*, 1998). Partly as a result of this, the 'cluster' remains something of a 'fuzzy' or alternatively 'chaotic' concept which lacks clarity and coherence (Markusen, 1999; Martin and Sunley, 2003). According to Steiner (1998, p. 1), "the still rather vague character of clusters poses problems of theoretically sound definition, of empirical measurement, of policy recommendation and evaluation". Fostering dialogue between academics and policy-makers is an important and worthy contribution (Benneworth *et al.*, 2003), but we would contend that the real test of the clusters approach will be its capacity to provide us with a better understanding of urban and regional growth processes that moves beyond previous approaches.

A number of important questions concerning the value of a clusters approach to urban and regional economic development and the actual operation of clusters can be identified. At a theoretical level, what is the added value of the cluster approach to existing theories of agglomeration? A second set of issues concerns how the key linkages and networks that underpin clusters function as channels of knowledge exchange and interactive learning. From a geographical perspective, the respective contributions of local and extra-local business networks are of particular interest (Bunnell and Coe, 2001; MacKinnon *et al.*, in press). Thirdly, the question of cluster evolution and renewal in terms of how constituent firms and organisations adjust to economic change has been neglected in much of the literature (Chapman *et al.*, forthcoming; Tichy, 1998). Fourthly, what scope is there for public policy to shape the development of clusters and how effective are local and regional initiatives in the context of national policy frameworks and wider processes of economic change? The issue of sectoral and spatial diversity runs across these questions, focusing attention on the scope and limits of the cluster concept beyond celebrated knowledge economy sectors such as biotechnology and a restricted number of innovative core regions such as Silicon Valley, the City of London and the 'Third Italy' (Martin and Sunley, 2003). In particular, serious questions need to be asked about the relevance of a clusters discourse for less favoured cities and regions that have been marginalised by processes of economic restructuring and globalisation. Whilst we do not claim to provide definitive answers to these questions, our purpose in putting together this special issue is to evaluate critically the concept, and to broaden the empirical base of cluster analysis through a number of detailed case studies.

Explaining the Rise of Clusters Research and Policy

An interest in the economics of spatial agglomeration can be traced back to the late 19th century and Alfred Marshall's observations about specialist industrial districts in the UK (Marshall, 1890). Whilst remaining relatively marginal to mainstream economic theory, agglomeration became a major theme of urban and regional studies. According to the traditional Marshallian conception, the advantages of agglomeration are rooted in the reduced costs that arise from the operation of three sets of 'localisation economies': the growth of various intermediate and subsidiary industries which provide specialised inputs; the development of a pool of skilled labour; and the establishment of a dedicated infrastructure and other collective resources (Malmberg and Maskell, 2002). These basic factors have remained prominent in much subsequent work on agglomeration, from the cumulative causation theories of the 1950s (Hirschman, 1958; Myrdal, 1957) to studies of industrial districts and new industrial spaces in the 1980s (Brusco, 1982; Scott, 1988; Storper and Scott, 1989). In the context of a purported shift towards a knowledge-based economy, however, more recent work has taken a different approach, placing particular emphasis on knowledge and information spillovers (Malmberg and Maskell, 2002; Storper, 1997).

On an empirical level, research demonstrating the continuing importance of processes of spatial agglomeration within a globalising and knowledge-based economy has served to focus attention on clusters. Whilst much research has consisted of case studies of successful clusters (Henry and Pinch, 2000; Keeble *et al.*, 1999; Kenney and von Burg, 1999; Saxenian, 1994), a number of studies by economists, geographers and others have pointed to high levels of agglomeration across a broad range of industries (Head *et al.*, 1995; Krugman, 1991; Malmberg and Maskell, 1997; Porter, 1998). These provide a considerable volume of evidence to support the observation that the geographical clustering of industries and firms represents a "fundamental fact of economic life" (Dicken, 2003, p. 75). At the same time, however, the apparent divide between intensive research based on single-cluster case studies, often interview-based, on the one hand, and extensive studies which deploy official statistics to measure the extent of clustering within particular economies, on the other, seems to have hindered understanding of agglomeration processes. It is in this context that the work of Porter (1990, 1998) has gained such influence, offering an accessible theory that emphasises the general importance of clustering as a phenomenon and outlines how the operation of particular clusters promotes growth and competitiveness.

Whilst Porter was originally concerned with the external conditions that support firm competitiveness at the national scale (Porter 1990), he has increasingly focused upon the importance of sub-national business clusters in underpinning competitiveness and innovation in modern economies (Porter 1998). In his earlier work, Porter developed his famous diamond model, arguing that national competitiveness was rooted in the relationships between four sets of factors: firm strategy, structure and rivalry, demand conditions, factor input conditions and related and supporting industries (Porter, 1990). More recently, he has argued that geographical concentration enhances processes of interaction within the 'competitive diamond' by increasing the static productivity of constituent firms or industries, stimulating higher rates of innovation and thereby productivity and encouraging high rates of business formation (Porter, 2000, p. 259). For Porter, clusters are defined as

> geographical concentrations of interconnected companies, specialised suppliers, service providers, firms in related industries, and associated institutions (for example universities, standards agencies and trade associations) that compete but also co-operate (Porter, 1998, p. 197).

Alongside Porter's promotion of clusters, other strands of economic and economic geographical research have also contributed to the renewed focus on spatial agglomeration. The two main approaches here are the work of economic geographers on innovation and learning processes in cities and regions and economists' concerns with the relationships between agglomeration, specialisation and trade. The rediscovery of geography by orthodox economists, most notably Paul Krugman and his colleagues, represents a particularly significant development (Fujita *et al.*, 2000; Krugman, 1991). This 'new geographical economics' has analysed processes of spatial concentration through the application of economic models, emphasising the importance of increasing returns to scale in underpinning localised advantages (Martin, 1999). What Krugman and colleagues have done is to deploy formal economic analysis, with the innovation of more advanced computational techniques, to highlight the continued significance of agglomeration in a globalising economy. Arguments long accepted by most geographers have been given a new stamp of legitimacy by the hegemonic social science discipline (see Fine, 1999; Henry *et al.*, 2001).

Whilst this 'new geographical economics' and economic geography share a substantive interest in spatial agglomeration, there is considerable divergence in the approaches and methods employed to study this issue. In contrast to the deductive methods of theorising and analysis employed by Krugman *et al.*, economic geographers tend to draw on more qualitative and open-ended methods in order to capture the contingency and embeddedness of economic processes (Amin, 1999; Cooke and Morgan, 1998; Henry and Pinch, 2000; Storper, 1997). Related strands of this latter type of research have been pulled together under the label of the 'new regionalism' by Lovering (1999) (see Cumbers *et al.*, 2003b; MacLeod, 2001), defined by an emphasis upon the renewed importance of regions as a key scale of economic organisation and policy intervention under late capitalism (Cooke and Morgan, 1998; Storper, 1997). There is considerable overlap between 'new regionalist' work and the clusters approach in terms of the focus on agglomeration and the advantages of clustering for regions that must compete in an increasingly global economy. Several studies have applied concepts associated with economic geography such as 'learning regions', 'innovative milieux' and 'institutional thickness' in order to understand the development of particular clusters (Henry and Pinch, 2000; Keeble *et al.*, 1999; May *et al.*, 2001). Where the two approaches diverge is in the new regionalism's more explicit focus on knowledge and learning together with its more theoretical orientation and style (Martin and Sunley, 2003). Put simply and briefly, a key argument within economic geography is that the increasing importance of knowledge-creating processes for competitive advantage in a global economy (see Lundvall and Johnson, 1994; Hodgson, 1998) is reinforcing the tendency towards urban and regional clustering (Cooke and Morgan, 1998; Malmberg and Maskell, 2002; Storper, 1997). Spatial proximity between specialist firms facilitates the creation and exchange of tacit knowledge, viewed as a crucial form of competitive advantage in a world in which codified knowledge is easily replicated and rendered ubiquitous (Cooke and Morgan, 1998; Maskell and Malmberg, 1999). The proposition that spatial proximity promotes innovation and learning through the rapid circulation of tacit knowledge has, however, been challenged by critics of clusters and learning regions who argue that such knowledge can also be transmitted through organisational and relational linkages between spatially distant actors (Amin and Cohandet, 1999; Bunnell and Coe, 2001; Gertler, 2001).

Whilst the work of geographical economists and economic geographers is also consistent with the prevailing direction of policy, particularly in terms of fostering the conditions for innovation, learning and entrepreneurship within regions, Porter's work on clusters has exerted by far the most direct influence. The proliferation of clusters policies in recent years highlights the attractions of the concept to policy-makers at regional, national and international scales. In many cases, a clusters approach has initially been developed by urban and regional agencies looking for initiatives to assist the efforts of leading firms and industries within their regions to compete in global markets before being taken up by national governments and international organisations such as the European Commission, OECD and World Bank (Lagendijk and Cornford, 2000, pp. 214–215). Porter's own Monitor consultancy performed an important role here in pioneering cluster studies and promoting the concept, undertaking mapping exercises in regions such as Scotland and the Basque Country, (Benneworth *et al.*, 2003, p. 517; Lagendijk and Cornford, 2000, pp. 214–215).[1] As Martin and Sunley (2003, pp. 8–10) outline, there are a number of reasons for the widespread adoption of the clusters approach by policy-makers, including how Porter has anchored it within a broader understanding of competitiveness as *the* imperative confronting economic development organisations, Porter's own status as an internationally renowned management theorist, coupled with his authoritative and accessible style, and the rather vague and generic nature of the cluster concept itself.

In geographical terms, the clusters approach has been applied across a range of scales from highly specialised craft districts within urban neighbourhoods (see—for example, Crewe, 1996) to concentrations of industry within particular regions (Trends Business Research, 2001), complexes of related industrial sectors at the national level

(Porter, 1990) and even interlinked activities in a group of neighbouring countries (Porter, 2000). As this suggests, clusters provide development agencies with a malleable and elastic tool that can be moulded to the particular circumstances of the industry or region in question. In a regional context in particular, a clusters approach seems to provide development agencies with a new and compelling rationale for both identifying a limited number of sectors for support—generally those that are deemed to have the highest growth potential—and defending and justifying this to those interests that are consequently excluded.[2]

These general observations are designed largely to provide a context for what follows, not to deny that particular development agencies have adopted or, indeed, rejected cluster polices on rational grounds related to their policy priorities and their interpretations of the economic circumstances and needs of their regions (Benneworth and Henry, this volume). Neither do we seek to deny that the clusters approach can succeed in helping to achieve higher rates of urban and regional growth and innovation. This is an empirical question that can only be answered through further research and observation. Nonetheless, it is our hope that the papers in this Review Issue offer a significant contribution towards the development of a firmer understanding of the value of the cluster concept as focus for urban and regional development research and policy. We now turn to outline the papers, which have been grouped into three sections: conceptualising clusters; clusters, knowledge and innovation; and from dynamic to mature clusters.

Conceptualising Clusters

The first set of papers engage with the theoretical literature from different standpoints. One of the most pronounced criticisms of the clusters debate regards the spatial elasticity of the term (Martin and Sunley, 2003). In this respect, the paper by Phelps provides a welcome analysis of the changing scale of agglomeration. As Phelps notes, there is a disjuncture in the literature whereby Marshallian insights about the operation of industrial districts, originally conceived of as highly localised districts or quarters within cities, are still the reference point for considering much larger agglomerations at the level of cities, city-regions or even nation-states in some cases. A key concern is whether a neo-Marshallian analysis which places the emphasis upon the role of external economies, whether of the traditional cost-based kind or the more relational kind (Storper, 1997) is relevant to understanding much larger agglomerations in which there is considerable internal dispersal of activities. Other concepts, such as 'borrowed size', might be more useful in assessing the evolution of clusters given the recent tendencies of firms to relocate to suburbs, enabling them to avoid the congestion and high costs of dense urban environments whilst remaining close enough to take advantage of agglomeration economies such as a skilled labour pool, specialist firms and local information networks. In short, Phelps raises some important questions about the ability of established theories of clusters to capture important changes in the geography of economic activity.

By contrast, the papers by Maskell and Lorenzen and Benneworth and Henry are more supportive of a cluster-based analysis. Rather than falling back upon the clichéd assertion that collective learning processes underpin clusters, Maskell and Lorenzen concentrate upon the individual firm and the advantages to be gained from different forms of market organisation. One reaction to market instability and uncertainty is for firms to organise themselves in vertical production networks where close relations are built up between customers and suppliers. However, where the market is characterised by rapid shifts in fashion and design where products have a short turnover time, and the requirement is for greater experimentation and flexibility, there are strong pressures towards geographical clustering. For evidence, Maskell and Lorenzen point to the organisation of the popular music industry, where the importance of being located near to a range of collaborators providing experience and finance suggests a different sort of agglomeration economics at work, founded on weak rather than strong ties. Similarly, the European furniture industry also shows tendencies towards clustering, although here it is the pressures to deliver a varied and rapidly changing range of products of a reasonably high quality to tight deadlines that encourage co-location and the ability to switch from one supplier to another.

Benneworth and Henry take a different approach, arguing that, whilst clusters can mean different things to different people, this can be a 'source of vitality' where the various discourses overlap. They argue that the concept has been valuable in opening up lines of conversation between geographers and economists and between academics and policy-makers. At the same time, they agree with Martin and Sunley (2003) that a central problem with clusters research is that too much of it is done badly. Recognising that the concept can be flexible does not excuse sloppy analysis and methodological rigour is required to ensure that research is consistent with the terms of the particular theoretical perspective taken. Nevertheless, drawing upon Trevor Barnes' recent critique of rationalist thought in economic geography, they argue that "multiple explanations can interact conceptually to provide a richer understanding of the situation than permitted by theoretically monistic approaches" (p. 1018). Their point is illustrated through the example of business services in London, where, using the work of other researchers, they uncover different layers of analysis that provide a deeper geographical knowledge of cluster dynamics.

Clusters, Knowledge and Innovation

Recent academic and policy discourses have stressed the advantages of clusters for cities and regions seeking to compete in an increasingly knowledge-driven global economy. The importance of locally specific forms of knowledge circulating through the labour market has been identified as a key feature of successful agglomerations such as Silicon Valley (Saxenian, 1994). Yet few detailed studies have sought to test this proposition empirically. In this respect, the paper by Power and Lundmark provides some powerful new evidence on labour market dynamics, using a unique longitudinal data-set from Sweden which contains detailed information on personal and occupational mobility for every resident. Drawing upon a case study of Stockholm's ICT cluster, they demonstrate a much higher level of mobility in clusters than for the urban economy as whole. This leads them to suggest that, contrary to the prevailing wisdom in the learning region literature, knowledge transfer occurs more through labour mobility than through informal exchanges outwith the workplace.

Harrison *et al.*'s paper is also concerned with the internal workings of clusters, although their focus is upon entrepreneurial rather than labour market dynamics. They are particularly concerned with the way in which new high-technology and successful businesses start up and develop. Drawing upon their own empirical study of the Ottawa technology cluster, they take issue with the emphasis in the literature upon the incubation effects of cluster organisations for new firm development. Whilst most entrepreneurs in the cluster have worked locally immediately prior to start up, a majority had "spent their formative years elsewhere and in most cases had come to Ottawa because of employment opportunities" (pp. 1059–1060). Their findings lead them to stress the importance of career histories and geographies in the development of entrepreneurs, whilst also highlighting the part played by teams—rather than sole entrepreneurs—who are able to draw upon broader and complementary experiences in starting up a new business.

The integration of clusters into wider spatial relations is a theme further developed by Wolfe and Gertler in their paper, which uses preliminary results from a cross-national study of Canadian clusters to engage with recent debates. Whilst Wolfe and Gertler accept that there are clear advantages to firms from clustering in specialist agglomerations, both from market-driven and institutional forces, they stress the importance of understanding clusters within broader geographical networks. Successful firms within clusters will make use of the relational assets available locally, but tend also to be plugged into non-local (and in many cases global) 'pipelines', which allow them to access key knowledge, information and skills. Wolfe and Gertler are also keen to stress the role of public infrastructure—particularly educational and training resources in supporting clusters—although once again they highlight the importance of relations operating at a series of different scales involving local, regional and national institutions and organisations. Their results do not provide easy answers for policy-makers as they conclude that cluster development is often due to a set of unique and path-dependent circumstances, which are not easily replicable elsewhere and in any case can take several decades to nurture.

The final paper in this section by Simmie provides further evidence to qualify simplistic analyses of the relationship between knowledge and clustering. Drawing upon recently published results from the UK Office of National Statistics' Community Innovation Survey, Simmie 'tests' Porter's recent assertion of the links between firm innovation and clustering. The data support Wolfe and Gertler's conclusions that national and international collaboration is more important for innovative firms than local linkages. There is little evidence that innovative firms are plugged into local supply chain or knowledge networks. In those cases where they use external support they are part of more dispersed communities of association. These findings, Simmie suggests, undermine notions of dynamic localised clusters at the heart of the knowledge economy, leading him to conclude that agglomeration is more likely to reflect more conventional urbanisation economies (Gordon and McCann, 2000).

From Dynamic to Mature Clusters

The final section of this Review Issue explores the cluster phenomenon through a range of empirical studies of clustering processes in specific 'new' and 'old' economy sectors. The papers identify the main influences on cluster development and assess how the key linkages between firms and institutions actually operate, focusing on clusters at different stages of development. Whilst the papers by Cooke and Leibovitz are concerned with biotechnology clusters, and Isaksen assesses the software cluster in Oslo—all of which could be viewed as relatively new and dynamic clusters with potential for further growth—Tödtling and Trippl examine the development of mature clusters in an old industrial region. Cooke demonstrates the tendency towards specialised regional clustering in the biosciences, arguing that the decline in research and development in large corporations and reductions in national public-sector funding for strategic science are leading to the development of regional centres of excellence around a new generation of specialist research firms and universities. A key factor is the emergence of transdisciplinary research networks involving key drug firms, 'academic entrepreneurs' and dedicated research firms whose co-location involves important knowledge spillovers. In this context, Cooke suggests that recent political devolution in the UK is providing opportunities for the new regional administrations in Wales and Scotland to develop their own regional science policies to capitalise on international expertise in their higher education institutions.

Cooke's paper is complemented by Leibovitz's, which explores the attempt to promote a biotechnology cluster in Scotland. He provides a rich empirical case study that seeks to embed cluster dynamics within broader processes of urban and regional change. Challenging the existing literature's emphasis upon the importance of localised interfirm relations, Leibovitz stresses the role played by path dependency, public-sector organisations, urban labour markets (in common with earlier contributions) and external linkages in promoting cluster development. Interview-based research reveals the important role of particular Scottish universities in providing a science-base and well-qualified graduates, although some firms expressed concern about obtaining more specialist staff in what is still a fledgling or embryonic cluster. In common with earlier work on another prominent Scottish cluster, Silicon Glen (Turok, 1993), Leibovitz also finds that local interactions between firms (whether buyer–supplier or competitor) and between firms and research institutions are surprisingly weak and that many of the key relationships operate at an international scale. Thus, Leibovitz's paper provides an important antidote to the overemphasis in the literature upon localised learning processes, suggesting that the most competitive firms within clusters will be those that also have well-developed external connections (see also Cumbers *et al.*, 2003a).

Isaksen examines the development of the software cluster in Oslo, emphasising the importance of close contact and interaction between consultancy firms, clients and suppliers. Echoing Maskell and Lorenzen's observations on pop music, Isaksen shows that the project-based nature of software consultancy fosters a need for face-to-face contact with a range of collaborators. Spatial proximity facilitates the exchange of complex bundles of specialised, tacit knowledge through collaboration between co-located firms. At the same time, it is important to situate the software cluster within the larger urban economy of Oslo, which provides access to a specialised labour market and advanced infrastructure. The Oslo software cluster can be seen as 'embryonic' in the sense of being at an early stage in its development, suggesting that there is substantial scope for further growth through the development of external economies and supporting institutions.

An important gap in the clusters literature is in its applicability to the experience of peripheral regions. Whilst the literature on the learning region has suggested that the economies of less favoured regions can be revived through appropriate knowledge and innovation policies (Cooke, 1995; Florida, 1995; Morgan, 1997), there have been few studies that have examined the possibilities for the renewal of mature industry clusters. The paper by Tödtling and Trippl makes an important contribution in this respect by examining the renewal of the automotive and metal clusters in the old industrial region of Styria in Austria. Using a regional innovation systems approach (see Braczyk *et al.*, 1998), they suggest that the problem for many older industrial regions is not an absence of 'institutional thickness' (see Amin and Thrift, 1994), but rather the legacy of a collective mindset oriented to older technologies and 'ways of doing'. This can be viewed as an expression of cognitive 'lock-in' in Grabher's terms (Grabher, 1993). In such cases, cluster renewal through the adoption of new and more innovative solutions requires a clean break from old habits or what Johnson (1992) has described as 'institutional forgetfulness'. Whilst they identify elements of successful restructuring in each case, the two clusters provide very different contexts for adaptation. In the automotive sector, a history of very limited interaction has been overcome through the efforts of the regional government to forge closer co-operation to address a highly competitive market environment. In the metals industry, however, progress has been slower, largely because of what they term an 'integration trap' where a close set of relations between state, former nationalised entities and trade unions has locked the cluster into an outmoded set of practices.

Conclusion

In summary, the papers in this Review Issue help to deepen our understanding of processes behind cluster formation and development. Whilst there is still a need for more empirically grounded but theoretically informed research that situates clusters within broader processes of spatially uneven development in the context of a changing global economy, the papers here have begun to address questions about the advantages of clustering in the context of a more knowledge-driven economy, the operation of key networks and linkages and the scale at which clusters operate. One important theme is that formal channels of labour mobility may be as (or more) important than informal conventions and relations in facilitating innovation and learning through the circulation of knowledge. At the same time, our attention is directed to the external connections that are critical to cluster development, whether this is of new sources of information, skilled labour, entrepreneurs or ideas. This is particularly true of more peripheral areas that lack the urbanisation economies of core cities and regions. But even in the more advanced regions, clusters cannot be regarded as self-contained assemblages of social and economic relations, as several of our contributors indicate. Hence, one of the most interesting issues for further research to explore concerns the linkages and connections between specific clusters and wider processes of information exchange and knowledge construction within international networks.

Notes

1. A number of prestigious consultancies have subsequently undertaken cluster studies across a range of locations.
2. Whilst advocates of clusters would strenuously disagree (Porter, 1990, 1998), in practice this approach does seem to amount to a form of 'picking winners', although the focus might be on business complexes or networks rather than specific firms or industries and the methods employed tend to emphasise indirect support and facilitation rather than direct subsidisation (Raines, 2002).

References

ALTENBERG, T. and MAEYER-STAMER, J. (1999) How to promote clusters: policy experiences from Latin America, *World Development,* 27, pp. 1693–1713.

AMIN, A. (1999) An institutionalist perspective on regional economic development, *International Journal of Urban and Regional Research,* 23, pp. 365–378.

AMIN, A. and COHANDET, P. (1999) Learning and adaptation in decentralised business networks, *Environment and Planning D,* 17, pp. 87–104.

AMIN, A. and THRIFT, N. (1994) Living in the global, in: A. AMIN and N. THRIFT (Eds) *Globalisation, Institutions and Regional Development in Europe*, pp. 1–22. Oxford: Oxford University Press.

BENNEWORTH, P., DANSON, M., RAINES, P. and WHITTAM, G. (2003) Confusing clusters? Making sense of the clusters approach in theory and practice, *European Planning Studies,* 11, pp. 511–520.

BOTHAM, R. and DOWNES, B. (1999) Industrial clusters: Scotland's route to economic success, *Scottish Affairs,* 29, pp. 43–58.

BRACZYK, H. J., COOKE, P. and HEIDENRICH, M. (Eds) (1998) *Regional Innovation Systems: The Role of Governance in a Globalised World.* London: UCL Press.

BRUSCO, S. (1982) The Emilian model: productive decentralisation and social integration, *Cambridge Journal of Economics,* 6, pp. 167–184.

BUNNELL, T. G. and COE, N. M. (2001) Spaces and scales of innovation, *Progress in Human Geography,* 25, pp. 569–589.

CHAPMAN, K., MACKINNON, D. and CUMBERS, A. (forthcoming) Adjustment or renewal in regional clusters? A study of diversification amongst SMEs in the Aberdeen oil complex, *Transactions of the Institute of British Geographers.*

COOKE, P. (1995) *The Rise of the Rustbelt.* London: UCL Press.

COOKE, P. and MORGAN, K. (1998) *The Associational Economy: Firms, Regions and Innovation.* Oxford: Oxford University Press.

CREWE, L. (1996) Material culture: embedded firms, organisational networks and the local economic development of a fashion quarter, *Regional Studies,* 30, pp. 257–272.

CUMBERS, A., MACKINNON, D. and CHAPMAN, K. (2003a) Innovation, collaboration and learning in regional clusters: a study of SMEs in the Aberdeen oil complex, *Environment and Planning A,* 35, pp. 1689–1706.

CUMBERS, A. MACKINNON, D. and MCMASTER, R. (2003b) Institutions, power and space: assessing the limits to institutionalism in economic geography, *European Urban and Regional Studies,* 10, pp. 327–344.

DICKEN, P. (2003) *Global Shift: Reshaping the Global Economic Map in the 21st Century*, 4th edn. London: Sage.

DTI (Department of Trade and Industry) (1998) *Our Competitive Future: Building the Knowledge Driven Economy.* Cm 4716. London: DTI.

EUROPEAN COMMISSION (2002) *Regional clusters in Europe.* Report to the Enterprise Directorate General by KPMG Special Services, EIM Business & Policy Research, and ENSR. Brussels: European Commission.

FELDMAN, M. (2000) Location and innovation: the new economic geography of innovation, spillovers and agglomeration, in: G. L. CLARK, M. FELDMAN and M. GERTLER (Eds) *Oxford*

Handbook of Economic Geography, pp. 373–394. Oxford: Oxford University Press.

FINE, B. (1999) A questions of economics: is it colonising the social sciences?, *Economy and Society*, 28, pp. 403–425.

FLORIDA, R. (1995) Towards the learning region, *Futures*, 27, pp. 527–536.

FUJITA, M., KRUGMAN, P. and VENABLES, A. (2000) *The Spatial Economy: Cities, Regions and International Trade*. Cambridge MA: MIT Press.

GERTLER, M. (2001) Best practice? Geography, learning and the institutional limits to strong convergence, *Journal of Economic Geography*, 1, pp. 5–26.

GLAESER, E. L. (2000) The new economics of urban and regional growth, in: G. L. CLARK, M. FELDMAN and M. GERTLER (Eds) *Oxford Handbook of Economic Geography*, pp. 83–98. Oxford: Oxford University Press.

GORDON, I. R. and MCCANN, P. (2000) Industrial clusters: complexes, agglomeration and/or social networks, *Urban Studies*, 37, 513–532.

GRABHER, G. (1993) Rediscovering the social in the economics of interfirm relations, in: G. GRABHER (Ed.) *The Embedded Firm: On the Socio–economics of Industrial Networks*, pp. 1–31. London: Routledge.

HEAD, K., RIES, J. and SWENSON, D. (1995) Agglomeration benefits and location choice: evidence from Japanese manufacturing investment in the United States, *Journal of International Economics*, 38, pp. 223–247.

HENRY, N. and PINCH, S. (2000) Spatialising knowledge: placing the knowledge community of Motor Sport Valley, *Geoforum*, 31, pp. 191–208.

HENRY, N., POLLARD, J. and SIDAWAY, J. D. (2001) Beyond the margins of economics: geographers, economists and policy relevance, *Antipode*, 33, pp. 200–207.

HIRSCHMAN, A. O. (1958) *The Strategy of Economic Development*. New Haven, CT: Yale University Press.

HODGSON, G. (1998) *Economics and Utopia*. London: Routledge.

JOHNSON, B. (1992) Institutional learning, in: B. A. LUNDVALL (Ed.) *National Systems of Innovation*, pp. 20–43. London: Frances Pinter.

KEEBLE, D., LAWSON, C., MOORE, B. and WILKINSON, F. (1999) Collective learning processes, networking and 'institutional thickness' in the Cambridge region, *Regional Studies*, 33, pp. 319–331.

KENNEY, M. and BURG, U. VON (1999) Technology, entrepreneurship and path dependence: industrial clusters in Silicon Valley and Route 128, *Industrial and Corporate Change*, 8(1), pp. 67–103.

KRUGMAN, P. (1991) *Geography and Trade*. Cambridge, MA: MIT Press.

LAGENDIJK, A. and CORNFORD, J. (2000) Regional institutions and knowledge: tracking new forms of regional development policy, *Geoforum*, 31, pp. 209–218.

LOVERING, J. (1999) Theory led by policy: the inadequacies of the 'new regionalism' (illustrated from the case of Wales), *International Journal of Urban and Regional Research*, 23, pp. 379–395.

LUNDVALL, B. A. and JOHNSON, B. (1994) The learning economy, *Journal of Industry Studies*, 1, pp. 23–43.

MACKINNON, D., CHAPMAN, K. and CUMBERS, A. (2004) Networking, trust and embeddedness amongst SMEs in the Aberdeen oil complex, *Entrepreneurship and Regional Development*, 16(2), pp. 87–106.

MACKINNON, D., CUMBERS, A. and CHAPMAN, K. (2002) Learning, innovation and regional renewal: a critical appraisal of current debates in regional development studies, *Progress in Human Geography*, 26, pp. 293–311.

MACLEOD, G. (2001) New regionalism reconsidered: globalisation, regulation and the recasting of political economic space, *International Journal of Urban and Regional Research*, 25, pp. 804–829.

MALMBERG, A. and MASKELL, P. (1997) Towards an explanation of regional specialization and industry agglomeration, *European Planning Studies*, 5, pp. 25–41.

MALMBERG, A. and MASKELL, P. (2002) The elusive concept of localisation economies: towards a knowledge-based theory of spatial clustering, *Environment and Planning A*, 34, pp. 429–449.

MARKUSEN, A. (1999) Fuzzy concepts, scanty evidence, policy distance: the case for rigour and policy relevance in critical regional studies, *Regional Studies*, 33, pp. 869–884.

MARSHALL, A. (1890) *Principles of Economics*. London: Macmillan.

MARTIN, R. (1999) The new 'geographical turn' in economics: some critical reflections, *Cambridge Journal of Economics*, 23, pp. 65–91.

MARTIN, R. and SUNLEY, P. (2003) Deconstructing clusters: chaotic concept or policy panacea?, *Journal of Economic Geography*, 3(1), pp. 5–36.

MASKELL, P. and MALMBERG, A. (1999) The competitiveness of firms and regions: 'ubiquitification' and the importance of localised learning, *European Urban and Regional Studies*, 6, pp. 9–25.

MAY, W., MASON, C. and PINCH, S. (2001) Explaining industrial agglomeration: the case of the British high-fidelity industry, *Geoforum*, 32, pp. 363–376.

MORGAN, K. (1997) The learning region: institutions, innovation and regional development, *Regional Studies*, 31(5), pp. 491–504.

MYRDAL, G. (1957) *Economic Theory and Undeveloped Regions*. London: Duckworth.

OECD (Organisation for Economic Co-operation and Development) (2002) *International Conference on Territorial Development. Local Clusters, Restructuring Territories*, Paris, January.

PORTER, M. (1990) *The Competitive Advantage of Nations*. Basingstoke: Macmillan.

PORTER, M. E. (1998) *On Competition*. Boston, MA: Harvard Business School Press.

PORTER, M. E. (2000) Locations, clusters and company strategy, in: G. L. CLARK, M. FELDMAN and M. GERTLER (Eds) *The Oxford Handbook of Economic Geography*. pp. 253–274. Oxford: Oxford University Press.

RAINES, P. (2002) Clusters policy: does it exist?, in: P. RAINES (Ed.) *Cluster Development and Policy*, pp. 21–33. Aldershot: Ashgate.

SAXENIAN, A. L. (1994) *Regional Advantage: Culture and Competition in Silicon Valley and Route 128*. Cambridge, MA: Harvard University Press.

SCOTT, A. (1988) *New Industrial Spaces*. London: Pergamon.

STEINER, M. (1998) The discreet charm of clusters: an introduction, in: M. STEINER (Ed.) *Clusters and Regional Specialisation*, pp. 1–18. London: Pion.

STORPER, M. (1997) *The Regional World: Territorial Development in a Global Economy*. London: Guildford Press.

STORPER, M. and SCOTT, A. J. (1989) The geographical foundations and social regulation of flexible production complexes, in: J. WOLCH and M. DEAR (Eds) *The Power of Geography: How Territory Shapes Social Life*, pp. 21–40. Winchester, MA: Unwin Hyman.

SWANN, G. M. P., PREVEZER, M. and STOUT, D. (Eds) (1998) *The Dynamics of Industrial Clustering: International Comparisons in Computing and Biotechnology*. Oxford: Oxford University Press.

TICHY, G. (1998) Clusters: less dispensable and more risky than ever, in: M. STEINER (Ed.) *Clusters and Regional Specialisation*, pp. 226–237. London: Pion.

TRENDS BUSINESS RESEARCH (2001) *Business Clusters in the UK: A First Assessment*. London: Department of Trade and Industry.

TUROK, I. (1993) Inward investment and local linkages: how deeply embedded is Silicon Glen?, *Regional Studies*, 27, pp. 401–417.

WORLD BANK (2000) *Electronic Conference on Clusters*. Washington, DC: World Bank.

Clusters, Dispersion and the Spaces in Between: For an Economic Geography of the Banal

N. A. Phelps

[Paper first received, February 2003; in final form, October 2003]

Introduction

The recent interest in clusters or agglomerations of economic activity can draw on a long analytical tradition in economics and geography.[1] However, although such clusters or agglomerations are enduring features of the industrial landscape and a perennial source of theoretical and empirical interest, there have been few explicit theoretical treatments of the changing geographical scale at which agglomeration is apparent.[2] The industrial districts of the late 19th century—the inspiration for Alfred Marshall's theory of agglomeration—were, in the main, self-contained singular industrial towns or else quarters of large cities (Hall, 1962; Wise, 1949). Today, however, the major examples of industrial agglomeration are of an entirely different geographical scale. Take, for example, the likes of 'Silicon Valley' near San Francisco (Saxenien, 1983), Britain's 'Motor Sports Valley' (Henry and Pinch, 1999), the neo-Marshallian nodes represented by global cities (Amin and Thrift, 1992) that draw together suburbs and satellite towns and cities into polycentric agglomerations (Hall, 1997) or 'scatterated' urban forms exemplified by Los Angeles (Gordon and Richardson, 1996, cited in Coffee and Shearmur,

2002). Such scatterated urban forms have European parallels in the dense networks of towns and cities of the Randstad (Batten, 1995) or the regionalised economy of the South East of England (Allen, 1992; Coe and Townsend, 1998) and have been likened to the extended metropolitan forms found in east Asia (Dick and Rimmer, 1998).

The emergence of these increasingly spatially diffuse forms of agglomeration raises questions for economic geography. First, they direct attention explicitly to questions of the spatial scale at which external economies and agglomeration processes are present. From a geographical point of view, then, explanation of the variable spatial scale of agglomeration remains a key question that, nevertheless, has rarely been addressed directly. As Gottdiener has suggested, there is a need to understand

> the processes that have worked and reworked settlement space by increasing its scale, fragmenting its communities and nodal points of interaction; by decentralizing its businesses and recentralizing them in new, functionally specialized nodes (Gottdiener, 2002, p. 178).

A few recent exceptions aside, geographers still have a tendency to interpret today's spatially expansive urban agglomerative forms in terms carried over from Alfred Marshall. For economists, cities, including the sorts of sprawling urban formations noted above, have been and continue to be taken as evidence of the importance of increasing returns to scale, the presence of 'market-size effects', 'urbanisation economies' or 'pecuniary' external economies. Hence, it is tempting to perceive invariant principles to the clustering or agglomeration of economic activity. However, is geographers' and economists' reliance on these invariant themes—of Marshallian externalities and of 'size'—too restrictive to capture important changes in economies over time? Can variations on these principles shed light upon the changing form and causes of agglomeration?

Secondly, these diffuse forms of agglomeration are notable for throwing up rather anonymous 'intermediate' locations or places that are nevertheless important in economic terms. This begs the question of whether, in considering contemporary manifestations of external economies, the clustering of economic activity and the spatial scale at which these occur, more analytical effort ought not to be directed at uncovering the economic dynamics of intermediate locations. Clearly, what count as intermediate locations have varied geographically and historically to include suburbs, edge cities or edgeless cities, but they might be defined as the relatively new locations that exist between established urban centres. In comparison with the enormous weight of theoretical and empirical interest in industry clusters, we know very little about the economic basis of these seemingly banal locations.

In this exploratory paper, I address these two questions. This in turn rests on a need for sensitivity to the type of externalities, the spatial scale at which they are present and their meshing together. Some types of externalities are more mobile than others. Moreover, the salience of different externalities to particular sectors of the economy has changed over time. It should be no surprise therefore that any renewal of the theory of agglomeration ought to move beyond general modelling of invariant principles of agglomeration (Krugman, 1996) to more geographically and historically context-sensitive explanations. Some of these historical contrasts in agglomeration have been drawn in very broad schematic terms in a previous paper (Phelps and Ozawa, 2003). This paper makes a further contribution to renewing the theory of agglomeration. It first considers the virtues of variations on the classical themes of agglomeration theory in explaining today's sprawling urban forms. In particular, the paper considers whether neo-Marshallian analyses and the idea of 'borrowed size' are able to capture adequately the tension between the mobility of pecuniary external economies and the relative fixity of technological externalities. It then points to the economic significance of intermediate locations and explores their economic basis in terms of the meshing of different external economies. In conclusion, it is suggested that analytical enquiry might more usefully focus on the meshing of different types of internal and external economies as these produce a 'banal' economic geography of intermediate locations.

External Economies, Agglomeration and Geographical Scale

The analysis of industrial agglomeration has proceeded from Alfred Marshall who deployed the concept of external economies.[3] The existence of external economies—the advantages that are open to and shared among a collectivity of businesses—promotes the geographical agglomeration of economic activity. These advantages are chiefly, especially in the geographical literature, discussed in relation to a given singular town, city or territorial scale. Specifically, Marshall identified a trinity of external economies: those connected with input–output transactions, those of labour market pooling and technological externalities.

Whilst drawing on the work of Alfred Marshall, from the work of Hoover in the 1930s until comparatively recently, one additional distinction which was made in geographical literature on agglomeration was that between 'urbanisation' and 'localisation' economies (see, for example, Karaska, 1969). In part, this distinction could be read as an early example of geographers' sensitivities to the expanding geographical scale over which external economies operate. However, since the instances of localised agglomeration being studied coincided with or were contained within single urban areas, it is less the geography of externalities than their type that geographers were drawing attention to. Here, urbanisation economies appear to equate to what economists refer to as pecuniary external economies (although these are not confined to urban areas) and localisation economies to technological externalities. Moreover, urbanisation economies appear to combine aspects of pure external economies of scale effects and the diversity of external economies of scope effects (Parr, 2002a, 2002b).

As the discipline of economics has evolved, it is the distinction between pecuniary externalities (market size or true external economies of scale effects) and technological externalities (Brown and Jackson, 1985; Harrison, 1992; Martin and Sunley, 1996; Skitovsky, 1954) that has emerged from Marshall's original concept. It is this distinction that we wish to work with in the remainder of the paper when considering the geographical scale at which forces of agglomeration have operated. This binary distinction cuts across the trinity of Marshallian externalities. The specialised input–output transactions fall fairly squarely into the category of pecuniary externalities (Phelps, 1992). The economies of labour market pooling, however, embody both pecuniary and technological effects. Although acknowledging the existence of very localised forms of agglomeration, the preference of economists has been to concentrate on pecuniary externalities not least because of the difficulties of incorporating the latter into formal equilibrium models. Finally, the technological externalities—the industrial atmosphere, as it is sometimes referred to—are those that have been and continue to be localised.

Working from their different perspectives and from different sides of this pecuniary–technological divide, economists and geographers have paid comparatively little attention in theoretical terms to the changing extent of industrial agglomeration apparent over time. Yet one inescapable essence apparent to those concerned with agglomerative processes is that, to quote a geographer, "agglomeration effects in large industrial cities can *not* be assumed to be geographically fixed" (Scott, 1982, p. 118; original emphasis). The same observation was made somewhat earlier by the economist E. A. G. Robinson. Commenting on the dissolution of Britain's industrial districts prior to the Second World War, Robinson drew a distinction between mobile and immobile external economies. It is this logic which led the economic geographer Richard Walker to suggest that agglomeration economies are an "historically contingent feature whose force has gradually diminished to be replaced by the economies of internal organisational scale open to large companies" (Walker, 1981, p. 385), although he has in subsequent work re-emphasised the persistence of localised agglomeration as part-and-parcel of urban and regional specialisation and differentiation (Storper and Walker, 1989).

In the remainder of this section, I want to consider the question of the geographical availability of external economies. Given the likes of major historical trends (such as in the technological and sectoral basis of economies), industry specifics, national institutional specifics and even ambiguities in the concept itself, this question is not readily reduced to associating particular types of external economy with fixed geographical scales. Instead, I approach the question indirectly, drawing attention to the relative mobility of different types of external economies.

The Immobility of Localised or Technological Externalities

The distinction between pecuniary and technological external economies has rarely been apparent within geographical work on agglomeration; however, until very recently, in the guise of 'linkage studies', much of the geographical research effort focused on urbanisation economies or pecuniary effects, rather than technological externalities *per se*. Most notably, a large body of linkage studies in the late 1960s and 1970s examined these pecuniary effects in promoting agglomeration (Hoare, 1985). These studies aimed to trace the significance of local linkages to the agglomeration of industry but instead appeared to indicate that agglomeration often existed without strong localised linkages (for example, Lever, 1972). Whilst linkages may have played an important part in Alfred Marshall's late 19th-century industrial districts, their role in the sometimes ailing agglomerations of late-Fordism was altogether more questionable (Phelps, 1992).

Somewhat curiously, then, interest in the connections between localised linkages, external economies and localised agglomeration was revived with a string of industry studies in the Los Angeles area—an area frequently referred to as the exemplar of urban sprawl and the dissolution of the classical monocentric city spatial structure. This work represented a deepening of our understanding of the role of input–output transactional external economies in promoting agglomeration. Here localised input–output external economies were produced through the vertical and horizontal disintegration of production (Scott, 1983, 1986, 1988) which in turn was driven by the wider market dynamics of a new 'post-Fordist' regime of flexible accumulation. However, in light of the findings of earlier linkage studies, the widespread relevance of input–output transactions to contemporary, 'post-Fordist', instances of agglomeration was soon questioned (Phelps, 1992; Storper, 1997).

In fact, until recent work on neo-Marshallian agglomerations, relatively little academic attention focused on two of Marshall's trinity of external economies—labour market pooling and technological externalities. Perhaps the most notable work that did tend to concentrate squarely on the technological aspect of localised technological externalities focused on the importance of informational linkages or contact systems (Thorngren, 1970; Tornqvist, 1970). The greater the information content and its strategic value, the greater the significance of face-to-face contacts in high-level business decision-making which in turn promoted highly localised forms of agglomeration—as seen, for example, in the head-office complexes of major city centres (Goddard, 1975).

The Mobility of Pecuniary Effects

Improvements in transport and communications technology and infrastructure have meant that the need to locate in or near major settlements as these represent major markets is diminishing for a growing number of manufacturing and service industries. This has been reinforced by the growing 'roundaboutness' of industry (Young, 1928) whereby ever-expanding chains of intermediate activities are implicated in the production of manufactured items and services. Krugman's argument that "the existence of cities themselves is evidently an increasing returns phenomenon" (Krugman, quoted in Boddy, 1999, p. 818) surely only holds for *some* industries. In contrast, then, a longer historical view has led Bairoch to argue that

> as a result of progress in technology the size of the required market for an industrial concern has increased at a much faster rate than the average size of cities ... Thus, except for the upper fringe of very large cities ... it seems unlikely that the city any more than before provided a market sufficient in itself to support industrial development (Bairoch, 1988, p. 344).

If the technical bases of economies of scale available at the scale of the city were soon exhausted, so too were the managerial bases to such economies (Chandler, 1966, p. 478). In other words, the need to locate in or near major settlements as these represent major sources of pecuniary effects has long-since been diminishing for a growing number of manufacturing and service industries. Thus, in general terms, input–output linkages which represent perhaps the major manifestation of pecuniary externalities, now play a less critical role in promoting the agglomeration of firms at the urban let alone more localised scales (Maskell *et al.*, 1998; Phelps, 1992).

The mobility or wide geographical availability of pecuniary effects appears to be what Robinson was referring to when noting that

> certain external economies, though by no means all, depend not on the size of that industry in one locality, but on the size of that industry in the world as a whole. We can, I think, say that the proportion of all economies which are of this international mobile type is steadily increasing, and that the advantage of a large industry concentrated in one country is steadily declining as the mobility of economies increases; that is that the optimum local industry is diminishing in size (Robinson, 1953/1931, p. 142).

This point appears to have been expressed more formally in the 'new economic geography' (for example, Krugman, 1996, 2000) where a distinction has been drawn between the widely available pecuniary externalities and less widely available technological externalities (see Boddy, 1999; Martin, 1999; Martin and Sunley, 1996). The new economic geography suggests that these pecuniary externalities tend to operate over very broad geographical areas (Martin, 1999), such as whole continents, so that peripherality is unlikely greatly to hinder access to markets and pecuniary externalities for a wide range of industries—especially manufacturing industries.[4] Little wonder then that external economies of scale, although of some significance, are not determinant in the formation and functioning of important contemporary instances of high-technology agglomerations (Pinch and Henry, 1999). Economies of scale, whether internal or external, can be secured up to a point but

> the larger an industry grows, the less are the economies to be secured by further growth ... And so we chase this will-o'-the-wisp of external economy through industry after industry, and we find it vanishing in the end or absorbed in the economies of firms or organisations below their optimum capacity (Robinson, 1953/1931, p. 138).

This passage evokes both the relatively limited value of the quantifiable pure pecuniary external scale effects to firms in many industries and the extent to which these may have to be pursued, as it were, over broad geographical scales such as whole continents or even globally.

Between Mobility and Fixity: Intermediate Locations and the Limits of Classical Agglomeration Theory

Contemporary instances of agglomeration

are of a different spatial order from those implicit in Marshallian analysis. Terms such as polycentricity (Suarez-Villa, 1989), scatteration (Gordon and Richardson, cited in Coffey and Shearmur, 2002), metaclustering (Bennett *et al.*, 1999) and regionalisation (Allen, 1992; Coe and Townsend, 1998) have all been used to try to capture the way in which external economies and the agglomeration they promote now appear to be available over much less localised geographical scales. In what follows, I suggest that, in recognising the growth of suburbs, edge cities or even more diffuse urban forms, we need to move away from classical agglomeration theory and the tendency to simply up-scale the territory over which an undifferentiated category of external economies is now perceived to be available.

Initial academic interest in polycentric urban forms tended to work broadly within the framework of Marshallian external economies. Thus there was a sense here in which the classical concepts of agglomeration might, in a modified form, be of use in analysing multinodal metropolitan areas. Such polycentric, polysectoral regional complexes are

> like a spatially extended agglomerative field based on a more elaborate social division of labour than would be found in a single sector in a dense urban industrial district. ... Nonetheless the analytical foundations for understanding such fields are very much the same (Storper and Walker, 1989, p. 141).

In the same vein, several authors have therefore analysed the multicentred form of Los Angeles in terms of the changing locus of external economies and in particular their suburbanisation (Scott, 1982, 1988; Suarez-Villa, 1989; Suarez-Villa and Walrod, 1997).

Central cities have been at the forefront of the structural transformation of advanced economies from industrial to post-industrial economies. The post-war period has, however, witnessed a greater specialisation within and between urban areas as centres of service employment. A complex pattern of city–suburb competition has evolved (Stanback, 1995) which should not be seen as simple 'suburban dependency'—as recent debate on the differential growth of US cities and suburbs demonstrates (see Hill *et al.*, 1995; Hill and Wolman, 1997; Savitch, 1995). A range of experiences is evident (Coffey and Shearmur, 2002; Hughes, 1993), including a degree of city-centre dependence on suburban growth. Indeed, taking up the latter point, there are growing signs of the economic performance of suburban areas leading, or being independent of, city centres. The growing disparities between the economic performance and conditions of suburbs and city centres has centred attention on the genesis of new suburbs and so-called edge cities.[5]

On the one hand, there are those who see such new urban centres as inextricably linked and part of a process of intraurban and interurban specialisation and differentiation.[6] According to one estimate, edge cities may account for only one-third of all non-downtown office floor space (Lang, 2003, p. 10). Instead, a growing proportion of economic activity, particularly in North America but also, we can suggest, elsewhere, centres on still yet more anonymous places or non-places that constitute what Kunstler refers to as the geography of nowhere typified by the urban sprawl of North America. "It is where most American children grew up. It is where most economic activity takes place. ... It is the geography of nowhere" (Kunstler, 1993, p. 15). Most recently, Lang has used the term 'edgeless cities' to describe "the unmarked phenomenon of the new metropolis" (Lang, 2003, p.5).[7] These diffuse and barely definable edgeless cities (which encompass edge cities) account for two-thirds of such office space (Lang, 2003, p. 10). Instead, the likes of edge and edgeless cities appear to "have replaced a segment of downtowns, which are now only one component of specialised regional economic landscapes" (Bingham and Kimble, 1995, p. 259).

On the other hand, there are those who see a much greater degree of autonomy from other nearby urban centres in the social, economic and political bases of edge and edgeless city developments. This point is emphasised by the Ghent Urban Studies Team (1999) who argue that the role of edge urban areas within urban systems is, if anything, more complex in the European setting than in the North American context.

> In some European cities, two ostensibly antagonistic processes even run side by side. ... The growth of the central city goes hand in hand with an increase in density within the periphery. A particular zone within the periphery (around the airport, for instance) can itself become the center of important sectors in the post-Fordist economy. *Ultimately, the periphery begins to interact not only with the central city, but with the entire region* (Ghent Urban Studies Team, 1999, pp. 37–38; emphasis added).

Peter Hall (1997, 2000) has perhaps gone further in stressing the novelty of new suburbs and edge cities within intrametropolitan and city-region-wide patterns of economic specialisation.

> The high level intelligence and control functions of the global cities are increasingly dispersed across a wide geographical area, limited only by certain geographical constraints of time-distance. Though traditional face-to-face locations retain their power, *they are increasingly supplemented by new kinds of node for face-to-face activity*. The resultant geographical structure is quintessentially polycentric (Hall, 2000, p. 73; emphasis added).

What we see here, as in the case of the South East of England, is a complex and selective process of economic dispersion that generates

> extensive networks of 'suburbs' or 'edge cities' and the emergence of what have been called non-places, each of which nevertheless has its own institutions of local governance and networks of local politics' (Charlesworth and Cochrane, 1994, p. 1725).

London's growth, for example, has not for many decades involved physical contiguity (Mogridge and Parr, 1997, p. 111). Such city-regions exist as highly functionally integrated regional systems which nevertheless are internally fragmented and discontinuous (Allen *et al.*, 1998).

In recognition of these empirical realities, there has been an assumption in some of the literature that such diffuse patterns of agglomeration can be explained in terms of simply recognising the greater—essentially regional—spatial scale at which an undifferentiated category of external economies operates. Thus Coe and Townsend explain the emergence of the software industry in the large swathe of territory within the South East of England adjacent to London in terms of the agglomeration economies that now operate at the regional scale. This resonates with an emphasis on the regional scale in some of the literature on North American patterns of urbanisation (Hise, 1997; Soja and Scott, 1996)—although it is Gordon and Richardson (1996) that have suggested specifically that such 'scatterated' urban forms are based upon agglomeration economies that operate at the city-region scale.

Partly in light of these recent developments and partly through a re-evaluation of the historical growth of multicentred urban forms, some recent work has moved away from the idea that classical analysis of agglomeration is sufficient to understand contemporary and historical instances of city-regions. This idea has been championed most explicitly by Walker and Lewis when commenting on historical patterns of urban and suburban growth in Montreal. They argue that

> the process of urban-industrial growth has another crucial dimension besides the outward flow and build-up of the city: the appearance of distinctive industrial districts within a multinodal metropolitan area. *Classic agglomeration theory does not explain this phenomenon* (Walker and Lewis, 2001, pp. 7–8; emphasis added).

Commenting on contemporary urban forms

that include the likes of edge cities, Krugman (1996, p. 88) notes that polycentricity is "not simply a matter of agglomeration economies producing agglomeration". When discussing today's spatially extended city-regions, it becomes clear, then, that economists and geographers are in agreement on at least one thing. Although relevant in part, traditional theories of agglomeration are insufficient to explain the increasingly spatially diffuse forms of service-centred agglomeration. I turn now to a discussion in which I consider the utility of variations on the themes of classical agglomeration theory—neo-Marshallian external economies and borrowed size—in analysing spatially diffuse forms of contemporary agglomeration.

Variations on Themes: Neo-Marshallian Externalities and Borrowed Size

If the use of existing theory by economists and geographers with a singular emphasis on external economies of scale (size), on the one hand, and technological externalities, on the other, provides increasingly inadequate means of understanding contemporary forms of urban economic agglomeration, perhaps variations on these themes may shed some light?

Neo-Marshallian External Economies

Recent research on the ingredients of localised agglomeration has returned to Marshallian external economies to concentrate less on interfirm material linkages and more on the externalities of labour market pooling and technological innovation (MacKinnon *et al.*, 2002). Such externalities remain highly localised and a determinant in contemporary instances of neo-Marshallian agglomeration (Amin and Thrift, 1992; Malmberg, 1997; Maskell and Malmberg, 1999) or clusters (Porter, 1990). Moreover, they can be important not only to technologically sophisticated manufacturing or service industries but also to the locational patterns of 'low-technology' manufacturing industries (Maskell, 1998). Yet, quite conspicuously, the recent literature on instances of neo-Marshallian agglomeration and clusters has avoided direct consideration of the questions following from the changing geographical scale at which such externalities operate.

Porter (1990) deployed the term clusters initially without explicit reference to the territorial scale at which concentrations of industries existed.[8] Indeed, there is considerable ambiguity as to whether such clusters or neo-Marshallian agglomerations are functionally or spatially defined. It has acquired multiple spatial connotations to become a 'ridiculously elastic' concept (Martin and Sunley, 2003). The weight of effort has concentrated instead on processes of innovation and learning. The major lines of inquiry have focused on establishing *inter alia*: the significance of learning to economic competitiveness; the various mechanisms by which learning processes occur; and, when and under what conditions localised learning takes place. These accounts have been important in establishing the localised nature of learning and hence confirming the intrinsically geographical nature of neo-Marshallian external economies.

However, a few notable exceptions aside (for example, Amin and Thrift, 1992), and in keeping with the Marshallian tradition, the recent literature on neo-Marshallian processes of learning assumes singular places and ostensibly locally bounded processes. Terms such as place, mileu, region, apparently are used interchangeably with little discussion, let alone definition, of variations in their spatial extent to denote a singular territory within which processes of innovation and learning are, in the main, bounded. There is, then, what Park and Markusen (1994, p. 86) refer to as an "assertion of endogeneity" apparent within much of the recent interest in neo-Marshallian agglomerations. As a result, few of these accounts have taken seriously the need to explore the geographical scale at which such localisation occurs. At the outset of this renewed interest in localised neo-Marshallian agglomerations, Storper and Walker were able to observe that there has been "a remarkable tendency to think in terms carried over from the days of simple commodity production, of manufacturing and the industrial revolution" (Storper and Walker, 1989, p. 127). In similar vein, but more recently, Hall has noted that existing theory concerning urban form is "set in a relatively self-contained world of towns and hinterlands, with hardly a hint of more complex larger scale relationships" (Hall, 1997, p. 315). The analytical powers of this literature are therefore limited "most particularly because of a tendency to cling to a model which is locally based and which does not therefore recognize the importance of emerging global corporate networks" (Amin and Thrift, 1992, p. 574).

One might argue that these criticisms are indeed valid, despite the significant theoretical effort devoted to detailing the ingredients and processes constitutive of neo-Marshallian external economies. Extralocality ties have been considered explicitly in some accounts. In light of the internationalisation of corporate networks, Amin and Thrift (1992) in particular draw attention to the meshing of global and local dynamics in neo-Marshallian nodes but without moving beyond this binary distinction in the scale-dependent nature of different externalities. Even here there is some debate as to whether technological innovation is stimulated by localised external economies or the industry diversity associated with urbanisation economies (Simmie and Sennett, 1999). Although noting that the Ruhr could not be considered as a Marshallian industrial district, following Granovetter, Grabher (1993) has highlighted the importance of 'weak', non-local, ties to the process of regional innovation and industrial performance. The significance of extralocal ties to the innovative activity of firms in the Aberdeen oil industry complex leads Cumbers *et al.* (2003) to question the emphasis upon bounded regions in much of the extant literature on localised learning and cluster formation.

Literature on localised processes of learning (Malmberg, 1997; Maskell and Malmberg, 1999) does set localised neo-Marshallian externalities within broader geographical contexts—especially nationally specific innovation systems. In this way, the spatial *cul-de-sac* of focusing only on locally bounded processes is ameliorated by an appeal to a more widely drawn territory in which external economies are seen to operate (Malmberg and Maskell, 2002). However, there is little serious engagement with the question of the extent to which there is meshing of these different scale-dependent processes of innovation.

Labour remains the most immobile of factors of production and therefore continues to promote quite localised forms of agglomeration. However, even here increases in personal mobility and daily commuting patterns beg important questions of the geographical scale at which agglomerations of industry now form. Earlier work on contact systems and face-to-face contacts has been recast by those noting the embodiment of specialised knowledge and information in the labour market. The continued vitality of many central-city agglomerations of offices reflects the continued importance of face-to-face contacts in specialised labour markets (Glaeser, 1998). Pursuing this theme, Duranton (1999) has suggested that personal contacts (in which we might include labour market dynamics) now represent *the* major regulating institution of agglomeration and help to reaffirm the economic viability of cities. Whether by face or other means, such contacts appear to promote agglomeration over extended metropolitan regions. Away from the possibly unique case of central cities, such labour-market-related untraded interdependencies are also considered to promote agglomeration more generally in localities such as Silicon Valley, Route 128 and the British Motor Sport Valley (Henry and Pinch, 1999). Yet the fact that these latter examples of agglomeration represent 'localities' of an altogether different spatial scale from those being considered previously is rarely given thought.

The fundamental question of the differences in the geographical scale at which localised processes of learning or innovation operate either presently or historically has, then, scarcely been addressed (Howells, 1999). Here, neo-Marshallian analysts themselves note that "as things stand today even neo-Marshallian nodes of global corporate networks are finding it difficult to retain their status" (Amin and Thrift, 1992, p. 585). Indeed, following this line of thought, Amin and Thrift have since gone on to argue that

> economic organisation now is irremediably distributed. Even when economic activity seems to be spatially clustered, a close inspection will reveal that the clusters rely on a multiplicity of sites, institutions and connections which actually constitute them (Amin and Thrift, 2003, p. 52).

This seems to concede that the classical *form* of agglomeration is an endangered species and with it, we might speculate, the relevance of classical agglomeration theory itself.

Borrowed Size

The concept of 'borrowed size' emerged fleetingly in the early 1970s as one very early variation on a theme and attempt to adapt theory to circumstances of urban zero population growth that had become apparent in the US by the 1960s. The term borrowed size was first briefly described by Alonso (1973). According to Alonso

> The concept of a system of cities has many facets, but one of particular interest ... is the concept of borrowed size, whereby a small city or metropolitan area exhibits some of the characteristics of a larger one if it is near other population centres (Alonso, 1973, p. 200).

As far as I am aware, Alonso did not subsequently develop this idea of borrowed size in any detail. It is intuitively appealing in its simplicity, but it is difficult to pin down analytically and its utility remains to be demonstrated. I attempt here to clarify its meaning and potential relevance.

In using the term borrowed size, Alonso was thinking of the tendency for people and businesses to retain the advantages of being based in smaller settlements (for example, less congestion, lower rents) whilst also being able to reap the advantages on offer in larger settlements (such as access to sizeable markets, business services and expertise, larger and more diverse labour markets and cultural amenities).[9] At the time, he suggested that the phenomenon "seems to account for the fact that in the large American metropolitan constellations, the smaller metropoles are growing more rapidly than the bigger ones" (Alonso, 1973, p. 200). He also suggested that borrowed size was visible in European urban patterns especially those in Germany and the Benelux countries.[10]

Interaction between firms and populations of individual settlements is both a measure of, and the mechanism enabling, borrowed size. The concept of borrowed size stresses the relational nature of external economies—the way in which pecuniary external economies can be spread across a whole series of urban areas and the way in which location in one town or city does not preclude access to externalities present in another. It appears to prefigure the concept of network externalities which suggests that "the greater availability of externalities that can be found in smaller centers is not uniquely generated within each of them. It is also the result of increased interaction between centers" (Senn and Gorla, 1999, p. 249). Bennett *et al.* (1999, p. 399) have therefore recently emphasised that "the potential for a cluster to develop in a given location thus depends not only on the local business base, but also on its location relative to other clusters". With this emphasis on interaction, the concept is closely related to earlier work on the 'non-place urban realm' (Webber, 1963, 1964) and city-systems (for example, Bourne, 1975; Pred, 1976). On the one hand, whilst in some instances (notably globally competitive industry clusters) specialised communities of business interests are highly localised, Webber's (1964) notion of the non-place urban realm is useful in stressing the fact such 'networking' can also involve important non-local ties such that a degree of decentralisation is possible. On the other hand, in contrast to the tendency for "accessibility to be freed from propinquity" made famous by Webber (1963, p. 109), the concept of borrowed size is a manifestation of the complementarity between the advantages firms draw from their locality and those they draw from other nearby settlements through their business transactions and interactions.

The concept of borrowed size speaks to an extension of the processes of suburban growth noted by Scott (1982) and Suarez-Villa (1989) in which the likes of edge cities and free-standing towns and cities are woven into polycentric urban forms. It derives from the tension between forces of agglomeration, on the one hand, and those of decentralisation on the other hand (Champion, 1988; Cheshire, 1995). As Parr observes

> For many firms the benefits of a metropolis no longer require a location within it or even in close proximity, but merely accessibility to it. What seems to be happening is that the incidence of agglomeration diseconomies is largely confined to the metropolis while the benefits generated within the metropolis are not really so locationally constrained as formerly (Parr, 2002b, pp. 728–729).

This brings us close to Krugman's observation that "the tension between centripetal and centrifugal forces ... suggests the possibility of an emergent pattern with multiple centres (Krugman, 1996, p. 89). If it is technological externalities that continue to act as a significant centripetal force drawing activity to relatively localised urban industrial agglomerations, and if it is the now generally widely available pecuniary externalities that tend to act as centrifugal forces permitting their dissolution, it is the relationship between pecuniary and technological externalities which is of importance for understanding today"s spatially diffuse, service-sector-based agglomerations.

However, instead, the idea of borrowed size conflates elements of both these types of external economy. There is the idea that it is indeed size, or rather the pecuniary effects associated with industry size, that is being borrowed in some way. So—for example, Krugman (1996, p. 91) argues that "we see how the mysterious centripetal and centrifugal forces of the edge city model could arise from pecuniary externalities, alias market-size effects". Yet unlike Krugman's modelling of polycentric urban forms for industry *in general*, it is possible to see that such polycentric urban forms may actually be closely associated with a *particular* subset of service industries for which important pecuniary externalities are available at the scale of major conurbations or even heavily urbanised regions. These would include non-specialist business services which are linked to important markets offered by the financial and head-office activities of global cities but also, for example, the likes of communications and health services for which urbanisation economies appear to be important (O'hUllachain and Satterthwaite, 1992). Here a diverse set of businesses benefiting from a diverse customer-base within a major city or city-region nevertheless also benefits from the economies of scale associated with an agglomeration of specialist suppliers in industries such as business services which themselves are dependent on a major concentration of customers represented by head-office functions. This situation in which there are wider benefits produced from the sorts of head-office producer service complexes that are apparent in major urban areas, appears to combine elements of the two slightly different meanings of size—as industry scale and as market diversity. It is an important example of, what Parr (2002a, p. 165) terms, reciprocity. It becomes apparent that elements of industry size and diversity are being borrowed by businesses in settlements accessible to major urban centres. In this connection, it is interesting to note that while we tend, following Garreau (1991), to think of edge cities as retail-office complexes, their economies are often as diverse or more so than the cities they surround—with some even having disproportionate shares of manufacturing employment (Bingham and Kimble, 1995).

In sum, Alonso's use of the term borrowed size prefigured other recent terminological innovations such as polycentric urbanisation and regionalisation in pointing to on-going trends of economic dispersal and the seeming dissolution of urban economic agglomerations. It also is notable for stressing the relational nature of external economies. Yet, in mixing an analysis of different types of external economies, problems of similar nature to using traditional agglomeration theory are created.

The Relational Nature of External Economies and the Economic Geography of Intermediate Locations

When one views processes of agglomeration in a longer-term historical frame, it becomes apparent that our thinking regarding external economies derives from possibly unique moments in national history. The brief and unique moment in the history of the first industrial nations when pecuniary and technological externalities were bounded and coincident at the geographical scale of the city was erected into a powerful theory and continues to colour analysis of the location of economic activity (Phelps and Ozawa, 2003). Yet prior to this, and certainly since this time, the spatial fields of different types of external economy have expanded and also diverged.

In reviewing the relevance of the new economic geography to the sub-discipline of economic geography, Martin has noted how

> although there is some suggestion in the 'new economic geography' literature that pecuniary (market-size) externalities operate over wider regional spaces than informational and technological externalities, which are more localised, how these different scale-dependent processes might intermesh and interact is not discussed (Martin, 1999, p. 78).

Yet this question—of the meshing of different scale-dependent external and internal economies—is perhaps the most fundamental to a richer understanding of the evolving nature of urban economic agglomeration and in particular the variable and changing geographical scale at which it manifests itself presently and over time.

Arguably, it is 'banal' intermediate locations—what at points in history have variously been regarded as suburbs, edge cities and edgeless cities—that best exemplify this meshing of different types of externality and upon which economic geography might refocus itself. Intermediate locations—what Gottman (1961) referred to as 'interurbia'—were already the norm for manufacturing industries by the 1960s in North America and are today for many firms in many industries, especially for most manufacturing industries (McCann, 1995, p. 574). Hence, one might pose the rhetorical question as to whether analytical attention might usefully be redirected from the hunt for, and documenting of, neo-Marshallian districts or clusters to the seemingly banal economic geographies of otherwise anonymous intermediate locations. Do we instead need to take seriously the likes of Croydon (Phelps, 1998), Reading (Raco, 2002), Swindon (Boddy *et al.*, 1997) and the towns and suburbs (Charlesworth and Cochrane, 1994) of a heavily urbanised region such as the South East of England, if we are to understand a large part of the spatial dynamics of the UK and other advanced economies? Should the edge and edgeless cities of North America, to date studied primarily for what they tell us about post-modern urbanism, be studied also for what they reveal about the contemporary economic geography of advanced nations? Curiously, despite their autonomous dynamics and their seeming importance to the functioning of heavily urbanised regions, industrial and service suburbs and edge cities have been perceived as banal economic locations and excluded from the mainstream of economic-geographical analysis. Whilst the growth of these places has been charted in aggregate geographical studies of employment decentralisation, the economic basis of these places *per se* has rarely been the explicit subject of study.

As early as 1899, suburban districts accounted for nearly half of all metropolitan employment (Lewis, 1999) yet suburbs *per se* have never occupied a central position within economic geographical analysis. If they had been, the fact that the earliest of industrial suburbs "depended on their relationship with other parts of the metropolitan area and with regional and national markets" (Lewis, 1999, p. 159) would surely have raised doubts over the ability of the concept of external economies to capture the simultaneity of meaningful economic interactions at multiple scales.

More recently, the burgeoning literature on edge cities offers surprisingly few insights into what economic advantages businesses derive from these new locations. Those that can be gleaned suggest that we are not dealing with the simple suburbanisation of external economies. So, for example, it has been observed that "many edge cities lack an agglomeration economy which might justify the cost of a massive retrofit of their infrastructure to accommodate greater density" (Lang, 2003, p. 97). Instead, whilst stressing the novelty and self-contained nature of these urban forms, Garreau (1991) himself also notes that they embody a "balance between individualism and face-to-face contact" (Garreau, 1991, p. 42). This balance has a significance in economic terms in that it does indeed suggest that the economies of many such edge cities are ensnared in a wider set of relations with other urban centres regionally, nationally and possibly internationally.

Some early clues to the economic basis of intermediate locations—such as edge cities—within megalopolitan forms were, as we saw earlier, provided in the concept of borrowed size. Further clues are provided by Gottman and Harper when revisiting the idea of megalopolis. They point out that the emerging patterns of urbanisation "must be understood as less localised, more *pluralistic* and more transactional than productive of manufactured goods, and more dependent on far flung urban networks" (Gottman and Harper, 1990, p. 249; emphasis added). Gottman's ideas have rightly fuelled research on the transactional economic activities of cities, suburbs and edge cities, but the idea of plurality has perhaps been overlooked. This plurality of economic processes is also a feature noted in Hise's depiction of the 'planning' of Los Angeles. Hise argues that

> What is needed is an elastic analytical and interpretative framework that can expand to the region and contract to the district or neighbourhood level and encompass all points in between. Elasticity is important because as the community builders realised, no event or intervention takes places at only one scale (Hise, 1997, p. 12).

In drawing attention to a plurality of economic processes, Hise and Gottman and Harper problematise the idea that any simple scaling-up of the territory over which external economies now operate is sufficient to understand the economic dynamics of diffuse urban forms. This notion of plurality has been taken up more explicitly in the most recent work on agglomeration and external economies.

If external economies as originally defined are the advantages shared among a group of firms within a single given territory, it is also true to see that these advantages are themselves now increasingly distributed among collectivities of linked towns or cities. This issue is more than just one of balance between undifferentiated categories of external economies and external diseconomies, or of the simple scaling-up of the territories over which external economies are available. More fundamentally, the relational nature of external economies and the sorts of patterns of agglomeration that can emerge rest on the common-place co-existence of different types of external (and internal) economies that are available over different geographical scales and are potentially open to firms operating in one location. Thus

> Externally-based agglomeration economies differ from each other not so much in terms of their content ... but with respect to the mode in which the firms gaining

these economies exist in relation to each other (Parr, 2002a, p. 158).

There are many instances of the agglomeration of businesses which are predicated on reasons that are nothing to do with the presence of other firms locally (McCann, 1995, p. 573). The meshing of the different external and internal economies with different geographical fields in this way highlights that even distinguishing between the potential contribution of different external and internal economies to firm location in isolation is insufficient. It then becomes important to consider aspects of plurality and reciprocity of these different economies apparent in the location of businesses (Parr, 2002b). These properties of reciprocity and plurality are also captured in the term 'network externalities' (Camagni, 1993, cited in Capello, 2000). The problem is that "if the theoretical framework around the concept of city networks is well structured, the same cannot be said for the empirical analysis around this concept" (Capello, 2000, p. 1926).

A plurality of external economies—for example, exists where a firm may have access to localised technological external economies as a result of being among like firms, but may also benefit from the pecuniary effects of urbanisation economies. Moreover, this plurality of external economies is likely to be reciprocal in nature especially in those primate and other nationally significant city-regions in which a large part of the national economy is concentrated. The best examples of this, as we saw earlier, are the head-office-producer service complexes of major city-regions—where businesses benefiting from the pecuniary effects of a diverse customer-base can also benefit from the technological externalities of specialist suppliers who themselves also draw upon such diversity. Given this, "it becomes apparent that for the individual firm at a particular location, the coexistence of one or more internally based agglomeration economies with one or more externally-based agglomeration economies can be regarded as the norm" (Parr, 2002a, p. 166). This combination appears to underpin the 'Milan innovation field' where, in one of the few empirical studies, it was found that "both the size of the firm and the location play a role in defining the sources of metropolitan advantages for firms' creativity" (Capello, 2000, p. 116).[11]

Moreover, some of the major instances in which there is a reciprocity and plurality of external economy effects that promote suburban and edge city-type developments within heavily urbanised city-regions are produced from government expenditures to support collective consumption (Gottdeiner, 2002; Walker, 1981). There are important differences in the dynamics underlying local and non-local expenditures and their associated external economies (Tiebout, 1956). Hence, the role of non-market provision for collective consumption is complex both in its simultaneously leading and shadowing market processes and in its scale-dependent nature.

Conclusion

The geographical availability of external economies has increased over time and this has to a greater or lesser extent been reflected in the economics and geographical literature on the subject of the clustering of economic activities. The concept of external economies itself tends to be locked into an idiographic realm of self-contained places defined at whatever scale. As a result, there is a tendency within the literature to scale-up an undifferentiated category of external economies from the district, to the city and onwards to the city-region, region or continental scale to try to capture the empirical realities of diffuse forms of urbanisation and economic location. However, today's spatially diffuse patterns of urbanisation and economic activity, and the economic basis of the intermediate locations such as suburbs, edge cities and edgeless cities of which they are composed, are not adequately explained in these terms. The external economies open to businesses in these intermediate locations are not the unique product of these places solely, but embody a meshing of potentially different types of external (and internal) economies.

It is consideration of these aspects of the meshing of different external economies that offers the greatest potential for a renewal of theory and empirical studies on this perennial topic of interest for economists and geographers. Here the concepts of 'network externalities' and aspects of plurality and reciprocity help to explain the economic basis of these seemingly banal, but important, intermediate locations. However, they remain to be subject to significant examination and refinement through empirical research.

All this leads on to broader questions regarding economic geographical analysis. For a start, it raises questions regarding the value of existing classifications of external economies (McCann, 1995; Parr, 2002a, 2002b). The on-going dispersion of economic activity, and its seeming acceleration at certain junctures, periodically has prompted a questioning of the value of the concept of external economies. The phase of rapid decentralisation of manufacturing from established metropolitan centres in the 1960s, 1970s and 1980s coincided with, as we noted earlier, the suggestion that external economies had been a temporary phenomenon now eclipsed by the economies internal to large corporations (Walker, 1981). More recently, in a period that has seen the acceleration of international economic integration, Amin and Thrift (2003) have distanced themselves fully from the vestiges of an idiographic approach implied within neo-Marshallian analysis to concentrate exclusively on the networks of relations associated with communities of practice. Thus, the concept of external economies today finds itself more fully exposed to the sort of creative tensions between place-based and space-based perspectives that have characterised human geography more generally (Taylor, 1999).

In this regard, we might want to consider how our understanding of agglomerations in terms of the meshing of different external and internal economies might be visualised and incorporated within a refigured relational economic geography (Thrift and Olds, 1996). Thrift and Olds depict the expanding sub-discipline of economic geography in terms of four 'topological presuppositions'.[12] The most recognisably economic processes within Thrift and Olds' refigured economic geography belong to the realm of bounded regions and networks. "Yet what will be most interesting is the kind of work that will be done at the boundaries of these presuppositions" (Thrift and Olds, 1996, p. 331). The thrust of this paper has been to suggest that any renewal of the theory of agglomeration must embrace the realms of these two topological presuppositions—placing it at a most interesting conceptual boundary in economic geography.

Notes

1. The focus of this paper is upon the theory of agglomeration and not on clusters as defined by Porter (1990) and pursued by others. For the purposes of this paper, clusters are treated as part of the broader phenomenon of the agglomeration of economic activity.
2. Some discussion of these issues is included in Lampard (1955), Scott (1982) and Walker and Lewis (2001).
3. The geographical tradition also drew significantly upon Alfred Weber's discussion of agglomeration factors (Phelps, 1992).
4. Prefiguring the sorts of conclusion being reached in the new economic geography, Robinson noted that

   > Where the costs diminish through external economies, the rate of diminution is likely to be even smaller. For in general the external economies apply to a very small proportion of the total costs of an industry

   and

   > Where ... the final product and the raw materials are alike easily transportable, the optimum industry is likely to be far greater, and the limit to its efficient size will be set, not by gradually increasing diseconomies, but by the increasing transport costs as more distant markets come to be served (Robinson, 1953/1931, pp. 138–140).

5. Cities, because they contain all the functions a city ever has, albeit in a spread-out form ... Edge, because they are far from the old downtowns where little save villages or

farmland lay only thirty years before (Garreau, 1991, p. 4).

6. An edge city: has 5 million square feet of office space; 600 000 square feet of retail space; more jobs than bedrooms; is perceived as a single place by local people; and was nothing like a city as recently as 30 years ago. Despite their novelty, the current growth of such edge cities is argued to come from the expansion of existing businesses (Garreau, 1991).
7. Edgeless cities are ... not even easy to locate because they are scattered in a way that is almost impossible to chart. Edgeless cities spread almost imperceptibly throughout metropolitan areas, filling out central cities, occupying much of the space between more concentrated suburban business districts, and ringing the metropolitan areas' built-up periphery (Lang, 2003, pp. 1–2).
8. Porter's cluster was intended as a way of decomposing national economies. It has further evolved to become

    highly generic in character, being ... sufficiently indeterminate as to admit to a very wide spectrum of ... demand–supply linkages, factor conditions, institutional set-ups, and so on, while at the same time claiming to be based on what are claimed to be fundamental processes of business strategy, industrial organisation and economic interaction (Martin and Sunley, 2003, p. 9).

9. To be precise, it is individuals and individual firms which are able to borrow size. We take Alonso's reference to settlements being able to borrow size to be a shorthand expression of the individuals and individual businesses of given settlements, treated collectively, being embedded in similar networks of interaction through which they borrow size.
10. It is precisely these countries that recent interest in the relational nature of external economies has turned to as examples (for example, see Batten, 1995; Kloosterman and Lambregts, 2001).
11. Parr also notes the potential for different categories of external economies to exert discrete or separate effects upon the clustering of economic activity. The effects of external economies can be said to be separate when one group of firms benefits from one type of externality and a different group of firms benefits from another type of externality.
12. These are: 'bounded regions', around which interest in international political economy continues to focus; 'networks', which are the focus of interest in the likes of business networks and learning regions; 'flows', upon which interest in cultural determinants of the economic are centred; and, the virtual economy of 'two places at once' associated with cyber geography (Thrift and Olds, 1996).

References

ALLEN, J. (1992) Services and the UK space economy: regionalization and economic dislocation, *Transactions of the Institute of British Geographers NS,* 17, pp. 292–305.

ALLEN, J., MASSEY, D. and COCHRANE, A. (1998) *Rethinking the Region.* London: Routledge.

ALONSO, W. (1973) Urban zero population growth, *Daedalus,* 102, pp. 191–206.

AMIN, A. and THRIFT, N. (1992) Neo-Marshallian nodes in global networks, *International Journal of Urban and Regional Research,* 16, pp. 571–587.

AMIN, A. and THRIFT, N. (2003) *Cities: Reimagining the Urban.* Cambridge: Polity Press.

BAIROCH, P. (1988) *Cities and Economic Development: From the Dawn of History to the Present.* London: Mansell.

BATTEN, D. (1995) Network cities: creative urban agglomerations for the 21st century, *Urban Studies,* 32, pp. 313–327.

BEAUREGARD, R. (1995) Edge cities: peripheralizing the center, *Urban Geography,* 16, pp. 708–721.

BENNETT, R. J., GRAHAM, D. J. and BRATTON, W. (1999) The location and concentration of businesses in Britain: business clusters, business services, market coverage and local economic development, *Transactions of the Institute of British Geographers NS,* 24, pp. 393–420.

BINGHAM, R. D. and KIMBLE, D. (1995) The industrial composition of edge cities and downtowns: the new urban reality, *Economic Development Quarterly,* 9, pp. 259–272.

BODDY, M. (1999) Geographical economics and urban competitiveness: a critique, *Urban Studies,* 36, pp. 811–842.

BODDY, M., LAMBERT, C. and SNAPE, D. (1997) *City for the 21st Century? Globalisation, Planning and Urban Change in Contemporary Britain.* Bristol: Policy Press.

BOURNE, L. S. (1975) *Urban Systems: Strategies and Policies for Regulation.* Oxford: Clarendon Press.

BROWN, C. V. and JACKSON, P. M. (1985) *Public Sector Economics*, 3rd edn. Oxford: Blackwell.

CAPELLO, R. (2000) The city network paradigm: measuring urban network externalities, *Urban Studies,* 37, pp. 1925–1945.

CAPELLO, R. (2001) Milan: dynamic urbanisation economies vs. milieu economies, in: J. SIMMIE (Ed.) *Innovative Cities*, pp. 95–128. London: Spon Press.

CHAMPION, A. G. (1988) The reversal of migration turnaround: resumption of traditional trends?,

International Regional Science Review, 11, pp. 253–260.

CHANDLER, A. D. (1966) *Strategy and Structure: Chapters in the History of the American Industrial Enterprise*. New York: Anchor Books.

CHARLESWORTH, J. and COCHRANE, A. (1994) Tales of the suburbs: the local politics of growth in the South East of England, *Urban Studies*, 31, pp. 1723–1738.

CHESHIRE, P. (1995) A new phase of urban development in western Europe? The evidence for the 1980s, *Urban Studies*, 32, pp. 1045–1063.

COE, N. and TOWNSEND, A. R. (1998) Debunking the myth of localized agglomerations: the development of a regionalized service economy in South East England, *Transactions of the Institute of British Geographers NS*, 23, pp. 385–404.

COFFEY, W. and SHEARMUR, R. (2002) Agglomeration and dispersion of high-order service employment in the Montreal metropolitan region, 1981–96, *Urban Studies*, 39, pp. 359–378.

CUMBERS, A., MACKINNON, D. and CHAPMAN, K. (2003) Innovation, collaboration, and learning in regional cluters: a study of SMEs in the Aberdeen oil complex, *Environment and Planning A*, 35, pp. 1689–1706.

DICK, H. W. and RIMMER, P. (1998) Beyond the Third World city: the new urban geography of south-east Asia, *Urban Studies*, 35, pp. 2303–2322.

DURANTON, G. (1999) Distance, land and proximity: economic analysis and the evolution of cities, *Environment and Planning A*, 31(12) pp. 2169–2188.

GARREAU, J. (1991) *Edge City: Life on the New Frontier*. London: Doubleday.

GHENT URBAN STUDIES TEAM (1999) *The Urban Condition: Space, Community and Self in the Contemporary Metropolis*. Rotterdam: 010 Publishers.

GLAESER, E. L. (1998) Are cities dying?, *Journal of Economic Perspectives*, 12, pp. 139–160.

GODDARD, J. B. (1975) *Office Location in Urban and Regional Development*. Oxford: Oxford University Press.

GORDON, P. and RICHARDSON, H. W. (1996) Beyond polycentricity: the dispersed metropolis, Los Angeles, 1970–1990, *Journal of the American Planning Association*, 62, pp. 289–295.

GOTTDIENER, M. (2002) Urban analysis as merchandising: the 'LA School' and the understanding of metropolitan development, in: J. EADE and J. MELE (Eds) *Understanding the City: Contemporary and Future Perspectives*, pp. 159–180. Oxford: Blackwell.

GOTTMAN, J. (1961) *Megalopolis: The Urbanized Northeastern Seaboard of the United States*. Cambridge, MA: MIT Press.

GOTTMAN, J. and HARPER, R. A. (1990) *Since Megalopolis: The Urban Writings of Jean Gottman*. Baltimore, MD: Johns Hopkins University Press.

GRABHER, G. (1993) The weakness of strong ties: the lock-in of regional development in the Ruhr area, in: G. GRABHER (Ed.) *The Embedded Firm: On the Socioeconomics of Industrial Networks*, pp. 255–277. London: Routledge.

HALL, P. (1962) *The Industries of London since 1861*. London: Hutchinson.

HALL, P. (1997) Modelling the post-industrial city, *Futures*, 29, pp. 311–322.

HALL, P. (2000) Global city-regions in the twenty-first century, in: A. J. SCOTT (Ed.) *Global City Regions: Trends, Theory, Policy*, pp. 59-77. Oxford: Oxford University Press.

HARRISON, B. (1992) Industrial districts: old wine in new bottles?, *Regional Studies*, 26, pp. 469–483.

HENDERSON, J. V. (2001) Urban scale economies, in: R. PADDISON (Ed.) *The Urban Studies Reader*, pp. 243–255. London: Sage.

HENRY, N. and PINCH, S. (1999) Spatialising knowledge: placing the knowledge community of Motor Sport Valley, *Geoforum*, 31, pp. 191–208.

HILL, E. and WOLMAN, H. (1997) Accounting for the change in income disparities between US central cities and their suburbs, *Urban Studies*, 34, pp. 43–60.

HILL, E., WOLMAN, H. and FORD, C. C. (1995) Can suburbs survive without their central cities? Examining the suburban dependence hypothesis, *Urban Affairs Review*, 31, pp. 147–174.

HISE, G. (1997) *Magnetic Los Angeles: Planning the Twentieth Century Metropolis*. Baltimore, MD: Johns Hopkins University Press.

HOARE, A. G. (1985) Industrial linkage studies, in: M. PACIONE (Ed.) *Progress in Industrial Geography*, pp. 40–81. Beckenham: Croom Helm.

HOWELLS, J. (1999) Regional systems of innovation?, in: D. ARCHIBUGI, J. HOWELLS and J. MICHIE (Eds) *Innovation Policy in a Global Economy*, pp. 67–93. Cambridge: Cambridge University Press.

HUGHES, H. (1993) Metropolitan structure and the suburban hierarchy, *American Sociological Review*, 58, pp. 417–433.

KARASKA, G. (1969) Manufacturing linkages in the Philadelphia economy: some evidence of external agglomeration forces, *Geographical Analysis*, 1, pp. 354–369.

KEEBLE, D. and TYLER, P. (1995) Enterprising behaviour and the urban–rural shift, *Urban Studies*, 32, pp. 975–997.

KLOOSTERMAN, R. C. and LAMBREGTS, B. (2001) Clustering of economic activities in polycentric urban areas: the case of the Randstad, *Urban Studies*, 38, pp. 717–732.

KRUGMAN, P. (1996) *The Self-organising Economy*. Oxford: Blackwell.

KRUGMAN, P. (2000) Where in the world is the 'new economic geography', in: G. L. CLARK, M. P. FELDMAN and M. S. GERTLER (Eds) *The Oxford Economic Handbook of Economic Geography*, pp. 49–60. Oxford: Oxford University Press.

KUNSTLER, J. H. (1993) *The Geography of Nowhere*. London: Touchstone.

LAMPARD, E. E. (1955) The history of cities in the economically advanced areas, *Economic Development and Cultural Change,* 3, pp. 81–136.

LANG, R. E. (2003) *Edgeless Cities: Exploring the Elusive Metropolis*. Washington, DC: Brookings Institution.

LEVER, W. (1972) Industrial movement, spatial association and functional linkages, *Regional Studies,* 6, pp. 263–280.

LEWIS, R. (1999) Running rings around the city: North American industrial suburbs, 1850-1950, in: R. HARRIS and P. LARKHAM (Eds) *Changing Suburbs*, pp. 146–167. London: Spon Press.

MACKINNON, D., CUMBERS, A. and CHAPMAN, K. (2002) Learning, innovation and regional development: a critical appraisal of recent debates, *Progress in Human Geography,* 26, pp. 293–311.

MALMBERG, A. (1997) Industrial geography: location and learning, *Progress in Human Geography,* 21, pp. 573–582.

MALMBERG, A. and MASKELL, P. (2002) The elusive concept of localisation economies – towards a knowledge-based theory of spatial clustering, *Environment and Planning A,* 34, pp. 429–449.

MARKUSEN, A. (1996) Sticky places in slippery spaces: a typology of industrial districts, *Economic Geography,* 72, pp. 293–313.

MARSHALL, A. (1920) *Principles of Economics*. London: MacMillan.

MARTIN, R. (1999) The new 'geographical turn' in economics: some critical reflections, *Cambridge Journal of Economics,* 23, pp. 65–91.

MARTIN, R. and SUNLEY, P. (1996) Paul Krugman's geographical economics and its implications for regional development theory: a critical comment, *Economic Geography,* 72, pp. 259–292.

MARTIN, R. and SUNLEY, P. (2003) Deconstructing clusters: chaotic concept or policy panacea, *Journal of Economic Geography,* 3, pp. 5–35.

MASKELL, P. (1998) Low-tech competitive advantages and the role of proximity: the Danish wooden furniture industry, *European Urban and Regional Studies,* 5, pp. 99–118.

MASKELL, P. and MALMBERG, A. (1999) Localised learning and industrial competitiveness, *Cambridge Journal of Economics,* 23, pp. 167–185.

MASKELL, P., ESKELINEN, H., HANNIBALSON, I. ET AL. (1998) *Competitiveness, Localised Learning amd Regional Development: Specialisation and Prosperity in Small Open Economics*. London: Routledge.

MASSEY, D., ALLEN, J. and HENRY, N. (1997) *Rethinking the Region*. London: Routledge.

MCCANN, P. (1995) Rethinking the economics of location and agglomeration, *Urban Studies,* 32, pp. 563–577.

MOGRIDGE, M. and PARR, J. B. (1997) Metropolis or region? On the development and structure of London, *Regional Studies,* 31, pp. 97–116.

OHULLACHAIN, B. and SATTERTHWAITE, M. A. (1992) Sectoral growth patterns at the metropolitan level: an evaluation of economic development incentives, *Journal of Urban Economics,* 31, pp. 25–58.

PARK, S. O. and MARKUSEN, A. (1994) Generalizing new industrial districts: a theoretical agenda and an application from a non-Western economy, *Environment and Planning A,* 27, pp. 81–104.

PARR, J. B. (2002a) Missing elements in the analysis of agglomeration economies, *International Regional Science Review,* 5, pp. 151–168.

PARR, J. B. (2002b) Agglomeration economies: ambiguities and confusions, *Environment and Planning A,* 34, pp. 717–731.

PHELPS, N. A. (1992) External economies, agglomeration and flexible production, *Transactions of the Institute of British Geographers NS,* 17, pp. 35–46.

PHELPS, N. A. (1998) On the edge of something big: edge city economic development in Croydon, *Town Planning Review,* 69, pp. 441–465.

PHELPS, N. A. and OZAWA, T. (2003) Contrasts in agglomeration: proto-industrial, industrial and post-industrial forms compared, *Progress in Human Geography,* 27, pp. 583–604.

PINCH, S. and HENRY, N. (1999) Paul Krugman's geographical economics, industrial clustering and the British motor sport industry, *Regional Studies,* 33, pp. 815–828.

PORTER, M. (1990) *The Competitive Advantage of Nations*. Basingstoke: Macmillan.

PRED, A. (1976) *City Systems in Advanced Societies*. London: Hutchinson.

RACO, M. (2002) Assessing the discourses and practices of urban regeneration in a growing region, *Geoforum,* 34, pp. 37–55.

RICHARDSON, R. and MARSHALL, J. N. (1996) The growth of telephone call centres in peripheral areas of Britain: evidence from Tyne and Wear, *Area,* 28, pp. 308–317.

ROBINSON, E. A. G. (1953 [1931]) *The Structure of Competitive Industry*. Cambridge: Cambridge University Press.

SAVITCH, H. V. (1995) Straw men, red herrings ... and suburban dependency, *Urban Affairs Review,* 31, pp. 175–179.

SAXENIEN, A. (1983) The urban contradictions of

silicon valley: regional development and restructuring of the semiconductor industry, *International Journal of Urban and Regional Research,* 7, pp. 237–262.

SCOTT, A. J. (1982) Locational patterns and dynamics of industrial activity in the modern metropolis, *Urban Studies,* 19, pp. 111–142.

SCOTT, A. J. (1983) Industrial organisation and the logic of intra-metropolitan location, i: theoretical considerations, *Economic Geography,* 59, pp. 233–250.

SCOTT, A. J. (1986) Industrial organisation and location: division of labour, the firm and spatial process, *Economic Geography,* 62, pp. 215–231.

SCOTT, A. J. (1988) Flexible production systems and regional development: the rise of new industrial spaces in North America and western Europe, *International Journal of Urban and Regional Research,* 12, pp. 171–186.

SCOTT, A. J. (1997) The cultural economy of cities, *International Journal of Urban and Regional Research,* 21, pp. 323–339.

SCOTT, A. J. and STORPER, M. (1987) High technology industry and regional development: a theoretical critique and reconstruction, *International Social Science Journal,* 112, pp. 215–232.

SENN, L and GORLA, G. (1999) Networking strategies as a factor in urban decentralization, in: A. A. SUMMERS, P. C. CHESHIRE and L. SENN (Eds) *Urban Change in the United States and Western Europe: Comparative Analysis and Policy,* pp. 243–262. Washington, DC: Urban Institute Press.

SIMMIE, J. and SENNETT, J. (1999) Innovative clusters: global or local linkages?, *National Institute Economic Review,* 170, pp. 87–98.

SKITOVSKY, T. (1954) Two concepts of external economies, *Journal of Political Economy,* 59, pp. 185–193.

SOJA, E. and SCOTT, A. J. (1996) Introduction to Los Angeles: city and region, in: A. J. SCOTT, and E. SOJA (Eds) *The City: Los Angeles and Urban Theory at the End of the Twentieth Century*, pp. 1–21. Berkeley, CA: University of California Press.

STANBACK, T. M. (1995) Putting city-suburb competition in perspective, in: J. BROTCHIE, M. BATTY, E. BLAKELY *ET AL.* (Eds) *Cities in Competition: Productive and Sustainable Cities for the 21st Century*, pp. 208–225. Melbourne: Longman Australia.

STORPER, M. (1995) The resurgence of regional economies, ten years later: the region as a nexus of untraded interdependencies, *European Urban and Regional Studies,* 2, pp. 191–222.

STORPER, M. (1997) *The Regional World: Territorial Development in a Global Economy*. London: Guilford Press.

STORPER, M. and HARRISON, B. (1991) Flexibility, hierarchy and regional development: the changing structure of industrial production systems and their forms of governance in the 1990s, *Research Policy,* 20, pp. 407–422.

STORPER, M. and WALKER, R. (1989) *The Capitalist Imperative: Territory, Technology and Industrial Growth.* Oxford: Blackwells.

SUAREZ-VILLA, L. (1989) Polycentric restructuring, metropolitan evolution, and the decentralization of manufacturing, *Tijdschrift voor Economische en Sociale Geografie,* 80, pp. 194–205.

SUAREZ-VILLA, L. and WALROD, W. (1997) Operational strategy, R&D and intra-metropolitan clustering in a polycentric structure: the advanced electronics industries of the Los Angeles basin, *Urban Studies,* 34, pp. 1343–1380.

TAYLOR, P. J. (1999) Places, spaces and Macy's: place–space tensions in the political geography of modernities, *Progress in Human Geography,* 23, pp. 7–26.

THORNGREN, B. (1970) How do contact systems affect regional development, *Environment and Planning A,* 2, pp. 409–427.

THRIFT, N. and OLDS, K. (1996) Refiguring the economic in economic geography, *Progress in Human Geography,* 20, pp. 311–337.

TIEBOUT, C. M. (1956) A pure theory of local expenditures, *Journal of Political Economy,* 61, pp. 416–424.

TORNQVIST, G. (1970) *Contact Systems and Regional Development.* Lund Studies in Geography (B) No. 35.

WALKER, R. (1981) A theory of suburbanization: capitalism and the construction of urban space in the United States, in: M. DEAR and A. SCOTT (Eds) *Urbanization and Urban Planning in Capitalist Societies*, pp. 383–429. London: Methuen.

WALKER, R. and LEWIS, R. D. (2001) Beyond the crabgrass frontier: industry and the spread of North American cities, 1850-1950, *Journal of Historical Geography,* 27, pp. 3–19.

WEBBER, M. M. (1963) Order in diversity: community without propinquity, in: L. WINGO (Ed.) *Cities and Space: The Future Use of Urban Land.* Baltimore, MD: Johns Hopkins University Press.

WEBBER, M. M. (1964) The urban place and the nonplace urban realm, in: M. M. WEBBER, J. W. DYCKMAN, D. L. FOLEY *ET AL. Explorations into Urban Structure*, pp. 23–56. Philadelphia, PA: University of Pennsylvania Press.

WISE, M. J. (1949) On the evolution of the jewellery and gun quarters in Birmingham, *Transactions of the Institute of British Geographers,* 15, pp. 57–72.

YOUNG, A. (1928) Increasing returns and economic progress, *Economic Journal,* 38, pp. 527–542.

The Cluster as Market Organisation

Peter Maskell and Mark Lorenzen

[Paper first received, April 2003; in final form, November 2003]

> In a world of strangers ... a new sucker arrives every minute ... The dealings of strangers are subject to social norms. In a world of Hobbesian asocial nomads the next stranger you meet would just as well shoot you as shake your hand. That is why the airlines are crowded with business travellers, on their way to making friends (McCloskey, 1994, p. 373).

1. Introduction

The aim of this paper is to investigate the spatial configurations of the economy that are particularly geared towards the creation, transfer and usage of knowledge. Knowledge as an economic input and as a commodity in its own right becomes crucial as the process of globalisation gradually renders most traditional factors of production increasingly ubiquitous—i.e. equally available to all competitors regardless of location (Maskell and Malmberg 1999, Maskell, 2001a). The paper pays special attention to the processes whereby the emerging knowledge economy translates into regional development patterns and patterns of interaction among business firms commonly associated with industrial 'clusters'.[1]

There are many competing schools of thought concerned with industrial clusters but they all agree that this real-life phenomenon has to do with the co-localisation of separate economic entities, which are in some sense related, but not joined together by any common ownership or management. In spite of this basic accord, no general understanding has yet emerged regarding the paramount reason why the separate entities became co-localised in the first place, what has made them stick together, what the effects may be and—at an even more basic level—why this matters at all.

The present paper addresses these fundamental questions related to industrial clusters by arguing that markets become organised in order to carry out certain kinds of economic activities efficiently. The paper proposes that the spatial arrangement of industrial clusters is in essence a market, organised in a particularly helpful way.

The paper is structured as follows. In section 2, we outline the two basic building-blocks of the modern exchange-economy—firms and markets—and their interaction. Section 3 investigates how markets become organised as rent-seeking firms build network relations to reduce uncertainty and create knowledge while keeping transaction costs at bay. Other arrangements are wanted when network-building is not a feasible strategy. The paper suggests that clustering provides such an alternative, coexisting with the non-spatial elements of network organisation. Section 4 focuses on the geographically delimited clusters and the kinds of highly flexible interfirm relations they support. Section 5 offers an empirical illustration of the paper's theoretical argument, sketching out the market organisation and spatial configuration of the furniture and pop music industries, respectively. While these industries are differently influenced by the emerging knowledge-based economy, they are also differently organised in terms of the importance of networks and clusters in general and the usage of various cluster benefits in particular. Section 6 concludes the paper by summing up the main arguments.

2. Firms and Markets

So commonly used, and so familiar to most of us, are the two concepts of firms and markets and so intimately are they jointly associated with the modern, globalising,[2] knowledge-based, exchange economy that we sometimes tend to forget that they are social innovations made centuries ago, independently of each other, and only recently combined. Markets seem to have emerged at least three millennia ago as temporary meeting-places for the exchange of gossip and produce among individuals (families, clans) but only within the past two centuries were they institutionalised and regulated with the explicit aim of enhancing competition. The existence of a market in this modern sense is by no means self-evident, as Williamson (1983) is sometimes taken to suggest.[3] Modern markets are, on the contrary, the result of careful deliberation and conscious effort combined with less rationally planned innovations (von Hayek, 1937, 1945; Polanyi *et al.*, 1957). Even the briefest spot transaction between partners at arm's-length normally takes place in markets that have nowadays become highly institutionalised.

Firms were conceived much later than markets. The word 'firm' was first recorded in 1574, deriving from *firmare* meaning 'to ratify by signature'. However, it appears that the lawyer Sinibald Fieschi, who in 1243 became Pope Innocent IV, was the first to use the phrase *persona ficta* when assigning rights and obligations to a strictly fictional entity. He thereby included the idea in Roman law that property rights could be executed, assets assembled, capital accumulated and contracts made by purely legal entities of limited liability and infinite lifetime (Maskell, 2001c). The subsequent dissemination of this legal innovation—and an innovation it certainly was—has had momentous consequences for the emergence of capitalism as the firm gradually became the main instrument and organisational form for co-ordinating an increasing part of all economic activity carried out by individual members of society.

Contemporary rent-seeking firms exchange plans and ideas as well as ownership of bundles of commodities in markets that provide the institutional framework—or setting (Loasby, 1994)—for such events of co-ordination and exchange.[4]

The boundary between the realm of the market and the realm of the firm remained obscure until the summer of 1932 when the later Nobel laureate Ronald Coase discovered why individuals sometimes prefer to utilise their skills under an employment contract rather than as free entrepreneurs (Simon, 1982; Loasby, 2000). The answer, Coase realised, was that there were costs when using the price mechanism of the market, and that these costs—later to become generally known as 'transaction costs' (Coase, 1937, 1992; Williamson, 1975, 1996)—sometimes exceed the costs of organising things under the authority of a single leadership within the framework of a firm.

His findings resulted in a steadily swelling interest in determining precisely which activities a firm may profitably combine under its managerial authority and which are best left to be produced by others and acquired through the market. While the questions of what to make and what to buy have become probably the most studied phenomena in all of economics during recent decades, far less attention has been paid to the related question of how activities divided between firms are subsequently combined to create something useful.

Hence our concern in this paper is exclusively with how firms maximise the benefits of external relations in general and, in particular, how firms through market relations:

(1) become more knowledgeable about present and future possibilities and requirements and transform this knowledge into product and process innovations (knowledge creation); and
(2) compensate for demand and supply fluctuations, thus helping to secure the efficient usage of resources that each firm has committed in production (resource efficiency)

While market relations offer opportunities for creating knowledge and compensating for fluctuations, these benefits may be offset by transaction costs. Such costs include costs related to the alignment of incentives and plans—i.e. search and information costs carried by the firm before any potential business partner is identified; the bargaining and decision costs before engaging in a transaction; and, the policing and enforcement costs to ensure that what was agreed on is actually delivered (Dahlman, 1979). They also include communication costs related to cognitive distance—i.e. the costs stemming from misunderstandings between and among firms with different mindsets or expectations, or the costs of bringing down such cognitive differences through negotiations and sharing of information (Lorenzen and Foss, 2003).[5]

To maximise one or both benefits of market relations while keeping transaction costs down, firms participate in the creation of institutions—they organise the market. The form of market organisation they will choose depends, *ceteris paribus*, upon the characteristics of the industry they are in.

Firms that operate in certain kinds of market, particular in standard goods, tend to prefer the freedom of spot-transactions, choosing different partners with each purchase based on competitive bidding when dealing with the 'buy'-related issues. However, in many other cases, firms find that more can be gained if the singular events are replaced in full or in part by commitments of repeated transactions.

Firms realising by conjecture or experimentation that there are disadvantages when playing the spot-market tend to choose between two exemplary strategies: they engage in network formation when their set of customers, suppliers and products is reasonably stable and in clustering when it is not.[6] The choice thus depends on the degree of expected relational stability that in turn reflects cardinal variance in industry uncertainty (Knight, 1921) in terms of unforeseeable changes in technology, supply and demand.[7] Regardless of which strategy is followed, the departing from spot-market interactions results in a structuring of the market into 'them' and 'us'. Firms become 'insiders' by investing in building or joining networks in cases of relatively low industry uncertainty, or cluster-building in cases of relatively high industry uncertainty.

This process, and the benefits associated with it, will be outlined in the following sections, starting with the building of networks.

3. Organisation of Markets through Network Building

If, and only if, the group of upstream and downstream partners is reasonably stable can firms hope to benefit from following a network strategy. Initiating 'dyadic' relations (Demsetz, 1968; Wilson, 1975) with a stable set of partners can reduce transaction costs to a degree where the benefits exceed the costs when attempting to achieve knowledge creation and/or resource efficiency.

If successful, such dyadic relationships are intensified as both parties contribute in strengthening the bilateral links and dilate their scope to involve several or all layers of the two organisations (Ford, 1990; Sabel, 1992). Each new link created, each new experience with the partner's peculiar ways and each new routine and convention (Egidi, 1995) facilitates future exchange and makes the interaction function easier. Through continuous interaction and information-sharing, former misunderstandings and suspicions are gradually eliminated and the interaction can encompass a still wider range of plan alignment and joint knowledge creation. Step by step, the cognitive distance is diminished as emerging 'code-books'—shared languages and ways of communicating and understanding information (Cowan *et al.*, 2000; Lissoni, 2001)—increase firms' cognitive abilities to co-ordinate activities and plans or share knowledge, even if (partly) tacit. Over time, the repeated interactions can also give rise to incremental learning of considerable significance for the overall competitiveness of the firms involved (Lundvall, 1985).

Once the element of relation-specific sunk costs is large enough,[8] a qualitative change takes place, as the scope for opportunistic behaviour (Williamson, 1975) becomes negligible. In a dyadic relation, the accumulated sunk cost makes the partners behave as if they *trust* each other (Ben-Porath, 1980).[9] As Granovetter shrewdly noticed

> Malfeasance is here seen to be averted because clever institutional arrangements make it too costly to engage in, and these arrangements—many previously interpreted as servicing an [other] economic function—are now seen as having evolved to discourage malfeasance. Note, however, that they do not produce trust but instead are a functional substitute for it (Granovetter, 1985, p. 489).

Trust will thus characterise a relation between business firms when each is confident that the other's present value of all foreseeable future exchanges exceeds the possible benefits of breaking the relation. The larger the sunk costs, the greater the confidence and the trust. And trust is a remarkably efficient lubricant to economic exchange, assuaging the friction for interaction, co-operation or exchange. Risks of committing resources inefficiently are reduced when crucial information about changes in demand and supply is volunteered. Sharing risks is especially important when confronted with the need to make great leaps into new business areas or when challenging the established ways of conducting business (Loasby, 1991). Moving forward together by sharing risks also secures against investing in procedures or technologies not acknowledged by others (Richardson, 1998). New markets can be developed and new solutions found when firms share information without fearing that they might never receive their fair share of a potential gain (Fukuyama, 1995).[10]

One important consequence of high levels of relation-specific sunk costs is that the flow of knowledge between the two business partners does not have to be strictly reciprocal or take place at precisely the same time. The overall exchange of knowledge is intensified and deepened when business partners believe that some piece of knowledge offered free of charge today will be repaid at some later moment in some way or another (von Hippel, 1987).

Networks may expand from dyads to encompass a growing number of partners. By utilising investments in already-existing relationships as channels to new partners (your-friend-is-my-friend), firms successfully minimise their search costs as well as other transaction costs when expanding their business network (Håkansson and Snehota, 1989). With the risk of severing carefully built bonds to intimate business associates if they misbehave, the participants in an expanding network are placed in a situation where any infringement of trust is so severely penalised that, in effect, malfeasance becomes a non-option (Roscher, 1989; Casson, 1991).

In the global arm's-length spot market for standard goods, where all customers and all suppliers can easily be substituted, an unsatisfied customer has no way of reaching all potential future buyers and opportunistic behaviour can therefore continue indefinitely. Not so in the business network, where any such wrong-doing will soon be known by all.

The collective awareness of this mechanism makes it possible to keep transaction costs down while attaining resource efficiency and exchanging knowledge to an extent no outsider can aspire to achieve (von Hippel, 1987).

Many of the benefits of networking may also become available to non-networking firms that are able to utilise a contrasting form of market organisation: the cluster. To this we shall now turn.

4. Organisation of Markets through Cluster Formation

Industry uncertainty implies that not all industries are characterised by reasonably stable sets of suppliers, customers and products. With high levels of uncertainty, it makes little sense for firms to engage in network-building with what will soon become yesterday's partners. Firms finding themselves in such circumstances tend instead to opt for a strategy of being a stakeholder in a *cluster*. Within a cluster, the structuring of markets usually takes place with the participation of more 'insider' firms and on a broader level than if embedded in a business network only. The extended range of 'insiders' with their own capabilities and resources allows for experimentation, flexibility and the use of shifting combinations of partners without carrying the full burden of spot-market transaction costs.

Firms opting for co-localisation strategies participate in building communities that share institutions, just like in networks, but the cluster institutions are often confined in scope (usually applicable to firms in one or a few related industries only) and always restricted in space (usually to be found on a local or regional level only).

If industry ambiguity[11] prevails and firms need regularly to redefine vital aspects of their product, project relations become the common form of interaction.[12] This is, for instance, the case in industries dominated by non-continuous production (such as construction), highly ambiguous consumer tastes (such as fashion, film, music) or customisation (advertisement, consultancy, etc.) where firms form projects to find solutions to a specific customer's demand within a definitive period (Goodman and Goodman, 1976; Ekstedt *et al.*, 1999; Davis and Brady, 2000; Hobday, 2000; Grabher, 2002; Engwall, 2002).[13]

In industries with less market ambiguity but moderate to high levels of industry uncertainty, firms frequently need to expand or reduce their output or adjust their product range to provide variety. In such industries, the common form of interaction is flexible relations where partners can be found or dropped according to the current market situation. The resulting freedom in choice of partners facilitates resource efficiency under industry uncertainty and is crucial for the competitiveness and survival of especially small firms faced with volatile markets and calls for perpetual variation. Along the vertical axis of the value chain, the clustered firms meet demand changes in the market by shifting between specialised suppliers and sometimes by using a new specialised supplier for a one-off delivery. Along the horizontal axis, firms may also level out the effects of a volatile demand volume by passing on tasks in case of excessive demand and supplying to others in times of low demand.

Table 1. Market organisation

Networks	Clusters
Institutional 'arrangement'	Institutional 'environment'
Firms as shareholders	Firms as stakeholders
Strong ties	Weak ties
Club institutions	Social institutions
Trust, sunk costs	Social trust, reputation
Codebooks	Social codebooks

Project relations and flexible relations among clustered firms both rely on collective institutions helpful in reducing transaction costs without imposing high switching costs precisely because they are based on social traits rather than on the idiosyncratic, partner-specific investments and dyadic sharing of information that characterise networks (as described above in section 3).[14]

Because the participants in projects and flexible relations originate from different firms and only work together temporarily, they have few incentives and little time to develop the kind of co-ordination mechanisms that emerge in stable business networks. However, the sheer clustering of firms and people favours the creation of 'weak ties' (Granovetter, 1973) that bring down information costs for the 'insiders' of the cluster by spreading information through meetings, gossip and direct observation. Over time, clustering supports cognitive alignment—i.e. 'social codebooks' of a communal social culture including collective beliefs, values, conventions and language, that significantly assist firms in obtaining and understanding information (Lorenzen and Foss, 2003). Clustering can also foster the additionally helpful collective institution of social trust (a general willingness to trust other 'insiders'), that is strengthened by the abundant information and the necessity of maintaining a high reputation of reliability to remain an 'insider' (Lorenzen, 2002).

Much exchange and creation of knowledge in a cluster will follow market relations along the vertical dimension of the cluster (Håkansson, 1987) or through external 'pipelines' (Bathelt *et al.*, 2002), but more is, perhaps, created along the horizontal dimension of clusters, consisting of firms with similar capabilities and performing alike tasks in parallel (Maskell, 2001b, 2001d). While it might be easy for firms to blame inadequacies on local or national market factors when confronted with the superior performance of competitors located far away, it is impossible to do so when the premium producer is located down the street. It is by watching, discussing and comparing dissimilar solutions that clustered firms become engaged in the process of continuous improvements on which their survival depends (Maskell, 2001d). Promising avenues identified by one firm become available to others without ever reaching the public domain (Marshall, 1919; Mansfield, 1985). Table 1 summarises the argument made so far.

5. Empirical Illustration: The Pop Music and Furniture Industries

The framework set out above allows us to explain the geographical organisation (i.e. tendency towards clustering) of different industries by investigating how industry uncertainty and ambiguity translate into firm strategies of participating to organising markets in clusters. To illustrate, this section presents empirical evidence from two clustered industries that are different in terms of uncertainty especially regarding the relative importance of resource efficiency vs knowledge creation: the EU pop music and furniture industries.

The two industries are both quite important in terms of employment and value added. In spite of the current alleged 'crisis', the global pop music industry has grown by 35 per cent during the past decade, with turnovers rising from US$27 billion to US$37 billion (IFPI, 2001), primarily due to technological and stylistic innovations and globalising markets. Even if a handful of major US-based record companies and publishing houses dominate the global industry, local firms in virtually all European countries are profit-earners in their own right and serve an important role by creating a continuous stream of artistic inputs to the global players (Power and Hallencreutz, 2002).

The EU furniture industry, accounting for just about half the global furniture sales, is one of the largest manufacturing industries in the EU and continues to enjoy high employment, outputs and exports (Lorenzen, 1998; Maskell *et al.*, 1998). In 2000, the EU furniture industry had a turnover of US$96.6 billion and the industry's almost 90 000 firms employed almost 900 000 persons in 1998 (UEA, 2003).

For a traditionally labour-intensive, low-tech industry, it is noteworthy that, even as the EU furniture industry now faces intense and growing price competition from countries in the former Soviet bloc and the Third World, its trade balance is still positive. In terms of absolute size, the major EU furniture exporters are Germany and Italy, but relatively small countries like the Netherlands, Portugal, Sweden and Denmark all have export rates well above 30 per cent (Lorenzen, 2004).

Below, we shall investigate the reasons behind clustering in the two industries, first by looking at how the two markets are organised.

The Pop Music Industry

The activities within the pop music industry are very diverse with high degrees of specialisation and disintegration as well as important differences in terms of competencies. Most important is, perhaps, the difference between 'artistic' and 'humdrum' competencies (highlighted by Caves, 2000), the first being populated with freelance artists and small 'artistic' firms, the second by firms sourcing, marketing and selling pop music consumer products like CDs, but also scores (sheet music) and tunes or jingles for mobile phones, etc. (Andersen and Miles, 1999). While focusing on the latter, 'humdrum' pop music firms,[15] we exclude from the analysis firms carrying out a range of 'input' activities (OECD, 1997; Castaner and Campos, 2002) where internal economies of scale and in-house skills are significantly more important for competitiveness than market relations.[16]

Even within the 'humdrum' segment, there are competence divides. In fact, the segment is broken up into a number of specialised firms with dense and complex vertical market relations, because sourcing, marketing and selling music require competencies that are not easily integrated. These activities usually involve (at least) event and concert firms, media firms, AD (art direction) firms, distributors, retailers, publishing houses, and financial and legal services, as well as record companies.

The market organisation of the industry reflects the high degree of uncertainty in the global demand for the main product: the pop music CD (Huygens *et al.*, 2002; Lopes, 1992). First, the product cycles for CD albums are usually very short (and even briefer for CD singles). Secondly, consumer tastes are highly unpredictable even for the most skilled and sagacious within the business (Shuker, 2001). Since nobody knows which CD is going to be a success, and since very few products sell on a large scale (Negus, 1992; Vogel, 1998), the strategy applied by end producers of music (i.e. record companies) is to ensure a steady stream of novel products. Any release of music CDs is in a sense an open-ended search process, where new products need to be tested on uncertain consumer markets, over limited test periods. The same goes, albeit to a lesser extent, for new national penetration efforts or marketing methods.

Product innovation in the pop music industry is a process of knowledge creation under uncertainty, through experimentation (Lorenzen and Frederiksen, 2003). In order to facilitate such knowledge creation, pop music firms use interfirm projects. Each new act of knowledge creation—i.e. each new experiment of producing, marketing and selling a new CD—is thus built on a new market relation, combining a multitude of partners, including at the very least a record company, one or more artists (some of whom are signed for only one CD), a publisher, an AD provider and often, in addition, media firms and event firms (some of which are also often used only once or twice).

The market relation is designed as a temporary project relation, with some partners (such as AD, media and event firms) participating only in parts of the process and others remaining at the heart of the relation for the project's entire lifespan. The artist remains in the project throughout, as (s)he is needed not only for creating the musical content of the CD, but also for marketing it through live and video performances. The publisher takes care of payments (royalties) to the artist and record company after the project is over, but may also sometimes be actively involved in signing artists and finding music content. The most important agent, however, is the record company, which establishes and co-ordinates the project and is involved in all aspects of it. The record company first signs artists—the source of musical content in CDs—and then 'pushes' the music through the other parts of the value-adding process by signing on firms with supplementary competencies and visions. Some record companies—particularly, the branches of the global majors (EMI, BMG, Sony, Universal and Warner)—do so by virtue of their dominating position in the project, but in other cases where power relations are more symmetrical, co-ordination also hinges on trust among project participants.

There are also often horizontal project relations *among* record companies, because some of them have superior marketing and distribution competencies in certain fields, while others often are more competent in discovering and signing artists. In many national markets, the branches of the major global record companies use smaller national independent firms as 'external R&D labs'. Such local firms may release national artists themselves, while artists perceived to hold great (global) sales potential are also licensed to major firms in order to utilise their larger marketing power and global distribution channels and sales networks (Darmer, 1999; Power and Hallencreutz, 2002; STEP, 2003).

Over time, project relations in pop music organise the market into relatively stable project clusters, where record companies, publishers, AD, media and event firms keep most project relations local when producing new CDs (Power and Lundequist, 2002; STEP, 2003). Whereas independent record companies, AD and event firms participate in the creation and marketing of a new CD and keep their relations within clusters, major record companies, while participating in clusters, also deal with global distribution and sales and hence often function as gatekeepers with respect to relations outside the clusters (Power and Hallencreutz, 2002; STEP, 2003).

Within such pop music clusters, new project partners can easily be found, because many have worked together in earlier projects. As a result of the high number of finished projects in pop music project clusters, the people who are skilled in co-ordinating music projects (i.e. the 'creatives' or persons responsible for 'A&R' (artist and repertoire) in the record companies, as well as a number of independent project co-ordinators) are all found within the cluster. Such experienced people, with know-how and know-who specialised in pop music projects, are central to the co-ordination (and sometimes also the initiation) of CD projects and are always in high demand.

Pop music clusters are typically found in the major cities of the world (Scott, 1999, 2000).[17] Here, we find national branches of major international record companies and publishers, the bulk of AD, media and event firms plus related legal and financial services, as well as many independent record companies and artists.[18] The record companies alone are often found within a few hundred metres of each other, in the city cores or in other high-prestige areas of the urban cluster (STEP, 2003).[19]

One effect of the clustering of project clusters within the pop music industry is of course that it lowers time costs when running projects. However, the major positive effect of the clustering of people with accumulated experience with project co-ordination is their many weak ties, which make information about people's and firms' skills and availability accessible to all local firms. Furthermore, a relatively high level of social trust,[20] facilitated by frequent interaction and information-sharing among people within and around the pop music industry, lowers transaction costs when new CD projects are initiated (Power and Lundequist, 2002; Lorenzen and Frederiksen, 2003).

The market organisation of pop music in clusters is characterised by many 'urbanisation economies' that supplement market relations in attracting pop music firms to cities. Artistic inspiration and stylistic information related to production and marketing of pop music heavily depend on global pipelines (Bathelt *et al.*, 2002) of people and information, and such pipelines are in practice unavoidably urban. Specialised educational institutions supplying new artistic talent, such as conservatories or management schools offering 'project management' or 'music management' courses targeted at the pop music industry, are only located in major cities. Furthermore, qualified labour is attracted by the diversity and global nature of large cities and its skills are enhanced there. As many project co-ordination competencies can only be acquired through hands-on experience, clustering of pop music labour markets in cities is important, as this allows people to acquire experience through moving between jobs in different local pop music firms (primarily, record companies).

However, pop music clusters also possess distinct 'locational economies' (Malmberg and Maskell, 2002), related directly to local clusters of firms with project relations—namely, weak ties among people and firms with experience and competencies—with distinct information and social trust effects. Such collective institutionalisation is the most important aspect of market organisation in pop music clusters.

The Furniture Industry

As in the music industry, it is furniture firms carrying out the end activities in the value chain[21] that have the highest tendency to cluster, but their reasons for clustering are very different from those of pop music firms. In spite of increased global competition, the size structure of the European producers of finished wooden or upholstered furniture and specialised components[22]—henceforth referred to merely as 'furniture' firms—seems relatively unaffected by the wave of vertical integration that has swept through other industries. In the furniture firms, internal scale and scope economies are still rather limited, even though a slowly growing number of European producers have managed to target export markets on a large scale with mass-produced (typically pine wood) furniture. The furniture segment hence continues to consist of SMEs: specialised suppliers and end-producers, with market relations along the value chain, as well as horizontally interwoven (Lorenzen, 1998; Maskell, 1998; Maskell *et al.*, 1998). To a very high degree, these market relations rest on the organisation of markets by clustering. To understand why, we need to look at the demand for furniture.

As in many other consumer industries, price matters. The increasing global price competition forces furniture firms to increase cost efficiency. At the same time, globalisation allows these firms to scan world markets for cheap suppliers of standard inputs (such as textiles and unspecialised components), which can be bought in bulk. For activities aiming at reducing production costs, the furniture firms use spot-market relations that are not localised, but include many far away producers. In order to escape the fiercest head-on price competition, the majority of EU furniture firms apply strategies that focus on less price-sensitive market segments (Maskell, 1998).

The export markets for EU furniture firms have a medium-level demand uncertainty in the guise of unpredictable volume fluctuations combined with a constantly increasing demand for product varieties. Furniture producers need to offer increasingly wide ranges of varieties of their models (in some cases, customise them to one or very few customers) and furthermore, have to be able to deliver quickly and efficiently. Average delivery times are forced down from months to weeks as retailers' and end-customers' tolerance towards delays rapidly evaporate. In other words, competitiveness is increasingly based on furniture firms' ability to deliver a growing range of product varieties, while maintaining high resource efficiency and time efficiency.

For the bulk of furniture producers, this is done by maintaining a portfolio of flexible relations to specialised suppliers. By being able to call on different suppliers at very short notice, end-producers can combine different inputs, allowing them to deliver furniture with quite different characteristics, at delivery times that are often considerably shorter than their large, integrated and automated competitors.[23] Vertical flexible relations are supplemented by less frequent horizontal flexible relations, where firms pass on excess orders to each other.

To maintain short delivery times and flexibility, firms depend on an organisation of the market in clusters, where furniture firms use each other again and again for both vertical and horizontal flexible relations, sometimes functioning as suppliers, sometimes as customers, sometimes giving favours, sometimes receiving them. Within such clusters of flexible relations, furniture firms also often help others to start up, creating new, specialised firms through spin-offs to the benefit of the entire cluster.

The EU furniture industry is highly clustered.[24] Producers/assemblers of finished furniture and components are often found in non-urban industrial districts, where a high rate of spin-offs from existing firms has created a significant co-location of producers, often in geographical areas spanning only tens of kilometres and a handful of villages. Some of these clusters are found in regions with long craft traditions, like in Italy (Bambi, 1998; Lojacono and Lorenzen, 1998), some in rather recently industrialising (but still predominantly rural) areas, like in West Denmark (Lorenzen, 1998).

In such furniture clusters, low time and transport costs of course ease both vertical and horizontal flexible relations ('just-in-time'), but the most important aspect of clustering is, as in the pop music industry, weak ties. Within furniture districts, being of small geographical size, weak ties (in professional associations as in civic life) are frequent among managers and workers. Apart from lowering search costs (updating local firms on the capacity and quality levels of all local suppliers of furniture components), the weak ties also create transaction-cost-reducing collective institutions, allowing firms to shift their relations quickly and cheaply.

Weak ties also facilitate knowledge creation—another important competitiveness factor in the furniture industry. Consumer markets (particularly, the medium- and high-end segments) demand ever more frequent product innovations. Most product innovations are incremental, with a genuinely novel model introduced only once every few years, subsequently followed by on-going adjustments and add-ons to the model design. Such adjustments and add-ons (as well as more genuinely new product designs) are often inspired by gossip and direct observation within clusters, facilitated by weak ties.

Hence, market organisation in clusters in the furniture industry rests on 'locational economies'. All firms in furniture clusters are specialised within a few and related economic activities and local collective institutions become specialised to support flexible market relations.[25]

Discussion

Above, we have illustrated the different market organisation of the pop music and the furniture industries. Indeed, large segments of these two industries are clustered, because the market relations used by firms in these segments are dependent on collective market organisation.[26] However, the markets for pop music and furniture become organised in clusters for different reasons. In the sourcing, marketing and distribution of pop music, clustering rests on a combination of urbanisation economies and locational economies supporting project relations; however, clustering in the production of finished furniture and components is mainly based on locational economies benefiting flexible relations. In both pop music and furniture, a crucial aspect of market organisation is the low cost of information, sustained by collective institutions that are predominantly found in clusters. In both industries, the community aspect of weak ties serves to bring down transaction and switching costs, through information and social trust effects. Hence, this collective institution, which depends on clustering, is a prerequisite for the flexibility upon which competitiveness within the two industries hinges.

Furthermore, the collective institutions of clusters serve to enhance knowledge creation, albeit in different ways. In the pop music industry, knowledge creation (being complex and demanding dedicated attention over a specified period of time) requires a dedicated market relation—namely, interfirm projects—while in the furniture industry (where product innovation is an incremental and relatively simple process and where imitation is abundant), flexible relations suffice to facilitate knowledge creation.

The industry examples have shown that an analysis of the market makes little sense without considering social institutions. While the literature refers to a range of beneficial effects stemming from the spatial clustering of firms,[27] the present paper has targeted the aspect of clustering of particular importance for the organisation of markets. As weak ties, compared with other types of relations among people and business firms, are highly sensitive to geographical distance, firms that depend on this aspect of market organisation tend to cluster spatially. We have empirically illustrated that, because knowledge creation and resource efficiency depend on weak ties, the degree of clustering is high in both the pop music and furniture industries. In other words, we have argued that the spatial arrangement of these industries in industrial clusters represents one particular form of market organisation that, over time and through market evolution, has proved to be advantageous for the performance of these specific kinds of economic activity.

6. Conclusion

The many competing schools of thought concerned with industrial clusters have at least one thing in common: they all agree that clusters are real-life phenomena characterised by the co-localisation of separate economic entities, which are in some sense related, but not joined together by any common ownership or management. So hierarchies they are certainly not.

Yet, it is usually taken for granted that clusters, almost regardless of how they are defined, all expatriate the 'swollen middle' (Hennart, 1993) of various hybrid "forms of long-term contracting, reciprocal trading, regulation, franchising and the like" (Williamson, 1991, p. 80) residing somewhere between hierarchies and markets. This fundamental (but usually implicit) assumption would, perhaps, be justified if markets could be reduced to the exchange of property rights, between "large numbers of price-taking anonymous buyers and sellers supplied with perfect information" (Hirschman, 1982, p. 1473), as they are commonly conceived in mainstream economics. One of the original attractions of neo-classical price theory was precisely that it promised a way of analysing the economy in general and market exchange in particular independently of specific institutional settings.

However, introducing transaction costs as more than fees paid to intermediaries leads inevitably to comparative institutional analysis and, not to be forgotten, to the perception of markets as institutions with specific characteristics of their own. Some sets of characteristics are so common that they represent a

Table 2. Benefits of market organisation

	Industry conditions	
Economic benefits:	High uncertainty or ambiguity	Low or modest uncertainty
Knowledge creation	Clusters of project relations	Networks
Resource efficiency	Clusters of flexible relations	

specific market organisation or market form. The cluster is one such specific market organisation that is structured along territorial lines because this enables the building of a set of institutions that are helpful in conducting certain kinds of economic activity.

This paper argues that clusters are markets where commodities, services and knowledge are traded in a notably efficient way among the insiders without restricting their abilities to build pipelines and to interact with suppliers and customers residing elsewhere. The institutions characterising this market form help to create an environment among insiders that reduces the barriers to acquiring and utilising knowledge produced or used locally.

Supported by the empirical illustrations of two selected industries the analysis undertaken in this paper has enabled us to make five important points (summarised in Table 2).

(1) The paper maintains that markets become organised by firms striving to reduce transaction costs in order to enhance the benefits of their market relations in general and knowledge creation and resource efficiency in particular.
(2) It concludes that firms tend to choose network formation if the set of dominant suppliers and customers is reasonably stable and co-localisation strategies of clustering when it is not.
(3) It is pointed out how flexibility and temporality in market relations reflect volatility in demand and that clustering often provides firms with tools to alleviate the consequences.
(4) A distinction is made between firms engaged in economic activities best conducted by project-based experimentation—where clusters constitute a competitive market organisation for knowledge creation when demands are ambiguous—and firms on highly volatile and unpredictable markets, where clustering can provide a competitive market form for resource efficiency by facilitating flexibility and adaptability.
(5) The paper asserts that cluster institutions can supplement firms' creation of knowledge when interacting with suppliers and customers along the vertical value chain; and, along the horizontal dimension, by enabling learning among rivals and competitors.

The empirical analysis has suggested one further aspect of clustering as market organisation. The evolution of each individual cluster is never premeditated but rather the complex result of an on-going interaction between external industry dynamics and the emergence of matching internal institutional settings helpful to firms harbouring divergent visions and pursuing different strategies.

Notes

1. Clusters are non-random (Ellison and Glaeser, 1994) geographical agglomerations of firms with similar or highly complementary capabilities (Richardson, 1972). The term 'cluster' is used in this paper in a generic sense including related concepts such as 'geographical agglomeration', 'industrial district', etc. For a further discussion, see Malmberg and Maskell (2002). The existence of clusters has been commented on by many scholars, earliest and most noticeably by Alfred Marshall (1890), (1919), and, more recently, by a swelling group including Beccatini (1990); Brusco (1986); Dei Ottati

(1994); Garofoli (1992); Gottardi (1996); Belussi (1999); Maillat (1991); Kirat and Lung (1999); Swann *et al.* (1998); Markusen *et al.* (1986); Saxenian (1994); Schmitz (1999); Arthur (1990); Staber (1994); Steiner (1998); Feldmann (1994); Phelps and Ozawa (2003); as well as the prevailing contribution by Porter (1998, 2000).

2. The process of globalisation brought about by the reduction of trade barriers, shrinking customer loyalty to local products, the efficiency of mass marketing and global supply chain management practices, etc. joins hands with recent advances in new communication technologies to reduce even further the friction of space.
3. Not least Brian Loasby (1994) who has pointed out that Williamson's famous dictum: "In the beginning there were markets" (Williamson, 1983, p. 20) should not be taken as a statement about the actual historical progression of economic life in the real world, but merely as the point of departure for a chain of arguments about the organisation of the economy.
4. This definition of markets is broader than the so-called Arrow–Debreu (1954) definition, commonly adopted in mainstream economics, where market interaction consists of *events* of exchange in which price signals are the only information-carrying device available to the participating firms. Loasby (1994), among others, argues that markets, as institutional *settings* where events of exchange take place, cannot meaningfully be reduced to standard price theory. Bowles (1998) stresses how markets do much more than allocate goods and services, while White (1981, 2001) contributes a coherent alternative from a sociological perspective. Rosenbaum (2000) attempts to build a basic classification of market approaches. See also Hodgson (1988, 1998), Hart (1995) and Foss (1999).
5. However, the transport costs of delivering physical goods and the time costs of interacting in a broad sense (Thompson, 1967; Grandori, 2000), often associated with the use of markets, are not specific to interfirm relations but will equally burden subsidiaries of a single firm.
6. The term 'exemplary strategies' denotes that while most communities are spatially defined in clusters some are not, and that although most networks seem to depend relatively little on spatial proximity, some are very geographically agglomerated.
7. In particular, high demand uncertainty raises a range of problems related to knowledge creation (such as how to identify and advance product and process innovations and improvements that can match unforeseeable market demands), as well as problems related to resource efficiency (such as how to respond efficiently to unforeseeable demand volumes, through varying production volumes and product varieties).
8. Relation-specific sunk costs are investments that cannot be recaptured even partly if the investing firm decides to leave the relationship (Baumol and Willig, 1981; Baumol *et al.*, 1982). High exit costs limit opportunistic behaviour and act as safeguards for exchange.
9. Following Glaeser *et al.* (1999), trust can be defined as the commitment of resources to an activity where the outcome depends on the co-operative behaviour of others.
10. The product innovation literature has convincingly established that firms learn from each other when interacting and that such interaction helps them to solve problems beyond the ability of each individual firm. See, for instance, Rosenberg (1972), Freeman (1982, 1991), Håkansson (1987), Kline and Rosenberg (1986), DeBresson and Amesse (1991), Hagedoorn and Schakenraad (1992), OECD (1992), DeBresson (1996).
11. Ambiguity differs from uncertainty in that it cannot be reduced by the collection of more facts.
12. Interfirm projects, being market-based, should be distinguished from the temporary *intra*firm task forces or 'project teams'.
13. Grabher (2002) uses the term 'project ecologies' about urban clusters of advertising firms with a high density of project relations. However, he does not discuss in any length the spatial clustering of such ecologies.
14. A cluster also provides an environment that curtails the costs and risks associated with establishing new firms through relocation or by spin-offs (Belussi, 1999; Klepper, 2002). By starting activities close to what is already going on in the cluster, new firms can skip the burdensome and costly process of gathering a lot of circumstantial knowledge about the business environment which would be crucial outside the cluster: "If it works for my neighbour, why shouldn't it work for me, too?" A favourable business environment includes a labour market with a sufficient supply of the most appropriate skills (to be hired if offering competitive wages), a competent capital market of bank managers accustomed to the business at hand or with particularly skilled venture capitalists or business angles, a ready-made access to suppliers, with the potential to reach customers (often supported by cluster-based branding or reputation ef-

fects) and a supportive infrastructure and institutional endowment.

15. Represented (albeit not perfectly) by NACE codes 22140 (publishers of sound recordings); 22150 (other publishers); and 22310 (reproduction of sound recordings).
16. We hence exclude from the analysis art making (songwriting, performance and studio production—represented by NACE codes 92310 (performing artists) and 92320 (theatre and concert halls, etc.) and support functions for music making (such as instrument making, equipment manufacture and software programming—represented (albeit poorly) by NACE codes 24650 (manufacture of prepared unrecorded media), 36300 (manufacture of music instruments) and 51433 (wholesale of CDs, tapes, records, videos). The artistic segment is typically constituted by freelance artists or one-man firms with little focus on profit and no clear localisation pattern. The support segment mostly consists of relatively vertically integrated, technically oriented firms, again with little obvious clustering compared with the end value chain segment subject to our analysis.
17. Outside national capitals, project clusters in the music industry are found in selected 'creative cities' that are able to attract specialised and highly qualified 'creative' labour (Florida, 2002).This pattern of urban clustering is something the pop music industry has in common with other so-called entertainment industries.
18. Many young artists want to live in the cities, but many of the most successful artists and/or songwriters dwell in the countryside, with no clear localisation pattern. These artists are more self-contained in their creative process and entertain fewer project relations (depending more on long-term network relations with record companies and publishers) and are consequently less dependent on urban location. By contrast, younger artists often have more diverse project relations, shifting between labels and bands, performing often and needing an abundance of weak ties to other artists to inspire their creative process.
19. For example, a recent study of the Scandinavian pop music industry showed that, when labour market data are used, in Denmark, pop music firms cluster in Greater Copenhagen, accounting for no less than 46 per cent of all Danish firms within pop music (as defined by the NACE codes specified above). This is significant, as the general concentration of firms—meaning, Copenhagen's share of all Danish firms—is only 19 per cent. A minor cluster of firms was also found around the second-largest Danish city of Aarhus, accounting for 12 per cent of all pop music firms (STEP, 2003).
20. The degree of formalisation is, however, generally higher in pop music than in furniture. This is due to the importance of intellectual property rights, which necessitate formalisation in contracts and stipulations to a higher extent than in more traditional industries, such as furniture.
21. Represented by NACE codes 361110 (producers of chairs, etc.), 361120 (upholsterers), 361200 (producers of office furniture), 361300 (producers of kitchens), 351410 (producers of wooden home furniture), plus (with some error) 203010 (producers of lists), 203020 (producers of construction wood) and 205110 (turners). The analysis in this section thus excludes production of raw wood, textiles, foam and standardised wood or metal parts (represented by NACE codes NACE 201010 (sawmills), 201020 (protection/treatment of wood) and 202000 (producers of chip and MDF boards). These industry segments are relatively consolidated, as their firms have for long enjoyed scale economies, and firms are randomly localised (or localised near sources of raw materials, such as forests)
22. It should be noted that many such firms produce both components and end-products—making the use of NACE codes even more problematic.
23. This is the effects of 'flexible specialisation', pointed to by Piore and Sabel (1984).
24. Similar clustering is also found in many other mature, low-tech European industries.
25. In clusters like industrial districts, we often also find specialised capital and local labour markets. While this may also be true for many furniture clusters, the industry is characterised by relatively low capital needs and opportunities to train many types of local labour to fit the activities of local firms. Hence, specialised financial institutions, and the ability to attract specialised labour, are factors that are not quite as crucial as they are for the pop music industry.
26. In the pop music industry, the upstream segments supplying artistic content and support functions are not clustered, because firms and agents focus on in-house artistic or technical skills; and in the furniture industry, the upstream segments producing raw materials and unspecialised inputs are not clustered, because firms focus on internal scale economies and resource efficiency.
27. These general advantages of clustering include transport and time cost reduction, training and attraction of specialised labour

and, in recent times, increased incentives for service providers and public policy-makers to target their offers at a few dominant economic activity areas.

References

ALLEN, R. C. (1983) Collective invention, *Journal of Economic Behaviour and Organization,* 4, pp. 1–24.

ANDERSEN, B. and MILES, I. (1999) *Orchestrating intangibles in the music sector: the royalties collecting societies in the knowledge based economy.* Paper prepared for the *CISTEMA Conference*, October.

ARROW, K. J. and DEBREU, G. (1954) Existence of an equilibrium for a competitive economy, *Econometrica,* 22(3), pp. 265–290.

ARTHUR, B. (1990) 'Silicon Valley' locational clusters: when do increasing returns imply monopoly?, *Mathematical Social Sciences,* 19, pp. 235–251.

BAMBI, G. (1998) The evolution of a furniture industrial district: the case of Poggibonsi in Tuscany, in: M. LORENZEN (Ed.) (1998) *Specialization and Localized Learning: Six Studies on the European Furniture Industry*, pp. 59–70. Copenhagen: CBS Press.

BATHELT, H., MALMBERG, A. and MASKELL, P. (2002) *Clusters and knowledge: local buzz, global pipelines and the process of knowledge creation.* DRUID Working Paper (available at: www.druid.dk).

BAUMOL, W. J. and WILLIG, R. D. (1981) Fixed costs, sunk costs, entry barriers, and sustainability of monopoly, *Quarterly Journal of Economics,* 96, pp. 405–431.

BAUMOL, W. J., PANZAR, J. C. and WILLIG, R. D. (1982) *Contestable Markets and the Theory of Industrial Structure.* San Diego: Harcourt Brace Jovanovich Publishers.

BECATTINI, G. (1990) The Marshallian industrial districts as a socio-economic notion, in: F. PYKE, G. BECATTINI and W. SENGENBERGER (Eds) *Industrial Districts and Inter-firm Co-operation in Italy,* pp. 37–51. Geneva: International Institute for Labour Studies.

BELLANDI, M. (1996) Innovation and change in the Marshallian industrial district, *European Planning Studies,* 4(3), pp. 357–368.

BELUSSI, F. (1999) Policies for the development of knowledge-intensive local production systems, *Cambridge Journal of Economics,* 23, pp. 729–747.

BEN-PORATH, Y. (1980) The F-connection: families, friends and firms in the organisation of exchange, *Population and Development Review,* 1(6), pp. 1–30.

BOWLES, S. (1998) Endogenous preferences: the cultural consequences of markets and other economic institutions, *Journal of Economic Literature,* 36(1), pp. 75–112.

BRUSCO, S. (1986) Small firms and industrial districts: the experience of Italy, in: D. KEEBLE and E. WEVER (Eds) *New Firms and Regional Development in Europe*, pp. 184–202. London: Croom Helm.

CARLSSON, B. and ELIASSON, G. (2001) *Industrial dynamics and endogenous growth.* Paper presented to the *DRUID Nelson and Winter Summer Conference*, Aalborg (available at www.druid.dk).

CASSON, M. (1991) *The Economics of Business Culture: Game Theory, Transaction Costs and Economic Performance.* Oxford: Clarendon.

CASTANER, X. and CAMPOS, L. (2002) The determinants of artistic innovation by cultural organizations: bringing in the role of organizations, *Journal of Cultural Economics,* 26, pp. 29–52.

CAVES, R. (2000) *Creative Industries: Contracts between Art and Commerce.* London: Harvard University Press.

CHANDLER, A. D. (1962) *Strategies and Structure. Chapters in the History of the International Enterprise.* Cambridge, MA: The MIT Press.

CHANDLER, A. D. (1977) *The Visible Hand: The Managerial Revolution in American Business.* Cambridge, MA: The Belknap Press of Harvard University Press.

CHANDLER, A. D. (1990) *Scale and Scope: The Dynamics of Industrial Capitalism.* Cambridge, MA: Harvard University Press.

CHEUNG, S. N. S. (1983) The contractual nature of the firm, *Journal of Law and Economics,* 26, pp. 1–21.

COASE, R. H. (1937) The nature of the firm, *Economica,* 4(16), pp. 386–405.

COASE, R. H. (1960) The problem of social cost, *Journal of Law and Economics,* 3, pp. 1–44.

COASE, R. H. (1992) The institutional structure of production, *American Economic Review,* 82(4), pp. 713–719.

COWAN, R., DAVID, P. and FORAY, D. (2000) The explicit economics of knowledge codification and tacitness, *Industrial and Corporate Change,* 9(2), pp. 211–254.

DAHLMAN, C. J. (1979) The problem of externality, *Journal of Law and Economics,* 22(1), pp. 141–162.

DARMER, P. (1999) *The Indie way: relationship between leisure and labour among Danish independent labels.* Paper presented at the *Leisure, labour and urban life conference*, Leicester University, March.

DAVIS, A. and BRADY, T. (2000) Organisational capabilities and learning in complex product systems: towards repeatable solutions, *Research Policy,* 29, pp. 931–953.

DEBRESSON, C. (1996) *Economic Interdependence and Innovative Activity*. Cheltenham: Edward Elgar.

DEBRESSON, C. and AMESSE, F. (1991) Networks of innovators: a review and introduction to the issue, *Research Policy,* 20, pp. 363–379.

DEI OTTATI, G. (1994) Co-operation and competition in the industrial district as an organisational model, *European Planning Studies,* 2, pp. 463–483.

DEMSETZ, H. (1968) The cost of transacting, *Quarterly Journal of Economics,* 82(1), pp. 33–53.

EGIDI, M. (1995) Routines, hierarchies of problems, procedural behaviour: some evidence from experiments, in: K. J. ARROW (Ed.) *Rationality in Economics*. London: Macmillan.

EKSTEDT, E., LUNDIN, R. A., SODERHOLM, A. and WIRDENIUS, H. (1999) *Neo-industrial Organizing: Renewal by Action and Knowledge Formation in a Project-intensive Economy*. London: Routledge.

ELLISON, G. and GLAESER, E. L. (1994) *Geographical concentration in the US manufacturing industries: a dartboard approach*. Working Paper No. 4840, NBER, Cambridge, MA.

ENGWALL, M. (2002) *No project is an island: linking projects and history and context*. Working paper, Stockholm School of Economics.

ESTALL, R. C. and BUCHANAN, R.O. (1961) *Industrial Activity and Economic Geography*. London: Hutchinson.

FELDMANN, M. P. (1994) *The Geography of Innovation*. Dordrecht: Kluwer.

FESER, E. J. (1999) Old and new theories of industrial clusters, in: M. STEINER (Ed.) *Clusters and Regional Specialisation: On Geography, Technology and Networks*, pp. 19–40. London: Pion.

FLORIDA, R. (2002) *The Rise of the Creative Class: And How It's Transforming Work, Leisure, Community and Everyday Life*. New York: Basic Books.

FORD, D. (Ed.) (1990) *Understanding Business Markets: Interaction, Relationships, Networks*. London: Academic Press.

FOSS, K. and FOSS, N. J. (2002) Organizing economic experiments: property rights and firm organization, *Review of Austrian Economics,* 15(4), pp. 297–312.

FOSS, N. J. (Ed.) (1999) *Theories of the Firm: Critical Perspectives in Economic Organisation*. London: Routledge.

FOSS, N. J. and LORENZEN, M. (2004) Analogy and the emergence of focal points: some suggestions for bringing cognitive coordination into the theory of economic organization, in: K. NIELSEN (Ed.) *Uncertainty in Economic Decision-Making: Ambiguity, Mental Models and Institutions*. Cheltenham: Edward Elgar (forthcoming).

FREEMAN, C. (1982) *The Economics of Industrial Innovation*. London: Pinter Publishers.

FREEMAN, C. (1991) Networks of innovators: a synthesis of research issues, *Research Policy,* 20(5), pp. 5–24.

FUKUYAMA, F. (1995) *Trust: The Social Virtue and the Creation of Prosperity*. London: Hamish Hamilton.

GAROFOLI, G. (1992) Industrial districts: structure and transformation, in: G. GAROFOLI (Ed.) *Endogenous Development and Southern Europe*, pp. 49–60. Aldershot: Avebury.

GLAESER, E. L., LAIBSON, C. L., SCHEINKMAN, S. J. A. and SOUTTER, C. L. (1999) *What is social capital? The determinants of trust and trustworthiness*. Working paper No. 7216, NBER, Cambridge, MA.

GOODMAN, R. A. and GOODMAN, L. P. (1976) Some management issues in temporary systems: a study of professional development and manpower—the theater cases, *Administrative Science Quarterly,* 21, pp. 494–501.

GOTTARDI, G. (1996) Technology strategies, innovation without R&D and the creation of knowledge within industrial districts, *Journal of Industry Studies,* 3(2), pp. 119–134.

GRABHER, G. (Ed.) (2002) Production in projects: economic geographies of temporary collaboration, Special Issue of *Regional Studies,* 36(3).

GRANDORI, A. (2000) *Organization and Economic Behaviour*. London: Routledge.

GRANOVETTER, M. (1973) The strength of weak ties, *American Journal of Sociology,* 78(6), pp. 1360–1380.

GRANOVETTER, M. (1982) The strength of weak ties: a network theory revisited, in: P. V. MARSDEN and N. LIN (Eds) *Social Structure and Network Analysis*, pp. 105–130. Beverly Hills, CA: Sage.

GRANOVETTER, M. (1985) Economic action and social structure: the problem of emdeddedness, *American Journal of Sociology,* 91, pp. 481–510.

HAGEDOORN, J. and SCHAKENRAAD, J. (1992) Leading companies and networks of strategic alliances in information technologies, *Research Policy,* (21), pp. 163–181.

HÅKANSSON, H. (Ed.) (1987) *Industrial Technology Development: A Network Approach.* London: Croom Helm.

HÅKANSSON, H. and SNEHOTA, I. (1989) No business is an island: The network concept of business strategy, *Scandinavian Journal of Management,* 5, pp. 187–200.

HART, O. (1995) *Firms, Contracts and Financial Structure*. Oxford: Clarendon Press.

HAYEK, F. A. VON (1937) Economics and knowledge, *Economica (New Series)*, 4(13), pp. 33–54.

HAYEK, F. A. VON (1945) The use of knowledge in society, *American Economic Review,* 35(4) 519–530.

HENNART, J.-F. (1993) Explaining the 'swollen middle': why most transactions are a mix of market and hierarchy, *Organizational Science,* 4(4), pp. 529–547.

HIPPEL, E. VON (1987) Cooperation between rivals: informal know-how trading, *Research Policy,* 16, pp. 291–302.

HIRSCHMAN, A. O. (1982) Rival interpretations of market society: civilizing, destructive or feeble?, *Journal of Economic Literature,* 20(4), pp. 1463–1484.

HOBDAY, M. (2000) The project-based organization: an ideal form for managing complex products and systems?, *Research Policy,* 29, pp. 871–893.

HODGSON, G. M. (1988) *Economics and Institutions.* Cambridge: Polity Press.

HODGSON, G. M. (1998) The approach of institutional economics, *Journal of Economic Literature,* 36(1), pp. 166–193.

HOOVER, E. M. (1948) *The Location of Economic Activity.* New York: McGraw-Hill Book Company.

HUYGENS, M., BADEN-FULLER, C., VAN DEN BOSCH, F. A. J. and VOLBERDA, H. V. (2002) Co-evolution of firm capabilities and industrial competition: investigating the music industry, 1877–1997, *Organization Studies,* 22(1), pp. 971–1012.

IFPI (The International Federation of Phonographic Industries) (2001) *The Recording Industry in Numbers.* London: IFPI.

KIRAT, T. and LUNG Y. (1999) Innovation and proximity: territories as loci of collective learning processes, *European Urban and Regional Studies,* 6(1), pp. 27–38.

KIRZNER, I. M. (1973) *Competition and Entrepreneurship.* Chicago, IL: University of Chicago Press.

KLEIN, B. (1983) Contracting costs and residual claims: the separation of ownership and control, *Journal of Law and Economics,* 26, pp. 367–374.

KLEPPER, S. (2002) *The evolution of the U.S. automobile industry and Detroit as its capital.* Paper presented at the *DRUID Winter Conference*, Denmark (available at www.druid.dk).

KLINE, S. J. and ROSENBERG, N. (1986) An overview of innovation, in: R. LANDAU and N. ROSENBERG (Eds) *The Positive Sum Game,* pp. 275–305. Washington, DC: National Academy Press.

KNIGHT, F. H. (1921) *Risk, Uncertainty, and Profit.* New York: Harper and Row.

KREBS, D. L. (1970) Altruism: an examination of the concept and a review of the literature, *Psychological Bulletin,* 73(4), pp. 258–302.

LANGLOIS, R. N. (1999) Scale, scope and the reuse of knowledge, in: S. C. DOW and P. E. EARL (Eds) *Economic Organization and Economic Knowledge: Essays in Honour of Brian J. Loasby, Vol. I,* pp. 239–254. Cheltenham: Edward Elgar.

LISSONI, F. (2001) Knowledge codification and the geography of innovation: the case of Brescia mechanical cluster, *Research Policy,* 30(9), pp. 1479–1500.

LOASBY, B. J. (1991) *Equilibrium and Evolution: An Exploration of Connecting Principles in Economics.* Manchester: Manchester University Press.

LOASBY, B. J. (1994) Organisational capabilities and interfirm relations, *Metroeconomica,* 45(3), pp. 248–265.

LOASBY, B. J. (1999) Industrial districts as knowledge communities, in: M. BELLET and C. L'HARMET (Eds) *Industry, Space and Competition: The Contribution of Economists of the Past,* pp. 70–85. Cheltenham: Edward Elgar.

LOASBY, B. J. (2000) *Organisations as interpretative systems.* Paper presented at the *DRUID Summer Conference*, Rebild, Denmark (available at www.druid.dk).

LOJACONO, G. and LORENZEN, M. (1998) External economies and value net strategies in Italian furniture districts, in: LORENZEN, M. (Ed.) (1998) *Specialization and Localized Learning: Six Studies on the European Furniture Industry,* pp. 71–94. Copenhagen: CBS Press.

LOPES, P. D (1992) Innovation and diversity in the popular music industry 1969 to 1990, *American Sociology Review,* 57(1), pp. 56–71.

LORENZEN, M. (Ed.) (1998) *Specialization and Localized Learning: Six Studies on the European Furniture Industry.* Copenhagen: Copenhagen Business School Press.

LORENZEN, M. (2002) Ties, trust and trade: elements of a theory of coordination in industrial clusters, *International Studies in Management and Organization,* 31(4), pp. 14–34.

LORENZEN, M. (2004) *Localized Learning and Cluster Capabilities.* Cheltenham: Edward Elgar (forthcoming).

LORENZEN, M. and FOSS, N. J. (2003) Cognitive coordination, institutions and clusters: an exploratory discussion, in: T. BRENNER (Ed.) *The Influence of Co-operations, Networks and Institutions on Regional Innovation Systems,* pp. 82–104. Cheltenham: Edward Elgar.

LORENZEN, M. and FREDERIKSEN, L. (2003) *Experimental music: product innovation, project networks and dynamic capabilities in the pop music industry.* Working paper, IVS/DYNAMO, Copenhagen.

LUNDVALL, B.-Å. (1985) *Product Innovation and User–Producer Interaction.* Industrial Development Research Series No. 31, Aalborg: Aalborg Universitetsforlag.

MAILLAT, D. (1991) Local dynamism, milieu and innovative enterprises, in: J. BROTCHIE, M. BATTY, P. HALL and P. NEWTON (Eds) *Cities of the 21st Century*, pp. 265–274. London: Longman.

MALECKI, E. J. (1991) *Technology and Economic Development: The Dynamics of Local, Regional and National Change*. Harlow: Longman.

MALMBERG, A. and MASKELL, P. (2002) The elusive concept of localization economies: towards a knowledge-based theory of spatial clustering, *Environment and Planning A,* 34(3), pp. 429–449.

MANSFIELD, E. (1985) How rapidly does new technology leak out?, *Journal of Industrial Economics,* 34, pp. 217–223.

MARKUSEN, A., HALL, P., GLASMEIER, P. A. K. (1986) *High Tech America*. Boston: Allen and Unwin.

MARSHALL, A. (1890) *Principles of Economics*. London: Macmillan.

MARSHALL, A. (1919) *Industry and Trade: A study of Industrial Technique and Business Organization and of their Influences on the Condition of Various Classes and Nations*. 5th edn (1927). London: Macmillan.

MASKELL, P. (1998) Successful low-tech industries in high-cost environments: The case of the Danish furniture industry, *European Urban and Regional Studies,* 5(2), pp. 99–118.

MASKELL, P. (2001a) Regional policies: promoting competitiveness in the wake of globalisation, in: D. FELSENSTEIN and M. TAYLOR (Eds) *Promoting Local Growth: Process, Practice and Policy*, pp. 295–310. Aldershot: Ashgate.

MASKELL, P. (2001b) Knowledge creation and diffusion in geographic clusters, *International Journal of Innovation Management,* 5(2), pp. 213–237.

MASKELL, P. (2001c) The theory of the firm in economic geography: or why all theories of the firm are not equally well suited for application within the conversation on space, *Economic Geography,* 77(4), pp. 329–344.

MASKELL, P. (2001d) Towards a knowledge-based theory of the geographical cluster, *Industry and Corporate Change,* 10(4), pp. 921–944.

MASKELL, P. and MALMBERG, A. (1999) Localised learning and industrial competitiveness, *Cambridge Journal of Economics,* 23(2), pp. 167–186.

MASKELL, P., ESKELINEN, H., HANNIBALSSON, I. *ET AL.* (1998) *Competitiveness, Localised Learning and Regional Development: Specialization and Prosperity in Small Open Economies*. London: Routledge.

MCCLOSKEY, D. N. (1994) *Knowledge and Persuasion in Economics*. Cambridge: Cambridge University Press.

NEGUS, K. (1992) *Producing Pop: Culture and Conflict in the Popular Music Industry*. London: Edward Arnold.

NELSON, R. R. and WINTER, S. G. (1982) *An Evolutionary Theory of Economic Change*. Cambridge, MA: Belknap Press.

OECD (ORGANISATION FOR ECONOMIC CO-OPERATION AND DEVELOPMENT) (1992) *Industrial Policy in the OECD countries. Annual Review*. Paris: OECD.

OECD (1997) *Content as a new growth industry*. Working Party on the Information Economy, Directorate for Science, Technology and Industry, Committee for Information, Computer and Communications Policy. Paris: OECD.

OECD (1999) *Boosting Innovation: The Cluster Approach*. Paris: OECD.

PHELPS, N. A. and OZAWA, T. (2003) Contrasts in agglomeration: proto-industrial, industrial and post-industrial forms compared, *Progress in Human Geography,* 27(5), pp. 583–604.

PIORE, M. J. and SABEL, C. F. (1984) *The Second Industrial Divide: Possibilities for Prosperity*. New York: Basic Books.

POLANYI, K., ARENSBERG, C. M. and PEARSON, H. W. (Eds) (1957) *Trade and Markets in Early Empires. Economies in History and Theory*. New York: Free Press.

PORTER, M. E. (1998) Clusters and the new economics or competition, *Harvard Business Review,* 11, pp. 77–91.

PORTER, M. E. (2000) Locations, clusters and company strategy, in: G. CLARK *ET AL.* (Eds) *The Oxford Handbook of Economic Geography*, pp. 253–274. New York: Oxford University Press.

POWER, D. and HALLENCREUTZ, D. (2002) Profiting from creativity? The music industry in Stockholm, Sweden and Kingston, Jamaica, *Environment and Planning A,* 34(10), pp. 1833–1854.

POWER, D. and LUNDEQUIST, P. (2002) Putting Porter into practice? Practices of regional cluster building: evidence from Sweden, *European Planning Studies,* 10(6), pp. 685–704.

RICHARDSON, G. B. (1972) The organisation of industry, *Economic Journal,* 82, pp. 883–896.

RICHARDSON, G. B. (1998) *Production, planning and prices*. DRUID Working Paper (available at www.druid.dk).

ROELANDT, T. J. A. and HERTOG, P. (Eds) (1999) *Cluster Analysis and Cluster-based Policy: New Perspectives and Rationale in Innovation Policy*. Paris: OECD.

ROSCHER, N. (1989) *Cognitive Economy: The Economic Dimension of the Theory of Knowledge*. Pittsburgh, PA: University of Pittsburgh Press.

ROSENBAUM, E. F. (2000) What is a market? On the methodology of a contested concept, *Review of Social Economy,* 58(4), pp. 455–483.

ROSENBERG, N. (1972) *Technology and American Economic Growth.* White Plains, NY: Sharpe.

ROSENBERG, N. (1992) Economic experiments, *Industrial and Corporate Change,* 1, pp. 181–203.

RUGGIE, J. G. (1993) Territoriality and beyond: problematizing modernity in international relations, *International Organization,* 47(1), pp. 139–175.

SABEL, C. F. (1992) Studied trust: building new forms of cooperation in a volatile economy, in: F. PYKE and W. SENGENBERGER (Eds) *Industrial Districts and Local Economic Regeneration,* pp. 215–250. Geneva: ILO.

SAXENIAN, A. (1994) *Regional Advantage, Culture and Competition in Silicon Valley and Route 128.* Cambridge, MA: Harvard University Press.

SCHMITZ, H. (1999) Collective efficiency and increasing returns, *Cambridge Journal of Economics,* 23, pp. 465–483.

SCHMITZ, H. and MUSYCK, B. (1993) *Industrial districts in Europe: policy lessons for developing countries?* Institute of Development Studies, Brighton (mimeograph).

SCOTT, A. J. (1999) The US recorded music industry: on the relations between organization, location and creativity in the cultural economy, *Environment and Planning A,* 31, pp. 1965–1984.

SCOTT, A. J. (2000) *The Cultural Economy of Cities.* London: Sage.

SHUKER, R. (2001) *Understanding Popular Music,* 2nd edn. London: Taylor & Francis.

SIMON, H. A. (1982) *Models of Bounded Rationality I: Economic Analysis and Public Policy.* Cambridge, MA: The MIT Press.

STABER, U. (1994) The employment regimes of industrial districts: promises, myth and realities, *Industrielle Beziehungen,* 1(4), pp. 321–346.

STABER, U. ET AL. (Eds) (1996) *Business Networks: Prospects for Regional Development.* Berlin: Walter de Gruyter.

STEINER, M. (Ed.) (1998) *Clusters and Regional Specialisation: On Geography, Technology and Networks.* London: Pion.

STEP (Centre for Innovation Research) (2003) *Behind the Music: Profiting from Sound: A Systems Approach to the Dynamics of Nordic Music Industry.* Oslo: STEP.

STORPER, M. (1995) The resurgence of regional economies, ten years later: the region as a nexus of untraded interdependencies, *European Urban and Regional Studies,* 2, pp. 191–221.

SWANN, P. G. M., PREVEZER, M. and STOUT, D. (1998) *The Dynamics of Industrial Clustering: International Comparisons in Computing and Biotechnology.* Oxford: Oxford University Press.

TEECE, D. J. (1980) Economies of scope and the scope of the enterprise, *Journal of Economic Behavior and Organization,* 3(1), pp. 223–247.

THOMPSON, J. D (1967) *Organizations in Action.* New York: McGraw-Hill.

TRIVERS, R. L. (1971) The evolution of reciprocal altruism, *Quarterly Review of Biology,* 46, pp. 35–57.

UEA (Association of European Furniture Producers) (2003) The EU furniture Industry, *Outlook* (www.ueanet.co./outlook.htm).

VOGEL, H. L (1998) *Entertainment Industry Economics.* Cambridge: Cambridge University Press.

WESSELING, H. L. (2003) The idea of an Institute of Advanced Study: some reflections on education, science and art, *European Review,* 11(1), pp. 2–21.

WHITE, H. C. (1981) Where do markets come from?, *American Journal of Sociology,* 87(3), pp. 517–547.

WHITE, H. C. (2001) *Markets from Networks: Socioeconomic Models of Production.* Princeton, NJ: Princeton University Press.

WILLIAMSON, O. E. (1975) *Market and Hierarchies–Analysis and Antitrust Implications. A Study in the Economics of Internal Organisation.* New York: The Free Press.

WILLIAMSON, O. E. (1983) Credible commitments: using hostages to support exchange, *American Economic Review,* 73(4), pp. 519–540.

WILLIAMSON, O. E. (1985) *The Economic Institutions of Capitalism.* New York: The Free Press.

WILLIAMSON, O. E. (1991) Strategizing, economizing, and economic organization, *Strategic Management Journal,* 12, pp. 75–94.

WILLIAMSON, O. E. (1996) Transaction cost and market failure, in: M. GOOLD and K. S. LUCHS (Eds) *Managing the Multibusiness Company: Strategic Issues for Diversified Groups,* pp. 17–40. London: Routledge.

WILSON, D. T. (1975) Dyadic interaction: an exchange process, in: B. ANDERSON (Ed.) *Advances in Consumer Research, Vol. 2,* pp. 394–397. Cincinnati: ACR.

Where Is the Value Added in the Cluster Approach? Hermeneutic Theorising, Economic Geography and Clusters as a Multiperspectival Approach

Paul Benneworth and Nick Henry

[Paper first received, March 2003; in final form, November 2003]

> There is no doubt that what it means 'to do' theory is quite different between the old and the new economic geographies, and in the process redefining the very discipline itself (Barnes, 2000, p. 19).

> Hermeneutic theorising ... is a much better description of the kind of theorising found in the new economic geography, and marked by an interpretive mode of inquiry that is reflexive, open-ended and catholic in its theoretical sources (Barnes, 2000, p. 546).

Introduction

The popularity of the idea of clusters in recent years has met marked ambivalence within the academic community of economic geography. Ultimately, academics have engaged, lured by renewed pressures to achieve policy relevance and tempered by an uneasy recognition of the concept's theoretical immaturity. Arguably rooted in the realm of stylised facts and thin abstractions (Clarke, 1998; Allen, 2002), cluster analyses have, nevertheless, seen the employment of a diversity of theoretical frameworks ranging from geographical econometric models to normative recipes for generalised economic success (Benneworth *et al.*, 2003; Gordon and McCann, 2000; Martin and Sunley, 2003).

In Martin and Sunley (2003), the authors deconstruct the concept to make better sense of the diversity of activity taking place under the rubric of 'clusters', although not, purportedly, with the wish to debunk it. Their 2003 *Journal of Economic Geography* paper "Deconstructing clusters: chaotic concept or policy panacea?" provides a powerful argument that the eclectic assembly of a diversity of perspectives is a "dubious endeavour" (p. 14). For them, the concept has become 'chaotic'

> in the sense of conflating and equating quite different types, processes and spatial scales of economic localisation under a single, all-embracing universalistic notion (Martin and Sunley, 2003, p. 10).

In this paper, we continue the process of theoretical reflection on the concept of 'clusters' by drawing on Barnes' hermeneutical theorising framework to explore whether this diversity has more beneficial implications for economic geography. Despite agreement with much of the detail of Martin and Sunley's critique, from this perspective we find deconstruction of the cluster concept to be a rather more positive experience. Opening the 'black box' of the cluster approach provides recognition of the (multiperspectival) range of academic and policy threads (both encompassed and evolving). Accepting these multiple perspectives permits the difficult act of "share(d) conversations in epistemology" (Haraway, 1991, p. 191, quoted in Barnes, 2000) at worst and, at best, the possibilities for theoretical, empirical and policy cross-fertilisation. It is in this sense, therefore, that a rigorous multiperspectival clusters approach (incorporating hermeneutic theorising) can add the value that Martin and Sunley are seeking.

Following this introduction, the paper sets out Barnes' (2000) argument for a new style of theorising in economic geography. This first section sets the epistemological context for a short review of the 'rise of clusters' in economic geography as perceived by Martin and Sunley (2003). In response, we argue for clusters as a multiperspectival approach and this approach is exemplified by reference to the variety of work that has been undertaken on London's business services agglomeration. The paper concludes with a discussion of the potentials and pitfalls of the conversations provided by a multiperspectival clusters approach engendered through hermeneutic theorising in economic geography.

Theorising Economic Geography

Barnes (2000) argues that there is a new form of theorising that characterises much of the contemporary work of economic geography. In contrast to the 'epistemological theorising' of the late 1950s and early 1960s—for example, he suggests that the very idea and practice of theorisation has shifted within economic geography towards an approach he labels as 'hermeneutic theorising'. Each of these approaches, epistemological and hermeneutic theorising, is an attitude to developing and using theory and each "implies a specific set of conditions that shape and constrain what counts as appropriate, novel and persuasive vocabulary in redescription" (Barnes, 2000, p. 5).

Central to the first of these, epistemological theorisation, is a belief that the aim of theory-building is to develop abstract vocabularies that 'mirror' (as best they can) an external and independent reality. In this sense, the aim is to use vocabularies with unambiguous meanings which invoke clear and determined relationships which exist between objects in an independent 'world-out-there'. The problem from this perspective is to find the vocabulary that most transparently represents the real relationships between objects. For Barnes, this theoretical perspective first came to the fore in economic geography in the 1950s in what others have termed the 'quantitative revolution'. This body of work used an innovative set of practices to drive a theoretical programme which redescribed the world using novel vocabularies. New techniques and technologies in computing and statistics allowed the expression and modelling of relationships using mathematics. This methodology shift had the effect of moving economic geography from the 'field' to the 'desk'. Social physics, morphological laws, the spatial logic of geometry and gravity models were just some attempts to use these innovations the better to mirror the world and underlying reality to improve geographers' capacities to make a difference in the world.

In contrast, Barnes argues that hermeneutic theorising takes as its starting-point that no vocabulary is perfect or final in its (re)description and, moreover, that there is no end of possible sources for vocabularies of theorisation. Diverse vocabularies allow for and engender theoretical conversations that hinge on interpretation and which never achieve perfect representation. Theorising is a creative and open-ended process of interpretation performed by community members. Barnes argues that this does not mean that 'anything goes' but, rather, that a process of critical scrutiny will take place in order to establish the 'usefulness' of a theory. Communities of practice will define any theory's utility around criteria including resonance with other theories, the convincingness of the argument, its rhetorical power, political sensibilities and the different kinds of action it empowers. Theorising is circular, reflexive, indeterminate and perspectival (Bohman, 1993, p. 116, quoted in Barnes, 2000). Hermeneutic theorising therefore explicitly recognises that knowledge (interpretation) is situated and partial (although certain vocabularies may hold sway as useful for long periods of time).

Accounting for Clusters

The Rise of the Ideas of Clusters

To make sense of the notion of clusters in economic geography, it feels intuitive to begin by noting the general resurgence of interest in "the region as a scale of economic organisation and political intervention" (MacKinnon *et al.*, 2002, p. 293). More particularly, this generic 'reassertion of location' (Martin and Sunley, 2003) has revived historical work on ideas of specialised industrial location, but reinterpreted in the context of contemporary structural economic challenges. Newlands (2003)—for example, outlines five different theoretical traditions drawn upon by the current industrial clusters literature

—standard agglomeration theory, from Marshall onwards;
—transaction costs: the 'California School';
—flexible specialisation, trust and untraded interdependencies;
—innovative milieux: the GREMI Group; and
—institutional and evolutionary economics.

We can add to this list of theoretical traditions the theory of 'clusters' promulgated by Porter (1998), which draws on his strategic management background. As Martin and Sunley (2003) outline, Porter's cluster theory stems from his work on international competitiveness and national competitive advantage in the late 1980s and early 1990s. His work was driven by a concern with improving US competitiveness in the context of the apparent rise of Japan and other newly industrialising Asian countries. He developed the concept of the 'competitive diamond', which separated out four sets of factors which determine the export success of a nation's firms. Porter went on to argue that the intensity of interaction within the 'diamond' is enhanced if the firms concerned are 'geographically localised' or 'clustered' (Martin and Sunley, 2003) and that a nation's most competitive industries are likely to be clustered. Porter's theoretical stance in 1990 was close to long-standing American interests in regional science (Gordon and McCann, 2000).[1] From this position, Porter built and popularised the idea of clusters in the first half of the 1990s and, through this 'brand building' process, he refined his ideas. A further critical point was the publication of *On Competition* in which he set out his regional cluster concepts (Porter, 1998). Moreover, Porter's success was advanced by geo-economic changes; Japan's stagnation and the US' long boom allowed Porter to position his clusters ideas as generic and transmissable wherever neo-liberal deregulation/reregulation processes unfolded (Peck and Tickell, 2002).

While Porter was busy refining his thinking in the late 1990s, others involved in regional development theory debates saw clusters as a useful tool and began to appropriate and develop their own versions of 'clusters'. Lagendijk and Cornford (2000)—for example, argued for a distinct deviation from Porter's original work with the subsequent fusion of clusters and networks literature within debates on the restructuring of old industrial regions (Boekholt and Thuriaux, 1999; Cooke, 1995; Morgan, 1997). Another example was the OECD, which saw clusters as a way of promoting innovative growth in its member countries. In this instance, the OECD focused on clusters through their longer-standing interests in concepts of national innovation systems (Bergman *et al.*, 2001).

Over time, a number of theoretical traditions have joined the debate on clusters, each bringing their own views to the concept (Newlands, 2003). This academic engagement must, however, be viewed within the context of the exceptional success of the clusters concept within the policy community (compare Benneworth *et al.*, 2003; Raines, 2002). As Martin and Sunley comment

> From the OECD and the World Bank, to national governments ... to regional development agencies ..., to local and city governments ..., policy-makers at all levels have become eager to promote local business clusters ... Clusters, it seems, have become a world-wide fad, a sort of academic and policy fashion item (Martin and Sunley, 2003, p. 6).

Indeed, for some, the clusters concept has almost been 'reverse engineered' into academic disciplines such as economic geography from the policy community (Benneworth *et al.*, 2003).

The Problem with Clusters

In many ways, Martin and Sunley's (2003) review paper on clusters has crystallised the dissatisfaction and concerns of many within and beyond economic geography with the notion of 'clusters'. We do not seek to dismiss these concerns; indeed, we agree with many of them and their underlying arguments. Much clusters work has been done badly—for example, methodological naivety, lack of theoretical reflexivity and cartographic rather than relational views of space. This process of doing clusters badly is actively hurting the development of theory. For Martin and Sunley (2003) this leads them to question the value of 'theorising clusters' and, ultimately, makes them sceptical that 'clusters' can be 'done well' at all and thus add value to existing theories. It is this final theoretical step and resulting conclusion, the concept's possible added value, which we focus on.

Martin and Sunley neatly summarise their critique in the question

> Why is it that Porter's notion of 'clusters' has gate-crashed the economic policy arena when the work of economic geographers on industrial localisation, spatial agglomeration of economic activity, and the growing salience of regions in the global economy, has been largely ignored (Martin and Sunley, 2003, p. 7).

For them, they see little more than a powerful brand in the concept of clusters. We suggest that this leads them to treat the idea of 'clusters' as if it is a single stable idea for which Porter is a gate-keeper. As interest in regional development (Lovering, 1999) has grown, the clusters approach has spread in so many directions that we think it is problematic to consider it a proprietary term, solely identified with the work of Porter.

Instead, by the late 1990s, and presumably the trigger for the Martin and Sunley piece, clusters thinking was a web of interdependent academic thinking, policy-making and consultants' work. The examples of the Rustbelt School and the OECD noted above highlight that clusters ideas have co-evolved with, rather than distinct from, a range of other regionalist perspectives. Similarly, as Newlands (2003) outlines, economic geographers' theoretical traditions in industrial localisation, spatial agglomeration and high-growth regions can, in fact, be found within, informing and part of, the many cluster studies that have peppered recent literatures in economic geography and cognate disciplines (compare the London example below). Indeed, their observation that "clusters, it seems, have become a worldwide fad, a sort of academic and policy fashion item" (p. 6) demonstrates to us that clusters thinking is simply too big to be contained within Porter's writings and 'brand'. Rather than regarding clusters as a stable and well-defined entity, perhaps it is better to consider clusters as a black box, in the language of Lagendijk and Cornford (2000), in which a range of evolving academic and policy threads have been placed together.

Indeed, at the heart of Martin and Sunley's critique of clusters lies just such an unresolved paradox. On the one hand, they identify the enormous success of the 'Porter brand' of cluster (and by comparison the 'failure' of the work of economic geographers). Yet, on the other hand, and as part of their deconstruction of the 'chaotic concept' of clusters, they highlight and lament the range of theoretical perspectives (many of which are derived from the work of economic geographers) encompassed within the brand. It is on this basis that their paper concludes by asking whether or not the clusters' approach can add value to these constituent elements. We believe that this key tension in their critique whether the clusters concept is merely what Porter writes about clusters—provides the means to resolve their question of the significance and added value of the concept of 'clusters'. For us, clusters is a concept which incorporates and extends well beyond the work of Porter and allows the possibility of theoretical debate and multidisciplinary cross-fertilisation.

Deconstructing the Martin and Sunley Critique

Martin and Sunley (2003) offer a highly informative table (Table 1, p. 12) summarising nine different definitions of clusters. This table highlights the theories that have been placed together in the cluster approach. These differing definitions have brought with them ambiguities and contradictions to the attempts to develop cluster typologies. As Martin and Sunley ask, why does the focus shift from "national groups of industries and forms" to "a local grouping of similar firms … within a highly spatially circumscribed area" (p. 11)? Why do "the social dimensions of cluster formation and cluster dynamics remain something of a black box in Porter's work" (p. 16)? And, ultimately, to what degree is it "possible to construct a universal theory of cluster formation … capable of covering the wide range of cluster types" (p. 14)?

We do not dispute the inconsistencies, difficulties and problems that Martin and Sunley highlight with the clusters concept and its associated literature. Nevertheless, it is worth making several points. First, we would agree that 'clusters' can appear to lack logical consistency, as writers jump between explanatory frameworks such as agglomeration, transaction costs, institutions and culture. If these analyses are written as if within a singular framework, then the analysis can appear unconvincing, jumpy and illogical (see Martin and Sunley, 2003). Yet, carefully treated as a portmanteau concept, analyses drawing on different case studies and theories can make sense if the component ideas are each recognised to evolve along their own internal and logically coherent pathway. If the condition of logical consistency is met for each strand, then the fact that "empirical methodologies and 'mapping' strategies vary considerably" (Martin and Sunley, 2003, p. 19) does not automatically invalidate the comparability of approaches.

As Martin and Sunley themselves outline in their Table 2 (adapted from Swann, 2002 and echoing McKendrick, 1999), different theoretical perspectives seek different forms of evidence and will do so through different sets of techniques. Gordon and McCann (2000, p. 528) reflect a wider consensus that a single clusters 'situation' can empirically demonstrate a range of explanatory theories—"actual clusters may contain elements of more than one type"—and Martin and Sunley (2003, p. 16) agree that in "reality, such co-existence is likely to be the rule". Of course, rigour must be maintained with a suitable marriage of theory and method *within* each perspective (Massey and Meegan, 1985) but, equally, vitality can arise from the overlapping of different perspectives and precisely not the paring down to a singular narrative. Indeed, Martin and Sunley's argument that "there is no agreed method for identifying and mapping clusters" (p. 19) sits rather uneasily with their view that "top–down mapping exercises at best only ... provide a shallow, reduced view of clusters" (p. 21). In essence, Martin and Sunley would seem to be demanding that we seek a singular geographical theory of clusters but our argument, drawn from hermeneutic theorising, would be to highlight that this demand is rooted in philosophical perspective.

Indeed, and secondly, in arguing that "the existence of clusters, appears then, in part at least, to be in the eye of the beholder—or should we say, creator", Martin and Sunley (2003, p. 11) merely highlight the critical issue of the situatedness of our knowledge so central to hermeneutic theorising. This familiar issue has been raised, and rightly so, around other favourite geographical objects of research such as the 'urban', 'rural', 'the region' or even 'the regional economy'. The pathway out of such quandaries is, precisely, to return to the question that is being asked (Allen *et al.*, 1998)—to the eye of the beholder and to recognise the theory-laden and political nature of any research question that will be asked, investigated and subsequently judged (Barnes, 2000).

Thirdly, Martin and Sunley begin their useful deconstruction addressing the clusters approach but their narrative quickly reverts to the brand à la Porter. The diversity of perspectives that Martin and Sunley do acknowledge are allowed agency only in the sense that these perspectives are placed within Porter's ever expanding theoretical model/brand. In turn, the other voices and agents sitting behind the diversity of perspectives are reduced to an overdetermined and centred community of 'brand merchants'.

By centring the debate around Porter and his brand, and thereby hiding the various other agents active in the current debate, Martin and Sunley obscure that each theoretical thread retains its own coherence, explanatory structure and *set of audiences*. Academics, for example, who have used the term cluster are assumed to have transformed themselves into Porter's acolytes, rather than the more sympathetic interpretation of themselves being engaged with moving theoretical debates forward. In particular, concern is reserved for "economic geographers [who] themselves started to use cluster terminology in preference to their own" (Martin and Sunley, 2003, p. 8). Similarly, policy-makers, who have clearly shaped the development of clusters ideas in their own interests as a policy tool (Benneworth and Charles, 2001), are scorned as "public policy-makers eager to enter the cluster promotion game" (p. 21) and are presented as passive recipients of received wisdom. This characterisation fits uneasily with analyses which have demonstrated that policy-makers have fitted cluster theories into their own policy needs (Peck and McGuinness, 2001; Dalsgaard, 2001) or alternatively, if appropriate to their own technical and political needs, have abandoned clusters entirely (Gilsing, 2001; Learmonth *et al.*, 2003).

As examples of the instrumental use of 'clusters policy', Catalonia, Flanders, Scotland and Quebec pioneered clusters ideas as an inexpensive way of opposing their federal centre's dominant industrial policies (which were perceived systematically to discriminate against them). These and other less favoured regions identified a tool to create their own industrial policy priorities, rather than falling into step with federal industrial priorities (Cox, 1998). 'Clusters' allowed them to raise their policy profile and demonstrate opposition and independence, without challenging the overall political legitimacy of the national developmental state. It is only subsequently that clusters have lost much of this subtly oppositional character as national government has adopted them as part of industrial and regional policy. The net effect of Martin and Sunley's determination to centre these heterogeneous views around Porter is to reduce what are in reality multiple and evolving conceptualisations into a singular argument about 'clusters'.

A Multiperspectival Approach to Clusters

Martin and Sunley outline how different people have theorised clusters in different ways. They read this as a devaluation of the integrity of the brand. Ironically, we would suggest that—possibly in their zest for its deconstruction—they overendow the global brand of the 'Porter cluster'. By highlighting different perspectives, and decentring the idea of clusters from Porter, we seek to raise the contrary position that 'clusters' are more akin to a series of proximate debates. The 'clusters' approach can be thought of as the act of holding together these dissonant threads in conversation. Disentangling the threads, and hence performing convincing 'clusters work' produces (the possibility of) new academic knowledges. Academics as cluster theorisers are drawing on a range of perspectives to gain multiple points of access into the same situation (O'Neill and Whatmore, 2000). This allows for the possibility of a broader analysis to be produced in which the total knowledge of the 'cluster situation' is greater than that of the component parts. Rather than regarding clusters as stable and a known entity or 'brand', it is more that the brand is polycentric and adopted as such by its diversity of users (academic and policy-maker alike). Multiple strands need not imply incoherence. Coherence is an emergent outcome of how effectively academics perform the theorisation process and is contingent on theorisation being done satisfactorily and convincingly.

The Multiplicity of Cluster Debates: A 'Work in Progress'

The idea that clusters are a multiperspectival concept is central to Gordon and McCann's 2000 *Urban Studies* paper, in which they argue that three separate disciplines—regional economics, business/ management and geography/sociology—have each developed their own theories of what is a cluster. Crudely put, these theories are agglomeration, supply chain co-ordination and embeddedness/institutional thickness respectively. Significant in the argument they put forward is that each of the approaches has a very different locus, a consequence of different disciplinary backgrounds, with different rules of evidence, proof and causality implicit in each. They argue—for example, that agglomeration theories make no claim about causality and institutional theories say nothing about generalisability. Each approach is therefore limited in terms of the claims that can convincingly be made, but cross-referencing these specific/limited/fragmentary claims allows a more coherent analysis, which they demonstrate by reference to the London producer services 'cluster'.

Gordon and McCann use the three approaches to generate evidence that they then stitch into a single narrative about the London business services cluster. The thrust of this narrative is that there are a limited number of very local clusters; some (media and consultancy) are driven by a need for information exchange and others (financial services) are driven by the location preferences of key decision-makers. These local clusters exist in a broader greater South East agglomeration, for which the motives and causality are less clear, but in which producer services are more competitive than elsewhere in the UK. Although they do not reflect on the implications of their methodology, Gordon and McCann adopt an approach in which the cluster is explicitly a 'construct', but is also a domain in which various theoretical perspectives have salience. Different points of entry generate knowledge about a situation; they compare that knowledge with other knowledges produced from other entry points into that situation and assemble those knowledges into an overarching analysis. By defining the boundary to what will be compared as the 'cluster', dissimilar methodologies can be integrated into a singular insight (Murdoch, 1997; Cox, 1998; O'Neill and Whatmore, 2000; Gibbs *et al.*, 2001).

Our argument in this paper is that the value added of the clusters approach (drawing on hermeneutic theorising) lies in, first, allowing for and explicitly promoting these theoretical conversations and, secondly, the potential this may afford in which multiple explanations can interact conceptually to provide a richer understanding of the situation than permitted by theoretically monistic approaches. Above, we highlighted how Gordon and McCann used three perspectives to capture what was 'going on' in the London business services cluster. Below, we exemplify further the argument for multiple perspectives by reviewing work on the London business services cluster, which has built on the analysis of Gordon and McCann. What other conceptualisations of the London business service agglomeration could combine into a cluster narrative?

The debate over the nature of the London business services cluster was stimulated by Allen (1992). His essentially theoretical argument was for the service sector as a growth dynamic in its own right (not as an adjunct to manufacturing) and, moreover, a regionally inscribed growth dynamic. One example was the financial and commercial services of the London city-region (incorporating elements of the South East planning region and beyond) which could be understood in terms of a 'regional mode of service growth'. Subsequently, Coe and Townsend (1998) used the concept of a 'regional mode' to frame a more detailed empirical investigation that concluded that agglomeration of service activity was, indeed, best characterised at the spatial scale of the Greater South East rather than localised London-scale service clusters. Nevertheless, a couple of years later, Gordon and McCann's paper highlighted that the concentration of advanced service firms in London is in fact both of these things and that different theoretical and methodological tools were necessary to discern each of them.

The most comprehensive investigation of these competing explanations has been published recently by Keeble and Nachum (2002). They investigate in great detail—and with great care—clustering processes amongst small consultancies by using samples drawn from the purported heart of the cluster—central London—and the decentralised locations of South West England and East Anglia. What is most instructive for our argument is how they build a more comprehensive explanation by bringing together knowledge produced from different departure points. Keeble and Nachum (2002, p. 68) ask "what the highly problematic notion of 'clustering' may mean in the context of business services". In operationalising the idea of a cluster, they draw, first, on Porter's broad definition based on geographical concentration and functional interconnection. Secondly, they define cluster existence based on quantitative concentration (i.e. central London business consultancies) and, thirdly, their investigation of interconnectedness is driven by a search for interfirm networking, collaboration and labour mobility.

From this overarching working definition, Keeble and Nachum adopt an innovative methodology of following each of these operational strands (plus one of decentralisation based on enterprising behaviour theory) to their logical conclusion, producing a set of analytical postulates which they then combine into a synthetic narrative about the 'London business services cluster'. Effectively, in each strand they begin from a particular theoretical perspective, operationalise what that perspective would mean in terms of a business services cluster and then collect and analyse their data-sets for that particular feature. Thus, their conclusions include

(1) the existence of marked relative and absolute concentrations of consultancy SMES (regional science/economics/economic geography; top–down SIC mapping);
(2) in a functionally integrated cluster (Porter; industrial complexes; interfirm collaboration);
(3) reinforced and made 'strong' by localised collective learning (economic geography/sociology/STS; personal networks; embodied expertise in a fluid labour market; high rates of spin-off);
(4) powerfully influenced not by supply-side focused 'agglomeration economy' processes but demand-led benefits of proximity and accessibility to clients (Porter; business studies; economic geography; networks, location prestige); and
(5) with accessibility incorporating national and global communications nodality and driving the cluster as a neo-Marshallian node (economic geography/sociology) (Keeble and Nachum, 2002, pp. 85–86).

We would suggest that, in this case, the result of this multiperspectival approach is a more confident identification of the existence of a cluster with a more holistic and deeper understanding of what is, or is not, driving its existence. Critically, the approach produces a highly geographical model of the relationships between agents which define the spatial extent and constitution of this particular form of regional development. Through a process of careful synthesis of (selected) different cluster perspectives, 'geography matters' (Samers, 2001) at a variety of scales within their analysis and conclusions.

Weaving the Strands: A Spun Thread or Ragged Mess?

The example of work on the London business services cluster exemplifies the multiple perspectives to be found within the clusters debate. Different perspectives might include agglomeration economies, industrial complexes and a variety of social network models. Even this list by no means exhausts the literature but merely exemplifies those perspectives most common in the geographical and regional science literatures. Strategic management—for example, offers other perspectives on the roles of clusters and particularly the (undertheorised in geography) connection between firm performance and cluster membership (see Tallman *et al.*, forthcoming; Pinch *et al.*, 2003). Indeed, geographical approaches have largely skirted Porter's own competitiveness-based analysis which, although hotly contested for its assumptions (Krugman, 1994), still retains the potential to generate insights into why particular groups of firms at local and national levels succeed.

Each of these different theories might imply a different understanding of a cluster, an accepted methodological foundation and 'rules of evidence and argument'. Gordon and McCann might regard these categories as ideal types, and Martin and Sunley suggest that elements of several are likely to be present in each real cluster. We regard each of these approaches as a lens with which to look at the same situation to produce knowledge which contributes to how we understand both 'that cluster' as well as 'clusters'. Whilst one lens might not be able to discern a cluster in a given situation, that is neither automatic proof that that situation is not a cluster, nor that it is not then valid to consider whether other lenses can see the 'cluster in a situation'.[2]

Following O'Neill and Whatmore (2000), each lens provides a separate point of entry to understanding clusters in a particular location. This moves beyond the mere 'identification' of a cluster and rather closer to identifying (making visible) the significant relationships those lenses reveal. For clusters, these relationships might be that an agglomeration is associated with higher productivity, or that some particular network is improving innovation performance. Our argument is that, by identifying the key relationships between agents and objects in particular situations, it is possible to generate a deeper knowledge of the cluster. This, in turn, can enable more precise and useful statements to be made about the nature of economic organisation and regional development in particular contexts or, alternatively, greater explicitness about the theory-laden process which has driven any piece of research on clusters.

The converse, however, is also true. If the opportunity is not taken and we do not theorise clusters or, worse, we theorise clusters badly, then no such better understandings will be generated. We argue that the work of Keeble and Nachum and Gordon and McCann is engaging precisely because they are 'good theorisations', which are convincing, coherent and satisfying. The challenge remains to take forward this multiperspectival method more generally along both tracks, doing good cluster studies which improve how we understand cluster relationships more generally. As Martin and Sunley note, for instance, Porter has never really developed a rigorous theory of social capital and networking, whilst many qualitative approaches begin from analyses of industrial concentration measured in employment or value-added terms. Interpreting and explaining uneven development requires comparison between situations and this approach helps to make explicit the conditions where transfer and generalisation are fair (Sayer, 1991).

This helps also to address the critique of clusters as an abstraction in which particularities of context are generalised into overdetermining categories (Sayer, 1989). By adopting a multiperspectival approach, cluster analyses are explicitly and

> inescapably partial, provisional and incomplete. Refusing any vantage-point that purports to take in the world at a glance, they are more modest in the claims they can, and want to, make (Whatmore, 2002, p. 7).

Furthermore, the accumulation of evidence from the application of different lenses is likely to guide an understanding of the weight and significance of any individual example. Whilst we do not argue for a primitive summary of cluster factors to gauge the significance of the 'cluster' (Massey, 1984), it is clear that Gordon and McCann's multivalent London producer service cluster is highly robust because of dense local linkages, a broader milieu effect and a national innovation system favouring London as a financial services centre. This makes it qualitatively much stronger than clusters which can only be seen through a single cluster lens, such as a micro-clustering network organisation (Lagendijk, 1999).

Conclusion

This paper began by outlining Barnes' claims for a new kind of theorising in economic geography: hermeneutic theorising. Its key characteristics include the recognition that theory is social practice engaged in by reflective and situated practitioners and that theorisation itself be recognised as a less formal activity open towards a diversity of sources. In turn, the ensuing 'set of narrative communities' (Thrift and Olds, 1996) and their conversations represent a constant theoretical 'work in progress' as a constellation of perspectives engage within an anti-foundationalist and anti-essentialist atmosphere. Within such a provisional arena of process, political discussion is to the fore as the relationship among different perspectives motivates study aimed, as much as epistemological theorising, at improving the world.

Using the framework of hermeneutic theorising, we have sought to contextualise the concept of clusters in economic geography and the particular concerns of Martin and Sunley around the 'chaotic' nature of this successful brand. In many ways, we agree with much of their critique of the concept of clusters. Where we differ is in our belief that this concept is very much 'work in progress', with potential utility as a technique to drive forward diverse theorising on industrial agglomeration. For us, the portmanteau concept of clusters represents a potential uniting thread to bring multiple perspectives to explain industrial agglomerations. To date, in some cases this has been done well, but we acknowledge that many other analyses have failed to convince. This mix of success and failure is not inconsequential but a hallmark of the creative process of theorisation in action. We would not yet equate bad cluster analyses with fundamental fault-lines in the concept, as Martin and Sunley seem to do. Instead, we would draw on hermeneutic theorising to note that the potentialities of the theoretical conversations opened up by the clusters approach are, at present, being corroded by poor operationalisation and analysis. This may be the portent of things to come, or clusters may join the more successful lexicon of economic geography, as academics use the conversations allowed by the concept to generate understanding within logically coherent, if theoretically eclectic, frameworks.[3] Within this paper, we believe the work we have highlighted on London's business services cluster reflects just that potential.

Our aim in this paper was not to review the clusters debate as a whole; rather, it was to explore the very interesting issues raised by Martin and Sunley. The inconsistencies and ambiguities of the clusters concept are part of the theoretical 'work in progress' around this immature, yet politically powerful, concept. We deem this 'work in progress' as a multiperspectival approach to clusters and believe that this approach's rationale, justification and overarching 'rules of the game' can be set by the epistemological position of hermeneutic theorising. In this paper, we have introduced and enabled this *possibility* within the clusters debate. It is beyond this paper's scope, however, to specify a particular pathway for the approach (such as—for example, dealing with incompatibility and methodological pluralism). We also acknowledge that it will require much 'good theorising' before the value of clusters will be realised. We contend that this value added demanded by Martin and Sunley will be delivered, if at all, precisely through its ability to allow for a multiperspectival approach. A significant amount of theorising remains to be done, however, if such conversations are to remake a more holistic understanding of clusters with geography at its theoretical heart.

Notes

1. These interests primarily were in statistical analyses of sub-national industrial concentration, beginning with Isard, and disseminated through the work and the journals of the Regional Science Association International, founded in 1953, and publisher of the *Journal of Regional Science* and the *International Regional Science Review*.
2. See, for example, Miller *et al.* (2001). This rigorous national cluster mapping exercise in the UK recognised that SIC codes would not identify all the likely cluster possibilities and incorporated an additional qualitative programme of on-the-ground interviews.
3. Indeed, theoretical eclecticism has been identified as one element of the vitality of the new economic geography (Thrift and Olds, 1996; Bryson *et al.*, 1999, ch. 2; Barnes, 2000).

References

ALLEN, J. (1992) Services and the UK space economy: regionalisation and economic dislocation, *Transactions of the Institute of British Geographers,* 7(2), pp. 292–305.

ALLEN, J. (2002) Living on thin abstraction: more power/economic knowledge, *Environment and Planning A,* 34(4), pp. 451–466.

ALLEN, J., MASSEY, D., COCHRANE, A. *ET AL.* (1998) *Rethinking the Region.* London: Routledge.

BARNES, T. (2000) *Retheorizing economic geography: from the quantitative revolution to the 'cultural turn'.* Paper presented to the *Annual Conference of the Association of American Geographers*, Pittsburgh, USA, April.

BARNES, T. (2001) Retheorizing economic geography: from the quantitative revolution to the 'cultural turn', *Annals of the Association of American Geographers,* 91(3), pp. 546–565.

BENNEWORTH, P. S. and CHARLES, D. R. (2001) Bridging cluster theory and practice: learning from the cluster policy cycle, in: E. M. BERGMAN, P. DEN HERTOG, D. R. CHARLES and S. REMOE (Eds) *Innovative Clusters: Drivers of National Innovation Systems*. Paris: OECD.

BENNEWORTH, P. S., RAINES, P., DANSON, M. and WHITTAM, G. (2003) Confusing clusters? Making sense of the cluster approach in theory and practice, *European Planning Studies,* 11(5), pp. 511–520.

BERGMAN, E. M. DEN HERTOG, P., CHARLES, D. R., and REMOE, S. (Eds) *Innovative Clusters: Drivers of National Innovation Systems*, Paris: OECD.

BOEKHOLT, P. and THURIAUX, B. (1999) 'Public policies to facilitate clusters: background, rationale and policy practices in international perspective' in OECD (Ed.) *Boosting Innovation: The Cluster Approach*, OECD, Paris.

BOHMAN, J. (1993) *New Philosophy of Social Science: Problems of Indeterminacy*. Cambridge, MA: MIT Press.

BRYSON, J. R., HENRY, N., KEEBLE, D. and MARTIN, R. (Eds) (1999) *The Economic Geography Reader: Producing and Consuming Global Capitalism*. Chichester: John Wiley.

CLARKE, G. L. (1998) Stylised facts and close dialogue: methodology in economic geography, *Annals of the Association of American Geographers,* 88(1), pp. 73–87.

COE, N. and TOWNSEND, A. R. (1998) Debunking the myth of localised agglomerations: the development of a regionalised service economy in South East England, *Transactions of the Institute of British Geographers,* 23(3), pp. 385–404.

COOKE, P. N. (1995) New wave regional and urban revitalisation strategies in Wales, in: P. N. COOKE (Ed.) *The Rise of the Rustbelt*. London: ICL Press.

COX, K. R. (1998) Spaces of dependence, spaces of engagement and the politics of scale or: looking for local politics, *Political Geography,* 17(1), pp. 1–23.

DALSGAARD, M.-H. (2001) Danish cluster policy: improving specific framework conditions, in: E. M. BERGMAN, P. DEN HERTOG, D. R. CHARLES and S. REMOE (Eds) *Innovative Clusters: Drivers of National Innovation Systems*. Paris: OECD.

DETR (DEPARTMENT FOR THE ENVIRONMENT, TRANSPORT AND THE REGIONS) (2000) *Planning for Clusters*. London: The Stationery Office.

GIBBS, D. C., JONAS, A. E. G., REIMER, S. and SPOONER, D. J. (2001) Governance, institutional capacity and partnerships in local economic development: theoretical issues and empirical evidence from the Humber sub-region, *Transactions of the Institute of British Geographers,* 26(1), pp. 103–120.

GILSING, V. (2001) Towards second-generation cluster policy: the case of the Netherlands, in: E. BERGMAN, P. DEN HERTOG, D. CHARLES and S. REMOE (Eds) *Innovative Clusters: Drivers of National Innovation Systems*, pp. 361–376 Paris: OECD.

GORDON, I. R. and MCCANN, P. (2000) Industrial clusters: complexes, agglomeration and/or social networks?, *Urban Studies,* 37(3), pp. 513–532.

HARAWAY, D. (1991) *Simians, Cyborgs and Women: The Reinvention of Nature*. London: Routledge.

KEEBLE, D. and NACHUM, L. (2002) Why do business service firms cluster? Small consultancies, clustering and decentralisation in London and southern England, *Transactions of the Institute of British Geographers,* 27(1), pp. 67–90.

KRUGMAN, P. (1994) Competitiveness: a dangerous obsession, *Foreign Affairs,* (2), pp. 28–44.

LAGENDIJK, A (1999) Learning in non-core regions: towards intelligent clusters; addressing business and regional needs, in: R. RUTTEN, S. BAKKERS, K. MORGAN, and F. BOEKEM (Eds) *Learning Regions, Theory, Policy and Practice*. London: Edward Elgar.

LAGENDIJK, A. and CORNFORD, J. (2000) Regional institutions and knowledge: tracking new forms of regional development policy, *Geoforum,* 31, pp. 209–218.

LEARMONTH, D., MONRO, A. and SWALES, J. K. (2003) Multi-sectoral cluster modelling: the evaluation of Scottish Enterprise cluster policy, *European Planning Studies,* 11(5), pp. 567–584.

LOVERING, J. (1999) Theory led by policy: the inadequacies of the 'new regionalism' (illustrated from the case of Wales), *International Journal of Urban and Regional Research,* 23(2), pp. 379–395.

MACKINNON, D., CUMBERS, A. and CHAPMAN, K. (2002) Learning, innovation and regional development: a critical appraisal of recent debates, *Progress in Human Geography,* 26(3), pp. 293–311.

MACLEOD, G. (2001) New regionalism considered: globalisation and the remaking of political economic space, *International Journal of Urban and Regional Research,* 25(4), pp. 804–829.

MARTIN, R. and SUNLEY, P. (2003) Deconstructing clusters: chaotic concept or policy panacea?, *Journal of Economic Geography,* 3(1), pp. 5–35.

MASSEY, D. (1984) *Spatial Divisions of Labour: Social Structures and the Geography of Production*. Basingstoke: Macmillan.

MASSEY, D. and MEEGAN, R. (1985) (Eds) *Politics*

and Methods: Contrasting Studies in Industrial Geography. London: Methuen.

MCKENDRICK, J. H. (1999) Multi-method research: an introduction to its application in population geography, *Professional Geographer,* 51(1), pp. 40–50.

MILLER, P., BOTHAM, R., MARTIN, R. L. and MOORE, B. (2001) *Business Clusters in the UK: A First Assessment*, London: Department of Trade and Industry.

MORGAN, K. (1997) The learning region: institutions, innovation and regional renewal, *Regional Studies,* 31(5), pp. 491–403.

MURDOCH, J. (1997) Towards a geography of heterogeneous associations, *Progress in Human Geography,* 21(3), pp. 321–337.

NEWLANDS, D. (2003) Competition and cooperation in industrial clusters: the implications for public policy, *European Planning Studies,* 11(5), pp. 521–532.

O'NEILL, P. and WHATMORE, S. (2000) The business of place: networks of property, partnership and produce, *Geoforum,* 31(1), pp. 121–136.

PECK, F. and MCGUINNESS, D. (2001) *UK competitiveness and the regional agenda: making sense of clusters in the North of England.* Paper presented to "European regions and the challenges of development, integration and enlargement" *Regional Studies Association International Conference*, Gdansk, September.

PECK, J. and TICKELL, A. (2002) Neoliberalising space, *Antipode,* 34, pp. 380–404.

PINCH, S., HENRY, N., JENKINS, M. and TALLMAN, S. (2003) *From 'industrial districts' to 'knowledge clusters': a model of knowledge dissemination and competitive advantage in industrial agglomerations.* Department of Geography, Southampton (mimeograph).

PORTER, M. E. (1998) *On Competition.* Boston, MA: Harvard Business School.

RAINES, P. (2002) Cluster policy: does it exist?, in: P. RAINES (Ed.) *Cluster Development and Policy*. Burlington, VT: Ashgate.

SAMERS, M. (2001) What is the point of economic geography?, *Antipode,* 33, pp. 183–193.

SAYER, A. (1989) The 'new' regional geography and problems of narrative, *Environment and Planning D,* 7(2), pp. 253–276.

SAYER, A. (1991) Behind the locality debate: deconstructing geography's dualisms, *Environment and Planning A,* 23(2), pp. 283–308.

SWANN, G. M. P. (2002) *The implications of clusters: some reflections.* Paper presented at the *Clusters Conference*, Manchester Business School, April.

TALLMAN, S., JENKINS, M., HENRY, N. and PINCH, S. (forthcoming) Knowledge clusters and competitive advantage, *Academy of Management Review*.

THRIFT, N. and OLDS, K. (1996) Refiguring the economic in economic geography, *Progress in Human Geography,* 20(3), pp. 311–337.

WHATMORE, S. (2002) *Hybrid Geographies: Natures Cultures Spaces*. London: Sage.

Working through Knowledge Pools: Labour Market Dynamics, the Transference of Knowledge and Ideas, and Industrial Clusters

Dominic Power and Mats Lundmark

[Paper first received, April 2003; in final form, September 2003]

Introduction

In recent years, the role of clusters and cluster-based policy approaches has become increasingly important in both the analysis of urban and regional economies, and in public and private economic development initiatives. In Sweden, the cluster approach has proved particularly attractive to policy-makers and a rash of cluster-based initiatives and cluster identification exercises has swept the country (Sölvell *et al.*, 1991; Jonsson, 1992; Ivarsson, 1999; Sandberg, 1999; Braunerhjelm *et al.*, 2000; Johanisson and Jonson, 2000; Sölvell, 2000; NUTEK, 2001; Söderström, 2001; Lundequist and Power, 2002; Hallencreutz and Power, 2004). The metropolitan region of Stockholm has attracted much of this attention with a number of high-profile clusters garnering both praise and support. It is against this background that the article explores a prominent cluster in the

Stockholm region: the information and communications technology (ICT) cluster.

Throughout the 1990s, the ICT sector in Stockholm grew rapidly. Centred on the high-profile Kista Science Park and a number of large Swedish and foreign firms, the ICT cluster thrived. Despite difficulties in measuring the exact parameters of a cluster and even with the usefulness of the concept itself (Martin and Sunley, 2001; Malmberg and Maskell, 2002), by the late 1990s both outside observers and the firms themselves were almost entirely convinced that a strong ICT *cluster* existed in the Stockholm metropolitan region. In other words, it became clear to many people that Stockholm was not merely home to a large number of unassociated ICT firms, but to a set of complementary and interlinked firms and institutions that had developed a shared consciousness and identity as an industrial cluster and system. Although with some misgivings, this is a view the present authors broadly share and thus the term cluster is used throughout the article. This development was fuelled by many different factors and inputs (both public and private) and the generally high levels of success enjoyed by the firms in the cluster, until the bubble eventually started deflating at the end of the 1990s, had a range of powerful effects on the city and region. Not least of these was an influx of people into the city and the growing sector, and an increasing dependence on ICT as the core driver of Stockholm's labour market and industrial profile (Birkinshaw, 1998, 2000).

In common with appraisals and evaluations of other prominent ICT clusters (from Silicon Valley to Silicon Glen), much of the commentary on the rise of the Stockholm ICT cluster has concentrated on the role of interfirm linkages, venture capital, supporting institutions, sophisticated consumers, infrastructure, etc. in helping to create a dynamic local milieu supportive of innovation and knowledge creation. In particular, much attention has been focused on the ways in which knowledge and innovation—what are now commonly considered to be the key bases for industrial competitiveness—are formed and transferred within diffuse social and interpersonal networks, milieus and cultures. Whilst these are undoubtedly crucial elements in a sector's cluster dynamics and industrial development, it appears—at least to us—that relatively little attention has been paid to the role of labour markets and labour mobility in shaping clusters' competitiveness and growth.

Labour Market Mobility

A high degree of mobility in the labour market is desirable for several reasons. First, an economy or a cluster cannot function without labour and history has shown that the availability of a young and growing population is a prerequisite for economic development (Malmberg and Sommestad, 2000). Furthermore, mobility functions as a lubricant for structural transformation between declining sectors and companies and expanding ones. It is important for companies' supply of competences and their adjustment to technical development and new demands from the surrounding world.

Labour mobility can also be assumed to have a large impact on the distribution of knowledge among companies (Almeida and Kogut, 1999; Dahl, 2002; Dahl and Pedersen, 2003). Above all perhaps, mobility can greatly effect an area's supply of tacit or non-standardised knowledge and the social capital based on people's network positions. Whilst theoretical notions of tacit knowledge, socially embedded knowledge, social capital and the like are often ill-defined, extremely diffuse and poorly substantiated, there exists a significant enough body of literature (and for our parts also an instinctive feeling) pointing to the fact that less codified and socially enacted and embedded forms of knowledge, capital and network relations have important consequences for industrial performance and innovative capacity. These rather hard to pin-down forms of knowledge and social capital ultimately rest with individuals; as indeed it must be noted does much of what is referred to as codified knowledge (Gertler, 2001, 2003).

The movement of people between labour markets, sectors and firms must therefore have important consequences for industrial functioning and innovation. For industrial sectors or clusters, such movements may be crucial for the binding together of firms into clusters (Basant, 2002) and the meeting and interaction of people in the workplace through which ideas can be negotiated, transferred or developed. Indeed, evidence exists to suggest that for firms whose operations are almost entirely dependent on the quality of human capital—such as those in sectors such as advanced business services, computing, ICT consultancy, etc.—a steady stream of incoming labour is crucial for development (Saxenian, 1994; Angel, 1991; Gustafsson, 2002).

For us, it appears that it is in the supply and quality of labour available to such an emerging cluster that the most important underlying factors are anchored (Krugman, 1991). Cultures, milieus, innovation systems and the like ultimately rest upon people who spend most of their time in their homes and workplaces. Thus rather than focusing on diffuse and vague notions that knowledge and innovation reside in the 'Bohemian' nature (Florida, 2002), 'in the air' (Marshall, 1920) or in the 'buzz' (Storper and Venables, 2002) of urban life, we believe it is equally interesting to focus on the firms and the workplaces where most people spend almost all of their working days. The underlying assumption here then is that knowledge and innovation most commonly develop through interaction located in the workplace itself. If it is in the firm and its various offices and factories that workers predominantly interact and form ideas and knowledge, then the flow of people in and out of such locations may be the most likely channels for local and extra-local sources of knowledge and ideas. Although our data-set does not include reliable indicators of innovativeness and competitiveness, we can assume from secondary sources that: the Stockholm ICT cluster has been a growing and innovative cluster; and, theoretical and empirical work has established the supposed benefits of mobility for industry and cluster development. To summarise roughly the theoretical and empirically based arguments for supposing that innovation and competitiveness are boosted by mobility, three clear suppositions seem especially prominent: that labour mobility is likely to speed up knowledge dissemination and thus learning processes; that labour mobility is likely to create new combinations of knowledge (ideas, methods, etc.) embodied in people; and that labour mobility is likely to create bonds/links between firms, workplaces and institutions and thereby be an active part of cluster and network building.

For many people, sectors and industries, such things as a cosmopolitan street life or accidental face-to-face encounters play relatively little part in the flow of experiences, knowledge and innovation. Rather, it is in the workplace that these exchanges and flows are located and it is thus through labour mobility that intracluster exchanges occur. In order to understand whether the transference and generation of knowledge in clusters are higher than in non-clustered activities, we think that perhaps a good first question to ask is whether labour mobility is higher in clusters and clustered firms (Totterdill, 1999). This does not, of course, mean that we reject the importance of informal contacts as one potential source of knowledge dissemination in the local milieu (see Dahl and Pedersen, 2003, for evidence supporting this idea) or that impulses emanating from more distant interactions than on the local labour market, often are crucial to knowledge building in clusters (Bathelt *et al.*, forthcoming; Malmberg and Power, 2004 forthcoming).

Although our data do not fully allow us to estimate the costs to firms of labour mobility, it is still important to note that mobility has its price. For individuals, costs are involved in moving house, etc. and even moving between employer or workplace can negatively affect individuals' investments in social relations and the like. For firms, the turnover of staff does not only mean the introduction of new competences and ideas, but also involves costly investments in training and education and adjustment periods that can be costly to on-going projects and existing workplace cultures. Studies exist that suggest that staff turnover is considerably higher in larger urban areas and that this can be a problem for certain types of firm that happen to be located in such areas (Orkan, 1972); this tendency has been noted to be especially true for ICT firms (Carnoy *et al.*, 1997; Lawton-Smith and Waters, 2003). Furthermore, it appears that costs and benefits arising out of mobility are different between interfirm and intrafirm mobility (Tomlinson and Miles, 1999; Tomlinson, 2002). Again, our data do not allow us to investigate fully these important differences, although we feel that further research in this area is long overdue.

For the reasons outlined above, a high degree of mobility, or flexibility, in the local labour market (LLM) can be seen as a prerequisite for the functioning of structural transformations within the economy. The mobility of labour, both within and between local labour markets, could be seen as a matching process between labour supply and demand where resources and competences—even during recessions such as Sweden saw in the early 1990s—are continuously transmitted from static to growing sectors in the economy. This matching process takes place within the framework of a gradually more segmented labour market—i.e. there is a large number of labour markets within a region wherein the supply and demand for labour are matched. This is the explanation of why we can see today that there is a shortage of certain types of labour and a surplus of others. These micro labour markets are to a large degree separated from each other, in so far as the mobility between them is considerably lower than within them. Behind the development of a more segmented labour market in Sweden, there is a long-term shift towards a more pronounced information and knowledge-base for the economy and society, where different kinds of specialist competences are becoming more important than labour with a set of generalist competences. This development is more accentuated in the larger cities and in the largest and broadest labour markets (Edvardsson, 2000).

Mobility within the Swedish labour market has decreased since the 1960s (NUTEK, 2000). During the late 1980s, a time of general prosperity, a certain degree of increased collective mobility was visible. This was followed by a more explicit decline during the first half of the 1990s (this article's study period). Variations in the levels of mobility are thus strongly coherent with fluctuations in the economy. However, there are other factors that affect mobility. In this respect, it is important to note the role of demography; the higher rates of labour mobility that tend to characterise younger generations have in Sweden been increasingly outweighed by a long-term ageing of the population. Another circumstance that has counteracted mobility, at least until the 1990s, is the increased number of households consisting of two people active in the labour market. The difficulties involved in finding employment for two persons within the same local labour market have, in particular, restrained geographical mobility. Furthermore, institutional elements such as the existence of relatively compressed wage structures and a strong progressive fiscal system may have decreased mobility. Mobility is, of course, affected by the labour market regulations in force, as well as by the design of labour market policy reforms and the design of unemployment insurance. In general, one can say that Sweden has traditionally had labour laws significantly less 'flexible' than those found in countries such as the US and the UK. While an active employment agency, subsidies to help with the costs of moving to a new area and 'start out on your own'/entrepreneurship campaigns have tended to increase geographical mobility, labour market training and public relief work have often had the opposite effect (Nutek, 2000, pp.11–12).

There is considerable variation in how Sweden's local labour markets function. The size of the labour market region has vital importance for these differences (Persson, 2002). A large and diversified labour market signifies good opportunities for qualified and highly educated people to make careers, which contributes to a generally high level of education in the larger labour markets. Since mobility is higher and the career opportunities better in the larger labour markets, they also attract younger generations of labour. A large labour market also better facilitates both partners in a household finding appropriate work. In general, we can notice that internal mobility (moves between sectors and employers within the same local labour market) is higher in Sweden's larger labour market regions, while external mobility (moving in and out of the local labour market area) is higher in the smaller regions.

It is against this background, and in line with the theoretical assumptions outlined further above, that we set out to investigate the degree and nature of labour mobility in Stockholm and in its ICT cluster. We first briefly describe our methods and data sources and then move on to present our empirical findings.

Methods, Data and Definitions

The article uses a uniquely detailed time-series data-set, which is based on official taxation and civil registration records and contains complete details on everything from education to career changes to income levels for *every* individual employee active in the Stockholm local labour market—*lokal arbetsmarknad* (Statistiska Centralbyrån (SCB) and Expertgruppen för forskning om regional utveckling, 1991) between 1990 and 1995 (in tables and figures below, the database is referred to as GEOMETRO). The detail of the individual records and the complete nature of the data-set mean that it offers a unique possibility to examine, on a large scale, the micro dynamics of individuals in the labour market and in clusters.

In total more than 1.1 million individuals are recorded in the data-set. Information has been gathered from a number of registers produced by Statistics Sweden—for example, the yearly register on employment (*Årsys*), the latest census conducted in Sweden (*Folk- och bostadsräkningen 1990*), the central register of income and capital assets, the education register, the total population register (*RTB*) and the international migration register. One important feature of the database is that every individual is linked to a localised workplace and to a range of standard industrially coded (the Swedish SIC system *Svensk näringsgrensindelning SNI-92* is used) data on the workplace. The workplace is in its turn linked to its economic or juridical unit—i.e. the firm or the organisation that the workplace is a part of. Since every individual has a unique identification number, it is possible to record movements of the workforce between workplaces and firms for the period 1990–95. Thus we do not talk only about mobility between different firms, but also mobility between different workplaces (i.e. physical offices and factories, etc.): this means that our measures also capture mobility within the very large firms that dominate the Swedish ICT cluster.

As mentioned earlier, the data are used to examine whether there have existed over time higher levels of labour market mobility into and within clusters than in the rest of the urban economy. With the level of detail contained in the data-set, we can go a long way towards actually seeing whether clusters are in fact better at encouraging labour mobility (and the skill and knowledge transference this brings with it) than the rest of the economy or area. In order to focus on this issue more clearly, we decided to focus particular attention on the upper strata of workers in both the cluster and the region. We acknowledge that so-called manual or low-skilled activities have an important determining effect on the milieu and its operation. Nevertheless, in a knowledge-based economy such as that of the Stockholm region, it is the most highly educated and skilled workers that can, perhaps, be seen to contribute most to innovation and competitiveness. We decided upon the use of two criteria to define this group: those with high levels of formal education; and those with relatively high levels of income. In operational terms, this group is defined as those having more than 3 years of third-level/university education *or* the top 20 per cent of incomes from paid work. Using two criteria is important, especially in the case of the ICT cluster, since formal education is not the only valid indicator of knowledge and skills: high pay is a relatively good indicator that a person is considered particularly valuable or successful in their work (Dahl and Pedersen, 2003).

As will become clear below, we used a set of 19 standard industrial classification codes in our definition of the ICT cluster. The codes that were used represent a distillation of those used in both other academic studies and industry/governmental studies that have attempted to define ICT clusters in Sweden. Whilst there are many problems associated with the use of SIC-coded data (Power, 2002), at present they represent the best method of carrying out large-scale industrial data analysis.

The Stockholm Region and the ICT Cluster 1990–95

The paper's study period (1990–95) contains the Swedish economy's worst years of recession since the 1930s. During the early 1990s, gross domestic product was negative for a couple of years, financial and real estate sectors came under tremendous pressure, a huge deficit in the state budget was accumulated and unemployment rose to (by Swedish standards) exceptionally high levels. In terms of unemployment, the Swedish metropolitan regions (including Stockholm) were hit more or less as hard as any other part of the country—a situation that had not been witnessed before. On the other hand, the data utilised in the present analysis reflect the start and end years of this extraordinary period and the situation on the labour market was, in fact, not at all that bad in those two particular years. The year 1990 was the end-point of a period of economic growth and high demand for labour, especially in the Stockholm metropolitan region. In 1995, the situation in the labour market had already started to improve considerably but the level of employment was still far below 1990 figures. As can be seen from Table 1, the 1990–95 period was characterised first and foremost by a 10 per cent decrease in employment. What can also be seen is that the working population is getting older, that educational levels are increasing and that fewer immigrants worked in Stockholm (of course, these trends may be accentuated by the fact that in recession older, highly educated and 'native' people tend to hang on to their jobs or are harder to fire).

Against this background, the ICT cluster continued to grow steadily and resist the region's predominant trends and, by the end of 1995, had risen to employ almost 61 000 employees. As can be seen from Table 2, the cluster—at least in the definition we use—consists of a wide variety of ICT activities. Perhaps unsurprisingly, it is the telecommunications category—within which much of the Swedish global giant Ericsson is contained (SNI-92 code 32200)—that accounts for almost a quarter of employees. This category is closely followed by computer consultants (20 per cent) and a wholesale office equipment category (19 per cent) which includes a series of large combined research and distribution units owned by global names such as Hewlett Packard, IBM and Apple. For the purposes of this article, these different activities are grouped together into three categories: manufacture of ICT products; wholesale and distribution of ICT products; and ICT services.

ICT services was characterised (as in general is the rest of the LLM) by small workplaces (76 per cent of workplaces had less than 5 employees) and the majority of firms were small to medium-sized enterprises. ICT manufacturing, on the other hand, was characterised by large shares of the employment concentrated in small workplaces (74 per cent were in workplaces of 10 or fewer persons) and a few large firms (70 per cent in firms with over 500 employees). The firm structure of ICT trade was positioned in between the two extremes, with an emphasis in terms of employment on the mid-sized firms (between 10 and 100 employees); although the vast majority of workers were also in

Table 1. The Stockholm metropolitan labour market, all individuals of working age, 1990 and 1995

Category	1990	1995	Absolute change	Relative change (percentage)	Share 1990 (percentage)	Share 1995 (percentage)
Sex						
Men	482 673	437 046	− 45 627	− 9.4	50.9	51.2
Women	466 046	416 462	− 49 584	− 10.6	49.1	48.8
Age						
Up to 35	410 917	324 564	− 86 353	− 21.0	43.3	38.0
36-55	432 909	426 783	− 6 126	− 1.4	45.6	50.0
55 and older	104 893	102 161	− 2 732	− 2.6	11.1	12.0
Educational level[a]						
Low	509 050	396 504	− 112 546	− 22.1	53.6	46.4
Medium	285 262	294 545	9 283	3.2	30.1	34.5
High	144 281	154 982	10 701	7.4	15.2	18.2
Unknown	10 126	7 477	− 2 649	− 26.2	1.1	0.9
Country of origin						
Nordic countries	861 193	780 401	− 80 792	− 9.4	90.8	91.4
Central and east Europe	20 551	16 709	− 3 842	− 18.7	2.2	2.0
Western Europe	21 478	16 750	− 4 728	− 22.0	2.3	2.0
Other countries	45 497	39 648	− 5 849	− 12.9	4.8	4.7
Total employment	948 719	853 508	− 95 211	− 10.0	100	100

[a]Low = *Folkskola* (elementary school), *grundskola* (the 9 years of compulsory schooling) or 2 years *gymnasium* (senior high school); Medium = 3 years *gymnasium* or 2 years post-*gymnasium* education; High = more than 3 years post-*gymnasium* education or postgraduate training.

Source: GEOMETRO.

Table 2. Employment (all types of employee) in different parts of the Stockholm ICT cluster, 1995

SNI-code and industry	Number of employees	Share of total ICT employment
ICT manufacturing		
30020 Manufacture of office machinery and computers	1 673	2.7
31200 Manufacture of electricity distribution and control apparatus	445	0.7
31620 Manufacture of other electrical equipment	592	1.0
32100 Manufacture of electronic components	2 685	4.4
32200 Manufacture of TV/radio transmitters, line telephony and telegraphy	13 995	23.0
32300 Manufacture of TV/radio receivers, sound/video apparatus, etc.	1 252	2.1
33200 Manufacture of appliances for measuring, checking, testing, etc.	3 312	5.4
33300 Manufacture of industrial process control equipment	118	0.2
Sub-total	24 072	39.5
ICT trade		
51434 Wholesale of electrical equipment	2 146	3.5
51640 Wholesale of office machinery and equipment	11 337	18.6
51652 Wholesale of computerised materials handling equipment	233	0.4
51653 Wholesale of telecommunication equipment and electrical components	4 055	6.7
Sub-total	17 771	29.2
ICT services		
72100 Hardware consultancy	249	0.4
72201 Software consultancy	12 404	20.4
72202 Software supply	2 131	3.5
72300 Data processing	2 842	4.7
72400 Database activities	550	0.9
72500 Maintenance of office accounting and computing machinery	634	1.0
72600 Other computer-related activities	280	0.5
Sub-total	19 090	31.3
Total ICT	60 933	100

Source: GEOMETRO.

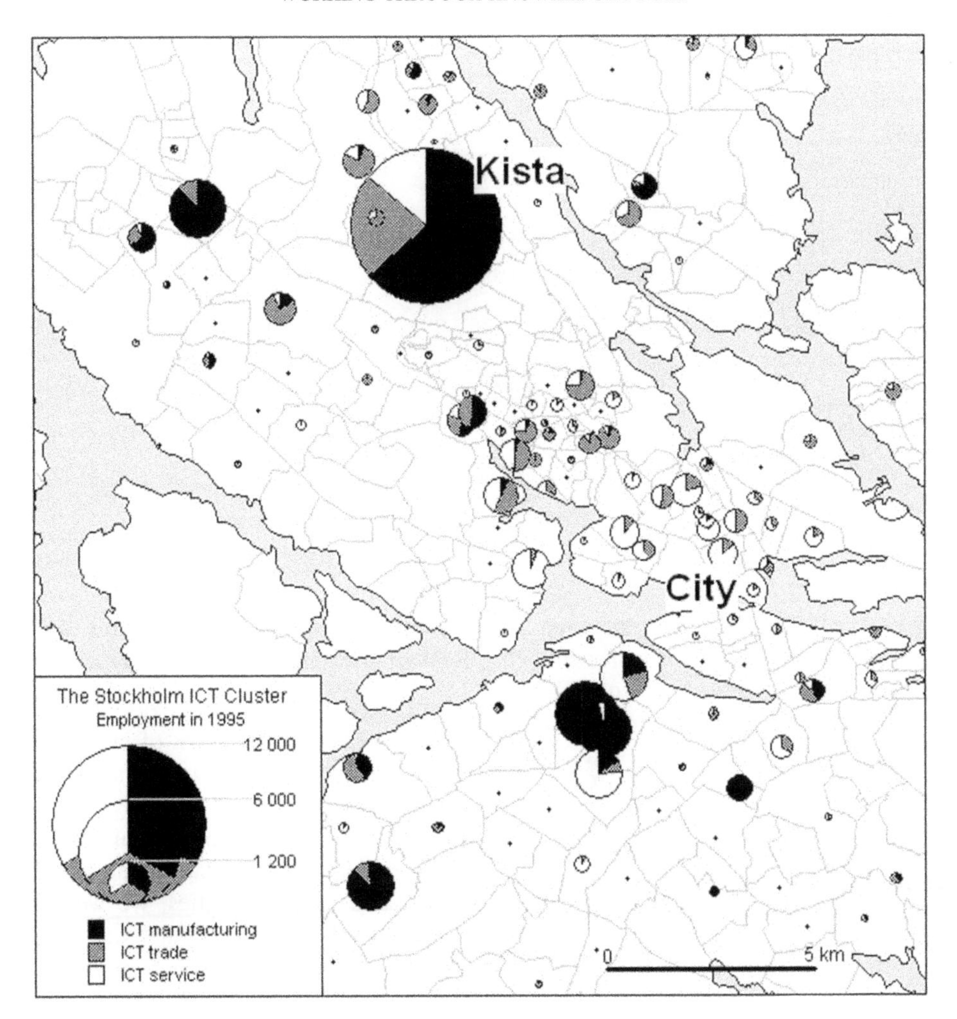

Figure 1. The Stockholm ICT cluster, 1995. *Note*: the map is based on small statistical areas (SAMS areas) that had at least 10 persons employed in the ICT cluster in 1995. Only the central part of the Stockholm LLM is shown on the map. *Source*: GEOMETRO.

smaller workplaces (76 per cent in workplaces of 10 or fewer persons). While each of these categories is heavily linked with the others (not least by strings of ownership and capital), we think that such a distinction is useful and says something important about the different components of the cluster: not least the different types of firm and workplace units that characterise each component, and the slightly different micro geographies and co-location tendencies that firms in each component appeared to share (see Figure 1).

As can be seen from Table 3, the ICT cluster performed considerably better in terms of employment growth than the rest of the Stockholm economy between 1990 and 1995. Whilst in general the labour market was in the grips of recession and shrunk by 10 per cent, ICT added 3 per cent to its employee base and in the case of ICT services grew by almost a quarter. This period indeed signalled a marked shift in the Stockholm ICT base away from manufacturing and distribution-type activities—mainly due to greater mechanisation and outsourcing abroad—towards the service-type activities that fuelled the bulk of the particular ICT bubble found in Sweden.

Table 3. Employment change in the ICT cluster and rest of labour market, all individuals of working age, 1990 and 1995

Category	1990	1995	Absolute change	Relative change
ICT manufacturing	24 841	24 072	− 769	− 3.1
ICT trade	18 896	17 771	− 1 125	− 6.0
ICT service	15 297	19 090	3 793	24.8
ICT total	59 034	60 933	1 899	3.2
Other	889 685	792 575	− 97 110	− 10.9
Total	948 719	853 508	− 95 211	− 10.0

Source: GEOMETRO.

It is not only, however, in the area of employment growth that the cluster proved to be at odds with the rest of the economy. It has a labour force that is predominantly male, younger, better educated and higher earning than the rest of the LLM (Table 4). When we look only at those with high education and/or high income (50 per cent of ICT employees, 30 per cent of rest of LLM's employees), we find that these differences are further accentuated (Table 4).

Finally, it was clear that the ICT cluster had a particular geographical distribution with a set of spatial agglomerations and centres standing out (Figure 1). Whilst it is true that many industrial activities also exhibited particular spatial patterns and centres, few other sectors, with the exception of newspapers and television and radio, demonstrated such strong spatial agglomerations. The map shows a concentration of ICT employment in the northern part of Stockholm where the Kista Science Park is located: the Science Park alone accounted for roughly 20 per cent of total employment in the ICT cluster in 1995. A distinct feature of the map is the locational separation of the different components of the ICT cluster: manufacturing activities were generally located further out from the city centre; computer services had a more central location; and wholesale trades were more often in semi-central locations.

Labour Flows into the Cluster and the Rest of the Regional Economy

It can be seen from the above, then, that the ICT cluster exhibited a set of characteristics and in particular growth rates that set it apart from the rest of the economy within the 1990–95 period. On the basis of such strong indicators as to the relative success of the cluster and its peculiarities, we were led to ask whether labour mobility was a further element that set it apart: and might perhaps help to explain some measure of its relative difference and growth.

Of course, the rather simplistic division between the cluster and the rest of the regional economy posited here in some way sets up a false contradiction: i.e. between a cluster and an amorphous non-clustered mass. In fact, the Stockholm region contains many other 'clusters', ranging from music (Power and Hallencreutz, 2002) to multimedia (Sandberg, 1999). Whilst we realise this is a problem, there is very little we can do about it: constructing a petri dish to be like a control economy is, of course, impossible. In our defence, we are not alone in doing this as researchers and regional development actors world-wide have posited similar divisions between a generalised regional economy and the cluster embedded within it.

As can be seen from Table 5, all the

Table 4. Labour characteristics for the ICT cluster and the rest of local labour market (LLM), all individuals of working age, 1990 and 1995

Category	ICT total	Rest of LLM total	ICT high education or income	Rest of LLM high education or income
Sex				
Men	70.7	47.7	81.4	63.2
Women	29.3	52.3	18.6	36.8
Age				
0–35	37.0	30.4	30.8	19.6
36–55	54.6	55.6	62.1	65.9
56 and over	8.4	14.0	7.1	13.2
Education[a]				
Low	32.1	49.9	19.9	25.3
Medium	44.4	31.9	44.5	29.5
High	23.2	17.8	35.4	45.0
Income[b]				
Mean	282	202	347	313
Median	253	184	291	259

[a] Low = *Folkskola* (elementary school), *grundskola* (the 9 years of compulsory schooling) or 2 years *gymnasium* (senior high school); Medium = 3 years *gymnasium* or 2 years post-*gymnasium* education; High = more than 3 years post-*gymnasium* education or postgraduate training.[b] Income from paid work in thousands of Swedish kronor (1995: US$1 = 7.13 SEK)

Source: GEOMETRO.

components of the ICT cluster had slightly lower levels of 'survivors'—those that managed to stay active in the LLM throughout the entire recessionary 1990–95 period—than the rest of the LLM and slightly higher inflows of workers new to the economy. ICT trade and services proved especially attractive it seems for people moving from other areas of the country and world into Stockholm (the in-movers category). ICT manufacturing by contrast attracted fewer in-movers but higher rates of workers new to the economy: which may reflect manufacturers' generally higher consumption of young workers in lower-grade assembly jobs and the like. It must be remembered, however, that during the period the rest of the LLM was shedding jobs at a much higher rate than the ICT cluster. With this in mind, it is clear that ICT firms were attracting higher rates of new labour and migrants/immigrants than is indicated by the figures above.

Labour Mobility between the Components of the ICT Cluster

Measuring basic rates of inflow into the economy and the cluster is interesting but tells us less of interest to the underlying theses of this paper than if we look at the rates of mobility between the components of the ICT cluster and between them and the rest of the LLM. Clear evidence suggests that mobility was much more intense within the cluster than in its relations with the rest of the LLM.

When applying a simple 'intensity measure', Figure 2 shows that mobility was more intense between the different components of the cluster than between the cluster and the outside world. In general, it is true to say that during the period the most intense labour mobility in the cluster involved people changing job within ICT by moving between the cluster's different components rather than

Table 5. Inflow of labour to the Stockholm ICT cluster and the rest of Stockholm LLM, all individuals of working age, 1990–95

	ICT manufacturing		ICT trade		ICT service		Rest of LLM	
	Number	Percentage	Number	Percentage	Number	Percentage	Number	Percentage
Survivors	18 423	76.5	14 109	79.4	15 068	78.9	637 047	80.4
New in the labour market	5 649	23.5	3 662	20.6	4 022	21.1	155 528	19.6
In-movers from another LLM in Sweden	1 407	5.8	1 308	7.4	1 458	7.6	50 880	6.4
Total	24 072	100	17 771	100	19 090	100	792 575	100

Source: GEOMETRO.

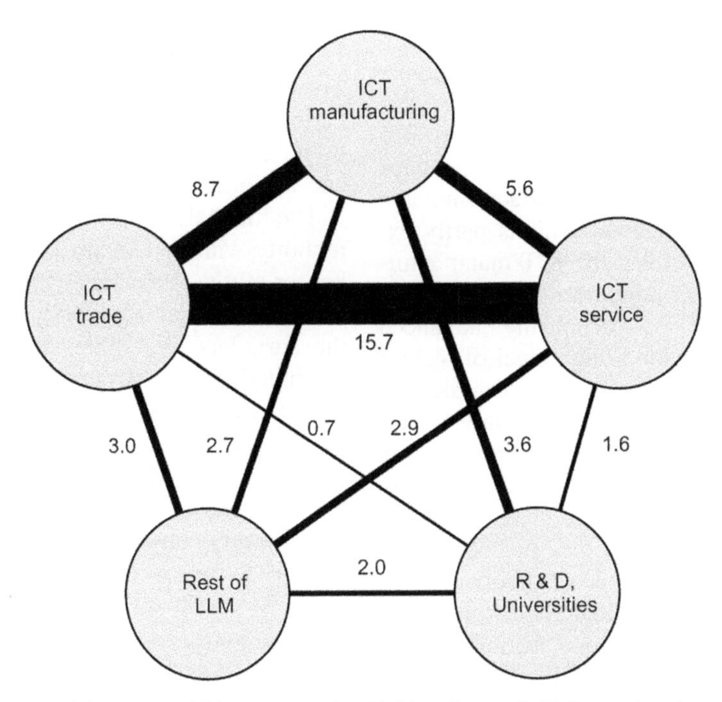

Figure 2. Intensity of labour mobility among the highly educated, high-earning 'survivors' between pairs of sectors, 1990–95. Notes: $I = (F_{ij} + F_{ji})/[(E_i + E_j)/2]*100$ where, I = intensity of mobility between sectors; F_{ij} = flow from sector i to sector j; F_{ji} = flow from sector j to sector i; E_i = total employment in sector i in 1995; and E_j = total employment in sector j in 1995. *Source*: GEOMETRO.

people moving from other sectors of the economy into ICT. Thus the cluster can definitely be said to have had something of an internal labour market that was relatively closed to individuals from outside the cluster's various parts: perhaps not surprising, bearing in mind the type of technical skills many ICT jobs require. Mobility was most intense between ICT trades and ICT services and mainly involved staff moving into the growing service firms; similarly, ICT services took in many more employees from ICT manufacturing than it sent. This may be an example of a matching process between the cluster's components, with labour mobility not only functioning to transfer skills, but also as an important element of industrial transformation. Whilst much of the mobility involved movements from the downsizing components (and the rest of the LLM) to the growing service component, this is not the entire story. Throughout the period, a significant minority of workers moved back and forth between the cluster's components: for example, leaving a large manufacturer for a job in a service firm for a short period and then going back to another manufacturer. Such back and forward movements undoubtedly have an effect both in linking the social worlds of different cluster components and in skills diffusion and knowledge transfer.

A large and indeed growing number of cluster and industrial innovation commentators and studies have suggested that a cluster or industrial system's relations to universities, higher education institutions and research and development centres/institutes are crucial to their industrial functioning, innovative capacity and long-term competitiveness (Cooke and Morgan, 1998; Porter, 1998; Lemarie *et al.*, 2001; Niosi and Bas, 2001; Santoro and Chakrabarti, 2001; Smith *et al.*, 2001; Acs *et al.*, 2002; Cooke, 2002, 2003; Furman *et al.*, 2002; Owen-Smith *et al.*, 2002). In Figure 2, we can see the intensity of labour flows, of the 'survivors', between the different components of the ICT cluster and the R&D/university sector. It is clear that it is only the manufacturing component of the ICT cluster that had a higher intensity of interaction with the research sector than did the rest of the LLM. In all, 275 survivors (an intensity of 3.6) moved back and forth between ICT manufacturing and research/higher education. This is partly explained by the fact that the ICT manufacturing component is dominated by firms very heavily focused on technically advanced products that demand a high level of technical and scientific training: for example, the mobile telecommunications manufacturers, such as Ericsson, that dominate this category. It may also be that the ICT trade and services components are much more heavily skewed towards small firms that may not be able to afford a full-time scientist of their own. Alternatively, it may be that Stockholm's research and higher education institutions are more useful or geared to ICT manufacturing than the services and trade components of the Stockholm ICT cluster.

It is unclear from our data whether such flows of full-time employees between the ICT cluster and the research and higher education sector are indicative of a further level of interconnections organised through—for example, research funding, personal contacts, advice-giving, etc. The data do indicate that labour mobility between the cluster and the research world outside does exist. However, the intensity of labour mobility is not significantly higher than it is for the rest of the LLM (although, of course, the rest of the LLM is full of highly knowledge-intensive industries) and does not come near to the intensity of intracluster and interfirm labour mobility. Whilst it is difficult to measure the other sorts of knowledge and idea flows that may link the cluster more strongly to the outside research environment, it appears that if labour mobility is an indicator of knowledge transference then this happens more intensely within the cluster itself and at firm and workplace levels. This may indicate that firms involved in ICT view interfirm and intrafirm co-operation and exchanges as more immediately fruitful than investing in deeper links with formal R&D or higher education units (de Propris, 2002).

Interfirm and Intrafirm Mobility

To what extent, then, are flows between the various components of the cluster matched by interfirm mobility within each component and by intrafirm mobility?

The first half of Table 6 shows the rates of mobility within elements of the labour market for all the 'survivors' (those active in the LLM throughout 1990–95): by focusing on the survivors, one filters out much of the 'mobility' arising from the arrival of employees new to the local labour market (such as school leavers, recent graduates, migrants). The levels of labour mobility seen for these survivors are interesting in that they reveal a very consistent pattern. In all but one case (ICT trade workers over 56), there were much higher levels of interfirm mobility in the ICT categories in comparison with the rest of the labour market and in comparison with selected other commercial activities in the same area of the production chain. Furthermore, these differences were extremely consistent across all the different age-groups. The ICT trade category was an exception to this general rule as interfirm mobility rates were only slightly higher and, in some cases, lower in comparison with circulation within other wholesale trade sectors. Again, the ICT service component led the way with higher interfirm mobility than all other areas of the cluster, other private-sector service firms and the rest of the LLM.

Due to the fact that many of the industry actors in both the region and the cluster are large firms—and if we think that the workplace itself is an important space for the exchange and creation of knowledge—it is important not to neglect mobility within these almost separate worlds. Movements of employees within the firms themselves are undoubtedly important to firms', and therein the cluster's, internal innovation processes and knowledge transfer. In Table 6, the intrafirm figures refer to movements between different departments and sections of firms and between different workplace locations owned by the same company. It must be remembered that some of these changes may

Table 6. The numbers of survivors (highly educated and/or high-earning) in 1995 who had been mobile during the period: had moved between firms (interfirm) and between workplaces (intrafirm) (figures are percentages of total employment in each category).

Category	All age-groups		Up to 35 years		Between 36 and 55 years		Older than 56 years	
	Interfirm	Intrafirm	Interfirm	Intrafirm	Interfirm	Intrafirm	Interfirm	Intrafirm
All individuals								
ICT manufacturing	51.7	5.5	60.7	4.7	49.1	5.7	42.5	6.0
Other manufacturing	33.5	3.9	42.5	3.3	32.3	4.1	23.4	4.1
ICT trade	41.4	5.0	53.4	3.1	36.9	5.8	23.9	6.1
Other trade	39.1	3.4	53.0	3.3	36.7	3.4	27.0	3.8
ICT services	67.1	1.6	73.5	1.0	64.9	1.8	53.3	4.3
Other business services	52.2	3.8	67.2	2.2	49.9	4.2	38.5	4.4
Rest of labour market	35.6	17.8	51.3	14.3	33.5	18.6	25.1	18.2
Individuals who only worked within one category								
Mobility within ICT cluster[a]	43.7	5.0	49.6	4.1	42.4	5.2	33.7	6.4
Mobility within other sectors[b]	36.0	14.0	50.4	10.8	33.9	14.9	25.7	14.4
Mobility within ICT manufacturing	40.6	7.3	44.2	7.3	39.9	7.4	37.4	7.0
Mobility within other manufacturing.	19.8	3.5	22.2	3.4	20.1	3.5	14.9	3.4
Mobility within ICT trades	23.9	6.0	30.4	4.4	22.1	6.7	13.0	6.8
Mobility within other trades	22.2	3.4	28.1	4.0	21.7	3.3	17.0	3.0
Mobility within ICT services	45.3	2.4	48.3	1.9	45.1	2.4	32.3	5.1
Mobility within other business services	34.1	4.7	46.5	3.4	32.6	5.0	24.4	4.8
Mobility within rest of LLM	31.7	18.8	45.9	15.9	30.0	19.6	22.8	18.8

[a]Mobility between firms and workplaces classified to the ICT cluster by individuals who were active both in 1990 and 1995.[b] Mobility between firms and workplaces outside the ICT cluster by individuals who were active both in 1990 and 1995.

Source: GEOMETRO.

reflect short-term company reorganisations, relocations due to temporary office renovations, etc. and therefore these figures should be taken as indicative and not as entirely precise figures.

Comparison with the 'rest of the labour market' figures is interesting as here we find consistently and significantly lower rates of *inter*firm labour mobility and consistently and significantly higher rates of *intra*firm labour mobility. Much of this difference is explained by the presence in this category of the public bodies that make up Sweden's extensive public sector. Many of these employers are large enough to form internal labour markets of their own and have tended to employ people on extremely long-term contracts that are almost impossible to break (for the employer, that is). Thus, this category includes large employers like the local council (*kommun*) and county council (*landsting*) which have a correspondingly large number of workplaces (schools, hospitals, etc.) in the Stockholm metropolitan area. Stockholm's position as capital city also means that large numbers of career civil servants work in the various state agencies and ministries located in the capital.

Between 1990 and 1995, intrafirm mobility was higher in other business services than in ICT services, but for both ICT manufacturing and trade intrafirm mobility was higher than it was for other private-sector manufacturing and trade firms. The lower rates of intrafirm mobility for ICT services reflect the fact that in other private-sector service activities, such as banking or business services, firms are generally much larger and have more workplaces. ICT services in this period were predominantly quite young firms that relied on as few as possible workplaces.

Interestingly, in the services and trade categories intrafirm mobility became even more pronounced with age than it was in other similar private-sector parts of the economy. Age was seen as more of a dividing factor than sex in determining one's mobility (Israelsson *et al.*, 2003). In general, the data show that, for individuals in all parts of the local labour market, interfirm mobility decreases sharply with age whilst intrafirm mobility slightly increases.

The lower half of Table 6 includes only those individuals that stayed during the period within the same category: for instance, a person who throughout the entire period only had jobs and workplaces in ICT manufacturing. This demonstrates that the patterns are similar for those people who throughout the entire period were dedicated to the one component of the ICT cluster. In short, the figures show that the higher interfirm and intrafirm mobility generally seen to have characterised the ICT cluster also hold true when one takes out the mobility resulting from people moving between the various components of the cluster and from the rest of the LLM. This means that cluster workers who have specialised in one particular activity area of the cluster have much higher rates of interfirm mobility. Furthermore, ICT workers had higher rates of intrafirm mobility than those in comparable private enterprises—although, like other types of ICT worker, they have generally lower levels of intrafirm mobility than workers in the rest of the LLM (which includes the public sector).

Conclusion

In conclusion, the analysis presented above provides evidence to support the hypothesis that a growing and innovative cluster is characterised by higher rates of labour mobility than in the rest of the region. All types of employees—short-term or longer-term members of the cluster, old and young, highly educated or high earners and normal workers—demonstrated significantly higher rates of mobility than their counterparts in the rest of the region's labour market. If we believe that labour mobility acts as a pipeline for the transfer of knowledge and new influences, then the higher rates of labour mobility seen in this cluster must surely have been beneficial to knowledge diffusion and creation in the cluster's components and firms. Furthermore, comparison of the three different ICT cluster components we constructed for the study confirms the idea posited in the introduction that labour mobility, within local labour markets and between sectors, could be seen as a matching process between labour supply and demand where resources and competences are continuously transmitted from static to growing sectors of the economy. The scale and nature of flows within and into the cluster strongly suggested that a matching process operated to supply the development of an increased service orientation in the ICT cluster.

Analysis of the intensity of labour flows also confirmed that transactions within the cluster were much higher than the cluster's relations to the surrounding labour market. Such intensified movement within the cluster may help to better bind together the cluster's different components and activities into an industrial system. In addition, the direction and nature of these flows indicate that something akin to a relatively closed internal labour market was at work within the cluster and that this phenomenon was a key indicator of the cluster's strength and borders. The fact that during the period the cluster also had a relatively small exchange of staff with R&D and higher education institutions further confirms this stand-offish picture.

In general, the ICT cluster had higher rates of interfirm labour mobility than the rest of the labour market and higher rates of intrafirm mobility than other comparable private-sector enterprises. However, it is clear that, for most of Stockholm's workers, intrafirm mobility was much higher than it was in the cluster. This can partly be explained by the existence of a number of large public bodies and private concerns with headquarters in Stockholm. However, this difference may also reflect an underlying tendency for ICT workers, especially the élite ones, to progress their careers through interfirm mobility and a tendency for ICT firms to prefer the influx of new staff and influences from the outside rather than from the inside. If these two tendencies are true, then they may explain much about the ways in which both workers and firms gain competitive advantage. To us, this suggests that two important areas for further research are: elaborated quantitative analyses of labour market mobility in other countries and labour market and regulatory contexts; and qualitative research that focuses on the actual mechanisms and causalities underlying the type of phenomena we have observed here.

Finally, we hope and believe that the dataset in itself makes this article an interesting contribution to the growing literature on the links between labour markets, knowledge flows and cluster development. There have been very few studies undertaken that are based upon a data-set as large as this one or using a set of data covering all individuals that have worked or earned money within a cluster and its surrounding milieu. Detailed analysis of individuals' working lives and trajectories, we argue, has much to offer geographers and others interested in regional economies, industrial dynamics, agglomerations, clusters, etc. A further implication of this study then is that analysing the intensity of labour mobility and linkages between activities is one of the more reliable ways of measuring, defining and delimiting both the existence of a cluster and the degree to which the supposed 'cluster' is functioning as an industrial cluster. The problem of how to define and measure clusters has continuously vexed analysts and a variety of methods have been used (Bergman and Feser, 1999; Bresson and Hu, 1999; Spielkamp and Vopel, 1999; Enright, 2000; Peters and Hood, 2000; Maskell, 2001; Malmberg and Maskell, 2002). On the basis of this study, one may say that using labour markets and individuals' labour mobility may offer a rewarding way forward in the empirical study of clusters and focus much-needed attention on the importance of labour mobility and labour markets in the formation and measurement of clusters and their underlying dynamics.

References

ACS, Z. J., FITZROY, F. R. and SMITH, I. (2002) High-technology employment and R&D in cities: heterogeneity vs specialization, *Annals of Regional Science,* 36(3), pp. 373–386.

ALMEIDA, P. and KOGUT, B. (1999) Localization of knowledge and the mobility of engineers in regional networks, *Management Science,* 45(7), pp. 905–917.

ANGEL, D. (1991) High-technology agglomeration and the labor market: the case of Silicon Valley, *Environment and Planning A,* 23(10), pp. 1501–1516.

BASANT, R. (2002) *Knowledge flows and industrial clusters. An analytical review of literature.* Indian Management, Ahmedabad, India (mimeograph).

BATHELT, H., MALMBERG, A. and MASKELL, P. (forthcoming) Clusters and knowledge: local buzz, global pipelines and the process of knowledge creation. *Progress in Human Geography,* 28(1).

BERGMAN, E. and FESER, E. (1999) *Industrial and regional clusters: concepts and comparative applications.* West Virginia University. The web book of regional science.

BIRKINSHAW, J. (1998) *The Information Technology Cluster in Stockholm: Prospects for Development, 1998/6.* Stockholm: Invest in Sweden Agency.

BIRKINSHAW, J. (2000) *The Information Technology Cluster in Stockholm: Changes from 1997 to 2000 and Prospects for Continued Growth.* Stockholm: Invest in Sweden Agency.

BRAUNERHJELM, P., CARLSSON, B., CETINDAMAR, D. and JOHANSSON, D. (2000) The old and new: the evolution of polymer and biomedical clusters in Ohio and Sweden, *Journal of Evolutionary Economics,* 10, pp. 471–488.

BRESSON, C. DE and HU, X. (1999) Identifying clusters of innovative activity: a new approach and a toolbox, in: OECD (Ed.) *Boosting Innovation: The Cluster Approach,* pp. 27–59 Paris: OECD.

CARNOY, M., CASTELLS, M. and BENNER, C. (1997) Labour market and employment practices in the age of flexibility: a case study of Silicon Valley, *International Labour Review,* 136(1), pp. 27–48.

COOKE, P. (2002) Regional innovation systems: general findings and some new evidence from biotechnology clusters, *Journal of Technology Transfer,* 27(1), pp. 133–145.

COOKE, P. (2003) Biotechnology clusters, 'Big Pharma' and the knowledge-driven economy, *International Journal of Technology Management,* 25(1/2), pp. 65–80.

COOKE, P. and MORGAN, K. (1998) *The Associational Economy: Firms, Regions and Innovation.* Oxford: Oxford University Press.

DAHL, M. (2002) *Embedded knowledge flows through labor mobility in regional clusters in Denmark.* Paper presented at DRUID's *New Economy Conference,* Copenhagen, June.

DAHL, M. and PEDERSEN, C. (2003) *Knowledge flows through informal contacts in industrial clusters: myths or realities?* Working Paper 03-01. DRUID, Copenhagen.

EDVARDSSON, I. (2000) *Competitive capitals: performance of local labour markets—an international comparison based on gross-stream data.* Working Paper 2000:7, Nordregio, Stockholm.

ENRIGHT, M. (2000) *Survey of the characterization of regional clusters: initial results.* Working Paper, Institute of Economic Policy and Business Strategy, Competitiveness Programme, University of Hong Kong.

FLORIDA, R. (2002) The economic geography of talent, *Annals of the Association of American Geographers,* 92(4), pp. 743–755.

FURMAN, J. L., PORTER, M. E. and STERN, S. (2002) The determinants of national innovative capacity, *Research Policy,* 31(6), pp. 899–933.

GERTLER, M. (2001) *Local knowledge: tacitness and the economic geography of context.* Paper presented at the *Annual Meeting of the Association of American Geographers,* New York, March.

GERTLER, M. (2003) Tacit knowledge and the economic geography of context, or the undefinable tacitness of being (there), *Journal of Economic Geography,* 3, pp. 75–100.

GUSTAFSSON, B.-Å. (2002) *Kreativa miljöer—Silicon Valley. Forum för småföretagsforskning.* Växjö: Växjö Universitet.

HALLENCREUTZ, D. and POWER, D. (2004) Cultural industry cluster building, in: P. OINAS and A. LAGENDIJK (Eds) *Proximity, Distance and Diversity: Issues on Economic Interaction and Local Development.* London: Ashgate Publishers (forthcoming).

ISRAELSSON, T., STRANNEFORS, T. and TYDÉN, H. (2003) *Geografisk rörlighet och arbetsgivarbyten. Ura 2003:1.* Stockholm: AMS utredningsenhet, Arbetsmarknadsstyrelsen.

IVARSSON, I. (1999) Competitive industry clusters and inward TNC investments: the case of Sweden, *Regional Studies,* 33, pp. 37–49.

JOHANISSON, B. and JONSON, M. (2000) *Partnership for regional development: an aluminium cluster in southern Sweden.* Paper presented at *C&C (Co-operation and Competition) Conference,* Växjö University, Sweden, November.

JONSSON, O. (1992) Det vårdindustriella klustret. Exempel på forsknings- och kunskapsintensiva klusters betydelse för Malmö/Lundregionens konkurrenskraft. Recept för Öresundsregionen?, in: C. WICHMAN-MATHISSEN and O. WÄRNERYD (Eds) *Öresundsbroförbindelsen,* pp. 55–60. Lund/Copenhagen: University of Lund and University of Copenhagen.

KRUGMAN, P. (1991) *Geography and Trade.* Cambridge, MA: MIT Press.

LAWTON-SMITH, H. and WATERS, R. (2003) *Rates of turnover in high-tech agglomerations: knowledge transfer in Oxfordshire and Cambridgeshire.* Paper presented at *Association of*

American Geographers Annual Meeting, New Orleans, March.

LEMARIE, S., V. MANGEMATIN and TORRE, A. (2001) Is the creation and development of biotech SMEs localised? Conclusions drawn from the French case, *Small Business Economics,* 17(1/2), pp. 61–76.

LUNDEQUIST, P. and POWER, D. (2002) Putting Porter into practice? Practices of regional cluster building: evidence from Sweden, *European Planning Studies,* 10(6), pp. 685–704.

MALMBERG, A. and MASKELL, P. (2002) The elusive concept of localization economies: towards a knowledge-based theory of spatial clustering, *Environment and Planning A,* 34, pp. 429–449.

MALMBERG, A. and POWER, D. (2004) On the role of global demand in local innovation processes, in: P. SHAPIRO and G. FUCHS (Eds) *Rethinking Regional Innovation and Change*. Dordrecht, Kluwer Academic Publishers (forthcoming).

MALMBERG, B. and SOMMESTAD, L. (2000) The hidden pulse of history: age transition and economic change in Sweden, 1820–2000, *Scandinavian Journal of History,* 25(1), pp. 131–146.

MARSHALL, A. (1920) *Principles of Economics*. London: Macmillan.

MARTIN, R. and SUNLEY, P. (2001) *Deconstructing clusters: chaotic concept or policy panacea?* Paper presented at *Regional Studies Association conference on "Regionalising the Knowledge Economy"*, London, November.

MASKELL, P. (2001) *Towards a knowledge-based theory of the geographical cluster*. Paper presented at *"Local Development: Issues of Competition, Collaboration and Territoriality": A joint conference of the IGU Commission on the Dynamics of Economic Spaces, and the IGU Study Group of Local Development*, Turin, July.

NIOSI, J. and BAS, T. G. (2001) The competencies of regions: Canada's clusters in biotechnology, *Small Business Economics,* 17(1/2), pp. 31–42.

NUTEK (NÄRINGS- OCH TEKNIKUTVECKLINGSVERKET/SWEDISH BUSINESS DEVELOPMENT AGENCY) (1998) *Kluster och klusterpolitik.* R 1998:29. NUTEK, Stockholm.

NUTEK (2000) *Arbetskraftens rörlighet : ett smörjmedel för tillväxt.* Rapport 2000:15. NUTEK, Stockholm.

NUTEK (2001) *Regionala vinnarkluster: en fråga om kompetensförsörjning, värdeskapande relationer och barriärbrytande visioner.* NUTEK Klustergruppen, Stockholm.

ORKAN, L. (1972) *Regionala variationer i företagens personalomsättning.* Avdelning för företagsekonomi, Ekonomiska institutionen, Umeå Universitet.

OWEN-SMITH, J., RICCABONI, M., PAMMOLLI, F. and POWELL, W. W. (2002) A comparison of US and European university–industry relations in the life sciences. *Management Science,* 48(1), pp. 24–43.

PERSSON, L. (2002) The Impact of Regional Labour Flows in the Swedish Knowledge Economy, in: H. HIGANO (Ed.) *The Region in the New Economy. An International Perspective on Regional Dynamics in the 21st Century,* pp. 221–237. Aldershot: Ashgate.

PETERS, E. and HOOD, H. (2000) Implementing the cluster approach, *International Studies of Management and Organization,* 30, pp. 68–90.

PORTER, M. E. (1998) Clusters and the new economics of competition, *Harvard Business Review,* 76(6), pp. 77–90.

POWER, D. (2002) 'Cultural industries' in Sweden: an assessment of their place in the Swedish economy, *Economic Geography,* 78(2), pp. 103–127.

POWER, D. and HALLENCREUTZ, D. (2002) Profiting from creativity? The music industry in Stockholm, Sweden and Kingston, Jamaica, *Environment and Planning A,* 34(10), pp. 1833–1854.

PROPRIS, L. DE (2002) Types of innovation and inter-firm co-operation, *Entrepreneurship and Regional Development,* 14(4), pp. 337–353.

SANDBERG, Å. (1999) The multimedia industry in Sweden and the emerging Stockholm cluster, in: H.-J. BRACZYK, G. FUCHS and H.-G. WOLF (Eds) *Multimedia and Regional Economic Restructuring*, pp. 238–251. London: Routledge.

SANTORO, M. D. and CHAKRABARTI, A. K. (2001) Corporate strategic objectives for establishing relationships with university research centers, *IEEE Transactions on Engineering Management,* 48(2), pp. 157–163.

SAXENIAN, A. (1994) *Regional Advantage: Culture and Competition in Silicon Valley and Route 128.* Cambridge, MA, Harvard University Press.

SMITH, H. L., KEEBLE, D., LAWSON, C. *ET AL.* (2001) University–business interaction in the Oxford and Cambridge regions, *Tijdschrift Voor Economische En Sociale Geografie,* 92(1), pp. 88–99.

SÖDERSTRÖM, H. Ed. (2001) *Kluster.se. Sweden in the New Economic Geography of Europe.* Stockholm: SNS Förlag.

SÖLVELL, Ö. (2000) *Sveriges Framtid—behovet av ökad klusterdynamik och förstärkta omvärldslänkar.* Rapport från Sveriges Tekniska Attacheer, Stockholm.

SÖLVELL, Ö., ZANDER, I. and PORTER, M. (1991) *Advantage Sweden.* Stockholm: Norstedt.

SPIELKAMP, A. and VOPEL, K. (1999) Mapping innovative clusters in national innovative systems, in: OECD (Ed.) *Boosting Innovation: The Cluster Approach*, pp. 91–123. Paris: OECD.

STATISTISKA CENTRALBYRÅN (SCB) AND EXPERT-

GRUPPEN FÖR FORSKNING OM REGIONAL UTVECKLING (1991) *Lokala arbetsmarknader och förvärvsregioner: Nya geografiska indelningar för regionala analyser*. Information om arbetsmarknaden 1991:7, SCB, Stockholm.

STORPER, M. and VENABLES, A. (2002) *Buzz: the economic force of the city*. Paper presented at *DRUID Summer Conference on "Industrial Dynamics of the New and Old Economy—Who Is Embracing Whom?"*, Copenhagen/Elsinore, June.

TOMLINSON, M. (2002) *Measuring competence and knowledge using employee surveys: evidence using the British skills survey of 1997*. Discussion Paper No. 50, Centre for Research on Innovation and Competition (CRIC), University of Manchester

TOMLINSON, M. and MILES, I. (1999) *The career trajectories of knowledge workers in mobilising human resources for innovation*. Proceedings from the Workshop on Science and Technology Markets. OECD, Paris.

TOTTERDILL, P. (1999) *Building communities of expertise: workplace innovation and regional development*. The Work Institute, The Nottingham Trent University, Nottingham.

Entrepreneurial Activity and the Dynamics of Technology-based Cluster Development: The Case of Ottawa

Richard T. Harrison, Sarah Y. Cooper and Colin M. Mason

[Paper first received, March 2003; in final form, October 2003]

Introduction

One of the most significant developments in economics, economic geography and management during the past decade has been the emergence of 'clusters' as a focus of academic research and debate. However, there is considerable diversity of approach. As Malmberg and Maskell observe

> Individual scholars and disciplines pursue different discourses and analyse the role of geographical space in the economic process as part of disparate wider agendas. Some frame their analysis in the context of specifying the role of local knowledge in a globalizing world economy, others relate such discussions to a broader transformation of capitalism, while yet others approach the role of localized clusters from the point of view of general business strategy, or attempt to bring economic geography into the core of mainstream economics (Malmberg and Maskell, 2002, p. 430).

Nevertheless, fundamental to this diverse body of research is the view that the locus of economic development in both advanced and developing economies is focused on the regional rather than the national scale, and that clustered, or locationally concentrated and integrated, economic development is central to innovation and competitiveness, to the development of the knowledge economy and to the process of economic growth and development (Krugman, 1991; Porter, 2000; Fujita *et al.*, 2000; Scott, 1998, 2001).

Although there are considerable doubts about the nature of the cluster development process (Martin and Sunley, 2003), industrial policy has increasingly been dominated by the cluster concept and governments in a wide range of economies have sought to adopt a cluster development policy for both traditional and high-technology industrial sectors (Doeringer and Terkla, 1996; World Bank, 2000; OECD, 1998, 2001; OECD–DATAR, 2001; Porter and Ackerman, 2001; Mytelka, 1999; Mytelka and Farinelli, 2000). Many of these policy applications, particularly those in the 'Porter tradition', assume some degree of universality in the cluster development process. In other words, the development of, in Porter's (1998) definition, a geographically proximate group of interconnected companies and associated institutions in a particular field linked by commonalities and complementarities, is assumed to follow common development paths and function in a common manner. However, even among advocates of cluster-based economic development policies, there is a growing recognition that not all clusters are the same (see, for example, Swann and Prevezer, 1996; Garnsey, 1998; Hendry and Brown, 2001): "clusters come in many forms, each of which has a unique development trajectory, principles of organisation and specific problems" (Mytelka and Farinelli, 2000, p. 11). In this paper, we focus on entrepreneurial dynamics as one key process in the origin and development of a technology industry cluster in Ottawa, Canada. The question "Where have the entrepreneurs have come from?" has received surprisingly little attention in prior studies of the evolution of technology-based clusters. Our distinctive contribution in this paper is to address this question. In doing so, we take a broader perspective on the origins of the entrepreneur than has hitherto been the case by considering their career and spatial mobility prior to start-up. The next section of the paper highlights the importance of entrepreneurial activity in cluster development. This is followed by a detailed examination of how the process of entrepreneurial activity has operated in Ottawa.

Entrepreneurial Dynamics and Cluster Development

Cluster definition is, as Martin and Sunley (2003) have recently pointed out, problematic. However, Mytelka and Farinelli's (2000, p. 7) definition of clusters as being "understood mainly in terms of spatial agglomerations of enterprises and related supplier and service industries" is as serviceable as any other. Swann (1998; Swann *et al.*, 1998) has provided a more specific interpretation of the cluster concept as covering a spectrum of definitions from the shallow (but easy to measure) aspects of co-location, co-location and technological proximity, and input–output table complementarities (which cover the key aspects of the Porter (1998) definition quoted above), to the rich (but difficult to measure) aspects of Marshallian externalities, explicit collaboration and informal knowledge spillovers. In this paper, we do not address specifically the definitional issue and work instead with an accepted identification of a specific technology cluster—in this case, Ottawa (Chamberlin and de la Mothe, 2003)—which is evident at the very least at the shallow level described by Swann (1998).

There have been a number of approaches to cluster research and resulting types of knowledge on spatial clustering (Malmberg and Maskell, 2002). First, there is a strong body of ideographic, historical work on the origin and development of various types of clusters. Secondly, there is more theoretically inclined research, which has dominated the recent debates on clusters, that tries to identify the mechanisms that give economic and other advantages to the individual firm located in a cluster, whether these reside in cost reduction (Marshall, 1890; Krugman, 1991; Scott, 1998) or in knowledge spillovers and learning and adaptation processes (Keeble and Wilkinson, 2000a, 2000b; MacKinnon *et al.*, 2002; Pinch *et al.*, 2003). Our approach in this paper is grounded in the ideographic and historical approach to cluster development. Specifically, we follow Wolfe in arguing that a limitation of much clusters research and policy prescription is

> A tendency to focus on the descriptive and analytical at the expense of the dynamic and explanatory level. What local economic authorities and policy-makers at all levels of government are most interested in is the process by which clusters take hold and expand in the context of local and regional economies (Wolfe, 2002a, p. 12).

In a similar vein, Feldman argues that, once established,

> clusters benefit from virtuous, self-reinforcing processes. A critical question is how these entrepreneurial processes begin, take hold and transform a regional economy (Feldman, 2001, pp. 861–862).

Key to this process is the role played by entrepreneurs, especially scientist- and engineer-turned entrepreneurs (Storper, 1993), who are already working in the local/regional economy. In what Malmberg and Maskell (2002) refer to as the *pioneering* stage of cluster development, these individuals use the expertise or ideas that they have developed in their employing organisation (the *incubator* organisation) to set up their own business to exploit this know-how (Keeble, 2000). This process then develops a momentum of its own as more and more entrepreneurs, either individually or in teams, leave their existing organisations to start their own businesses—what Malmberg and Maskell (2002) term the *imitation* stage.

The literature on cluster dynamics emphasises that the new venture creation process is local, with entrepreneurs starting their new businesses in the same geographical area as that of their immediate past employer (Cooper and Folta, 2000). This feature is illustrated in the company genealogies of various technology clusters, which 'map' the relationship between incubator organisations (i.e. the companies which entrepreneurs left to start their own companies and spin-offs) and those subsequent entrepreneurial start-ups (see, for example, Segal Quince and Partners, 1985; Lawton-Smith, 1991; Neck *et al.*, 1999; US Small Business Administration, 2000). These genealogies reveal a variety of different types of incubator organisations (Miller and Côté, 1987), including both large and small businesses (including the spin-offs of the pioneering era) and also universities and research organisations. A consistent feature of these studies is that they reveal a small number of very significant incubator organisations that have spawned large numbers of entrepreneurs.

For entrepreneurs, there are a number of advantages to remaining in the same area in which they were working and living (Cooper and Folta, 2000). It enables them to exploit the knowledge, networks and reputation that they have acquired in their previous employment. This provides significant economic benefits in terms of resource acquisition (for example, suppliers, finance, employees) and identification of market opportunities (although these are often likely to be non-local). By the time that the imitation stage has been reached there will also be agglomeration and localisation economies which reinforce the economic advantages of remaining local (Parr, 2002), notably cost-sharing for collective resources, the development of a local labour market for specialised skills, reduced interfirm shipment and transaction costs, and knowledge spillovers, learning and adaptation (Malmberg and Maskell, 2002). Personal considerations also play an important role (Garnsey, 1998). Regional collective learning theory emphasises the importance of individuals who leave existing firms to start their own businesses as being a crucial mechanism on account of the technological and management know-how and 'embodied expertise' which they carry (Keeble and Wilkinson, 2000a).

However, this paper argues that previous studies of technology clusters have overemphasised the significance of the incubator organisation—that is, the *immediate past employment* of the founder of the new business—in the new venture creation process. This focus on the immediate past employment of entrepreneurs has created a particular view of the entrepreneurial dynamics of technology clusters which overemphasises the locally embedded nature of the process. Entrepreneurs are shown to have worked locally immediately before starting their own firm. It is then inferred from this that the spin-off process involves the local diffusion of technological knowledge from the incubator organisation to the spin-off companies. The studies of high-technology clusters that comprised the Keeble-and-Wilkinson-led (1999, 2000b) project on 'Networks, Collective Learning and Research and Technology Development in Europe' are typical in this respect. As a consequence, the role of national and global links is downplayed.

This paper seeks to demonstrate the limited perspective of previous studies of the entrepreneurial dynamics of technology clusters by means of a study of the 'geographies and histories' of entrepreneurs in Ottawa's technology cluster. It makes three main points. First, the concept of a single incubator organisation for a new business is problematic in a context where most new technology-based firms have multiple founders. Secondly, characterising entrepreneurs as being 'local' is misleading: although their immediate past employer is almost invariably local most have exhibited considerable spatial mobility in their careers. This points to the need to situate technology clusters in the context of national and international scientific and professional labour markets. Specifically, the study highlights the role of 'magnet organisations' in attracting talented individuals to the Ottawa region, a proportion of whom subsequently started their own businesses. Thirdly, and following from the previous point, entrepreneurial learning is not confined to the incubator organisation nor is it necessarily confined to the cluster. Most entrepreneurs have worked for more than one organisation prior to starting their own business and their new venture draws upon their whole career experience, rather than just that of their immediate past employer.

The Ottawa Context

Ottawa—Canada's capital city and the country's fourth-largest metropolitan region with a population of just over 1 million at the 2001 census—is often thought to be simply a government town. In fact, in recent years it has become a world-class technology centre which is dominated by ICT industries—notably, telecommunications equipment (the oldest and largest sub-sector), semiconductors, software and an emerging photonics cluster. In addition, Ottawa also has an emerging life sciences sector, specialising in biotechnology and biomedical devices (Wolfe, 2002b; Chamberlin and de la Mothe, 2003). Self-identified as 'Silicon Valley North', it features in most listings of 'silicon valleys' around the world. At the peak of the technology boom in 2000, Ottawa had more than 1200 technology companies which collectively employed 85 000 people, exceeding (for the first time) the federal government workforce. The world-wide technology downturn, which had its biggest impact on the telecommunications sector, resulted in a significant fall in employment, with more than 20 000 technology jobs lost in 2001/02 as a result of redundancies by leading employers such as Nortel, Alcatel and JDS Uniphase and failures in the small company sector. Thus, by mid 2003, employment in the technology sector had fallen to 64 500.[1] Nevertheless, there has been no major population migration from the city and house prices have remained stable and have even appreciated (Chamberlin and de la Mothe, 2003). Indeed, there has been an increase in the number of technology companies to around 1500 in 2003 (*Ottawa Business Journal*, 23 June 2003), suggesting that many technology workers have opted to start their own companies rather than chasing possibly illusory job opportunities in other cities (Chamberlin and de la Mothe, 2003). An increase in the federal government's workforce in Ottawa to over 100 000 has also served to soften the impact of the job losses in the technology sector.

Ottawa has a variety of world-class R&D facilities and capabilities which has created a large community of scientists and technologists. Over 90 per cent of Canada's R&D in industrial telecommunications is undertaken in the Ottawa region. It is the primary centre for the federal government's spending on science and technology, which is conducted through such agencies as the National Research Council (NRC), Communications Research Centre (CRC), Atomic Energy of Canada Ltd and by major government departments. The Canadian Photonics Fabrication Centre has recently been developed in Ottawa by the NRC. Ottawa is also the location of Communications and Information Technology Ontario (CITO), a provincial research establishment. In addition, it is home to many leading private-sector technology companies including Nortel Networks (formerly Northern Electric), Newbridge Networks (acquired by Alcatel in 2000), Corel Corporation, JDS Uniphase, Mitel Corporation and Tundra Semiconductors. Nortel undertakes a large share of its world-wide research in Ottawa. In recent years, the recognition of Ottawa as an important centre for telecoms technology has led to global companies such as Cisco Systems, Nokia, Cadence Design Systems and Premisys Telecommunications seeking a presence in the region either through greenfield-site investment or acquisition of local companies. In the life sciences sector, there are 20 research institutes, notably the NRC, Ottawa Health Research Institute and the University of Ottawa Heart Institute.

The NRC was founded in 1916 as government's primary adviser and supporter of scientific and industrial research. The foundations of the NRC's growth into a world-class research institution were laid during the Second World War when it played a central role in research in a wide range of fields. The origins of Ottawa's technology cluster can be traced back to the early post-war period with the founding of Computing Devices of Canada Ltd in 1948 as a spin-out from NRC telecommunications laboratory to produce military computer hardware. It was taken over by Control Data Corporation of the US in 1969. Both NRC and CDC have been the origin of many other spin-out companies since then. A further significant building-block was the decision of Northern Telecom (the forerunner of Bell Northern Research and later Nortel Networks) to open an R&D facility in Ottawa in 1961.[2] Within five years of its opening, it had 800 employees (Chamberlin and de la Mothe, 2003). This facility has become one of the largest and most innovative telecommunications research centres in the world and has been instrumental in creating a pool of highly skilled knowledge workers in Ottawa. It too has been a major source of spin-out companies. Microsystems International—a subsidiary of Northern Telecom—was one of the earliest developers of semiconductor technology. It was closed in the mid 1970s when the chip business took a temporary downturn. Microsystems had attracted a large number of highly skilled IT engineers and scientists to Ottawa. Following its closure, many of these staff went to work for Nortel Networks but some started their own companies. Indeed, more than 20 start-ups can be attributed to former Microsystems employees.

The movement of firms into the cluster from elsewhere, to capitalise on the localisation, agglomeration and knowledge spillover benefits of cluster location—a process which is central to much of the debate on the evolution of clusters (see Porter, 1998; Krugman, 1991; Swann, 1998; Beaudry and Swann, 2001)—has been relatively unimportant in the Ottawa case. Although Ottawa contains several branch operations of multinational enterprises, its emergence as a major technology cluster is largely a 'home-grown' phenomenon, attributable to the start-up and growth of entrepreneurial companies over the past 30 years.[3] The first wave of start-ups—in the 1950s and 1960s—were mainly defence-oriented businesses; other waves of start-ups occurred in the 1970s and 1980s in such sectors as telecommunications (for example, Mitel), computing equipment (Cadence, Gandalf Technologies, Norpak), lasers (Lumonics) and software (Cognos), while the latest wave of start-ups, which started in the late 1990s, is primarily telecoms-related (such as photonics, optical networking, wireless), with an emerging biotech/life sciences cluster becoming a further source of new ventures.[4] Ottawa now has several entrepreneurs on their third or fourth start-ups. For example, Terry Matthews who, with Michael Copland started Mitel, which was sold to British Telecom in the early 1980s, went on to start Newbridge Networks and most recently March Networks, while Copland, who founded Corel, now runs ZIM Technologies. Another prominent entrepreneur, Antoine Paquin, started Skystone Systems and Philstar Semiconductors. These, and other serial entrepreneurs, have also been involved in financing new businesses, both directly as business angels, and also indirectly by establishing or investing in local venture capital firms (Mason *et al.*, 2002). However, on the negative side, critics point to the paucity of fast-growing, sustainable high-technology companies to have emerged from the most recent waves of start-ups. This can be attributed, at least in part, to the tendency of local firms to pursue a 'grow-to-sell' strategy where they sell out to multinational companies rather than striving to become global enterprises themselves (Chamberlin and de la Mothe, 2003).

Methodology

Researchers have employed a range of methodological approaches to investigate technical entrepreneurship. Several early US studies employed quantitative approaches to gather information on hundreds of firms (A. C. Cooper, 1973; Roberts, 1991) which provided data on aggregate trends but little anecdotal or contextual information. Bygrave (1988) considers that the entrepreneurial process is best understood through softer, qualitative investigation of the entrepreneur's background, permitting exploration of internal and external influences. Thus, for this research, a combination of survey techniques involving both quantitative and qualitative approaches was used. Aggregate data on entrepreneurial activity were gathered through a fax survey conducted among firms in the Ottawa region and trends identified during the large-scale survey were explored in detail during follow-up interviews with a small sample of entrepreneurs. Given the focus on 'entrepreneurial histories and geographies', we sought to explore the significance of employment prior to the incubator organisation in the start-up process—for example, as a source of expertise, ideas, motivation or resources. Hence, the fax survey collected data on the founders' employment in the immediate pre-start-up stage. It also investigated a range of business issues, including venture formation, composition and experience of the start-up team and the source of the technology on which the firm was based. The personal interviews[5] provided an additional dimension by probing the career histories of a smaller sample of entrepreneurs and, if relevant, the circumstances under which those entrepreneurs who were in-migrants to Ottawa came to the region. Case studies are presented on the genesis of three firms to illustrate career and migratory patterns which were apparent within the wider sample. The paper also draws on information from a parallel survey of venture capitalists who have invested in Ottawa's technology sector (Mason *et al.*, 2002). All of the surveys were undertaken in 2000, close to the peak of the technology boom.

The questionnaire was faxed to the CEOs of 423 technology product firms and 343 technology services firms in the Ottawa region. Firms were identified from the *Ottawa Business Journal* Electronic Directory (pro-

Table 1. Year of formation, by type of firm

Formation date	Technology product firms		Technology service firms		Total	
	n	Percentage	*n*	Percentage	*n*	Percentage
Pre-1988	49	44.1	30	40.0	79	42.5
1988–96	44	39.6	35	46.7	79	42.5
Post-1996	18	16.2	10	13.3	28	15.1
Total	111	100	75	100	186	100

Source: fax survey.

duced on CD-ROM by the *Ottawa Business Journal* in partnership with Ottawa Economic Development Corporation). The questionnaire was accompanied by a one-page outline of the research project and a letter from the research team seeking the co-operation of the CEO. A follow-up letter was faxed to non-respondents two weeks after the original fax was sent and, in turn, this was followed by a telephone call several days later. In all, 111 useable responses were obtained from technology product firms (32 per cent response rate) and 75 responses were received from technology service businesses (19 per cent). Respondent firms represented the broad spectrum of technologies present in the Ottawa region.

Entrepreneurs in 20 of the technology firms from which data were collected via the fax questionnaire then participated in in-depth interviews to explore a range of issues associated with the entrepreneurial process. The interviews were generally with the lead entrepreneur who also provided information on the career histories of the other members of the founding team.[6] These 20 firms had a total of 43 founders. All of the interviewed firms were founded in the Ottawa–Carleton region, with 19 of the 20 established as independent ventures. The final firm was established as a Crown Corporation by the federal government and was included because it was one of the region's early technology firms. The 20 firms were drawn from contrasting sectors of biotechnology (4), software (8) and electronics (8). This decision was informed by previous research on firms in different new-technology sectors that have revealed some sectoral variations in the characteristics of founders (S. Y. Cooper, 1996). The inclusion of biotechnology firms within the interview sample was prompted by the trend for increasing convergence between technologies.[7]

Temporal Trends in Technology Start-ups in the Ottawa Region

Our ability to comment on temporal trends in business start-ups in Ottawa is constrained by the survivor and non-response biases in our fax survey data. With this important *caveat*, it would appear that the formation of technology firms in Ottawa has taken place in a number of waves, with periods of expansion being followed by consolidation and then subsequent growth in the number of ventures established. Data from the fax survey on the start-up date of firms indicated that 43 per cent of respondent firms were established prior to 1988, the same proportion in the period 1988–96, with the remaining 15 per cent formed between 1997 and 1999. The balance of product and service firms formed in each period differs (Table 1). Proportionately larger numbers of technology product firms were established in the earliest and most recent periods, while greater numbers of technology service firms were formed between 1988 and 1996. The dominance of technology service firms in the middle phase is likely to reflect a growing recognition of Ottawa as the home to innovative and leading-edge technology firms which stimulated the formation of service firms to support on-going and emerging technology development activities.

The information in Table 2 on firm employment levels indicates that some of the largest firms were those established in the period immediately prior to the survey (1997–99). Particularly marked is the proportion of technology product firms employing 50 people or more that were started during this period of rapid new technology firm formation. These firms made a significant contribution to the increase in the region's technology employment during the latter half of the 1990s.

The vast majority of firms in both the product and service categories were established as independent firms: only 8 firms (4.3 per cent) were established as subsidiaries. Of the 5 technology service firms established as subsidiaries, 3 had a foreign parent compared with 2 of the 3 technology product subsidiaries. In addition, a number of firms had been acquired subsequent to formation so that by the time of survey, 39 firms (21.3 per cent) were no longer independent. Acquisition levels were the same amongst both technology product firms and those in the technology service category. This highlights the importance of entrepreneurial exit through acquisition (Storper, 1993; Garnsey and Cannon-Brookes, 1993; Lindstrom-Dahlstrand, 2000) and entrepreneurial recycling (Bahrami and Evans, 1995; Mason and Harrison, 2003) as important processes in the dynamics of technology clusters. However, there is also a potential danger, as noted earlier, that the continued growth of the technology cluster will be threatened unless some local firms achieve the transition from SME to large firm, rather than selling out to multinational companies (Chamberlin and de la Mothe, 2003).

Just over 1 in 10 firms did not originate in Ottawa (22 firms, 11.7 per cent) but had either relocated or been established as a subsidiary of a business based elsewhere. A larger number of technology service (12 firms, 15.8 per cent) than technology product firms (10 firms, 8.9 per cent) were non-local in these senses. The market opportunities created by the extensive technology firm base in Ottawa offered a prime attraction for technology service firms seeking to relocate.

The Entrepreneurial Team

With its emphasis on a single incubator organisation, the technology clusters literature implies that each spin-off company is started by a 'lone wolf' entrepreneur. The reality is that multifounder start-up teams are the norm amongst technology ventures both in Ottawa and elsewhere. Entrepreneurial teams were responsible for the formation of more than three-quarters of firms in the technology product ($n = 85$, 77 per cent) and technology service ($n = 57$, 76 per cent) categories. Two-person teams started 43 per cent of the technology product firms (47 firms) compared with 37 per cent of the supplier category (28 firms). The average team size was 2.6 founders for the product technology sample and 2.5 for the technology services firm group. Other studies have also noted the prevalence of founding teams in technology start-ups (A. C. Cooper, 1986; S. Y. Cooper, 1996; Roberts, 1991), typically accounting for between two-thirds and three-quarters of all firms (S. Y. Cooper, 1996; Oakey *et al.*, 1990). New technology ventures that are founded by teams have a greater variety and depth of talent and experience (often combining technology and managerial experience); the founders work more total hours and are able to devote more time to sales and marketing and start with greater initial capital. Moreover, because venture capital funds prefer to invest in businesses with large management teams, they are better positioned to raise external finance (Roberts, 1991). Given these attributes of team starts, it is not surprising that other studies have found that technology start-ups that are started by teams are more successful than those started by single entrepreneurs (for example, Oakey, 1995). Roberts observes that

> one of the strongest and most consistent correlates of later success of the new technology enterprise is the number of

Table 2. Year of formation and employment by type of firm

Number of employees in 2000	Technology product firms: year of formation				Technology service firms: year of formation			
	Pre-1988 *n* %	1988–96 *n* %	1997–2000 *n* %	Total *n* %	Pre-1988 *n* %	1988–96 *n* %	1997–2000 *n* %	Total *n* %
< 10	2 (10.0)	3 (10.3)	1 (9.1)	6 (10.0)	6 (28.6)	11 (33.3)	5 (62.5)	22 (35.5)
10–24	1 (5.0)	9 (31.0)	0 (0)	10 (16.7)	6 (28.6)	8 (24.2)	1 (12.5)	15 (24.2)
25–49	3 (15.0)	4 (13.8)	3 (27.3)	10 (16.7)	2 (9.5)	5 (15.2)	0 (0)	7 (11.3)
50 +	14 (70.0)	13 (44.8)	7 (63.6)	34 (56.7)	7 (33.3)	9 (27.3)	2 (25.0)	18 (29.0)
Total	20	29	11	60	21	33	8	62

Source: fax survey.

> original company founders ... [no matter how] success is measured (Roberts, 1991, p. 258).

Both the fax (Table 3) and interview surveys found that in the majority of multiple founder businesses at least some, and in many cases all, of the founders had worked together previously and, therefore, shared in common part of their career history. However, they had not spent all the time working together in the same department or company and therefore brought different experiences to the team. In some cases, the founders had worked together in the incubator organisation immediately prior to the formation of the new venture. In other cases, they had worked together either at an earlier stage in their careers or at both an earlier stage in their careers and also in the incubator organisation. This is illustrated in the case of Routes AstroEngineering (Figure 1)—a start-up which was founded by entrepreneurs who had worked together earlier in their careers. Thus, previous workplaces play an important role as settings in which future business partners meet and work together. In the cases where the team members had not worked together, they were normally friends or acquaintances.[8]

The interview survey confirmed that in most cases the teams had complementary skill-sets. Indeed, there were only 3 team starts that were clearly unbalanced: in each case, all of the founders had technical expertise. In all of the other cases, team members contributed a complementary blend of technical and business skills. In one instance, 2 members of the 4-man team had over 30 years each of experience in manufacturing environments, experience that complemented the technical and financial expertise of the remaining team members and was essential for the new venture. In 13 cases, the principal founder took up a technical or research-based functional role, with the other founders taking on business/management functions. None of the principal founders took on a role in finance while in 5 firms one of the co-founders assumed that role. However, founding teams are dynamic, and just under 30 per cent of the founding entrepreneurs were no longer with their firm by the time of the survey in 2000, most having moved on to other activities or, in a minority of cases, had retired or died.

The Incubator Organisation

Entrepreneurs in Ottawa came from a wide range of incubator organisations (Table 4). The largest single source of entrepreneurs is small firms. This confirms the conventional wisdom that small firms are better 'training grounds' for entrepreneurs than large firms, both for technology firms (A. C. Cooper, 1970, 1973) and in general (Cross, 1981; Fothergill and Gudgin, 1982; O'Farrell, 1986). Contrary to other studies (for example, Miller and Côté, 1987; Lawton-Smith, 1996), federal government laboratories have also been a relatively significant source of entrepreneurs.

One of the most significant aspects of the incubator organisation is that it is generally the setting in which the decision to start a new business is made. In other words, the incubator organisation is very often the source of the 'displacement' or 'trigger event' (which can either be positive or negative) which sets in motion the business start-up process (Shapero, 1984). Each of the firms in the interview survey was started because of factors related to the employment situation of one or more of the founders immediately prior to start-up. In half of these cases, the founder(s) had identified a market opportunity which arose from their knowledge of the technology, customer or supplier environments of the incubator organisation. In a further three cases, the firms were created to exploit technology which had been developed by the incubator firm, but which its senior management had decided not to exploit commercially. One of these entrepreneurs had been responsible for establishing a new facility on behalf of his employer to develop a new product line. Having set it up he was "basically asked to knock it down again because of a change in

Table 3. Incidence of founders working together previously

Incidence of founders working together	Technology product firms		Technology service firms		Total	
	n	Percentage	*n*	Percentage	*n*	Percentage
All worked together	54	63.5	30	55.6	84	60.4
Some worked together	13	15.3	11	20.4	24	17.3
None worked together	18	21.2	13	24.1	31	22.3
Total	85	100	54	100	139	100

Source: fax survey.

Figure 1. Routes AstroEngineering. *Source*: interview.

Routes Astro Engineering was established in Kanata in 1988 by three engineers, Tony Payne, Ron Buckingham and Geoff Barnes. Payne was raised in Alberta and attended high school in British Columbia before moving to eastern Canada to undertake his undergraduate studies in maths and physics and postgraduate Masters and PhD studies in aerospace engineering in Quebec and Ontario. His first post on graduation was that of sub-systems supervisor on a satellite project at the Communications Research Center (CRC), in Kanata, Ottawa. During his five years at the CRC, Payne saw the entire flow of the spacecraft programme and established a strong network of contacts in the aerospace sector. It was while working at the CRC that he met Ron Buckingham. Ron is a New Yorker who had undertaken undergraduate and Masters studies in engineering at New York University (NYU) before moving to Los Angeles, California, to work for a major aerospace company as a mechanical thermal designer on the Apollo programme. After about eight years on the west coast, he moved to Kanata to work at the CRC, where he met Payne.

In the late 1970s, Payne and Buckingham had the opportunity to join a young local start-up venture. Canadian Astronautics had been established by three founders two years earlier; Payne and Buckingham joined as the sixth and seventh members of the team. Over the next 10 years, they saw the firm grow to employ around 300 people, by which time Payne had risen to the position of Vice President of Advanced Systems. As a result of changes which occurred within Canadian Astronautics, Payne and Buckingham recognised that greater opportunities existed outside the company. Payne commented that he and Buckingham, "were getting a little bit away from the kind of area that we had our best expertise in ... so we returned to out roots, to do space systems design work which is what the two of us had been doing for some years in the past ... the two of us were becoming somewhat redirected into things like military electronics and things like that which were not really our areas of expertise or interest". Geoff Barnes, a personal friend of Payne and his neighbour for a number of years, was the third member of the team. Geoff was the only one of the trio to have been brought up in Ottawa, although he left the city to attend Queens University in Kingston where he studied civil engineering. After graduation, he worked on a number of major civil engineering projects including Expo 67. Eventually he joined a local Ottawa company, before leaving to establish his own practice in the city. The fourth partner joined a little bit later. According to Payne, "he left [Canadian Astronautics] for essentially the same reasons. He was into an engineering function where he was doing an awful lot of paper handling and specification brochures. He is a designer and that is what he wanted to do". Routes AstroEngineering had the opportunity to undertake work for the Canadian Space Agency, much of which had previously been performed by Canadian Astronautics; due to changes in direction Canadian Astronautics no longer did such work.

To a certain extent, Payne and Buckingham had parallel careers prior to establishing Routes AstroEngineering. The time that Buckingham spent in California on the Apollo programme enriched the networks and competences of the team. Important to the formation and development of Routes were the complementary skills, expertise and personal networks which the team members brought to the table. Payne and Buckingham brought technical and management knowledge and a rich network to the start-up; high-level contacts in NASA and the European and Canadian Space Agencies facilitated the establishment of commercial relationships essential in building the engineering design company. Barnes' particular strength was in strong project management and contract negotiation skills which were critical to Routes, given the contract-based nature of much of its work. Payne and Buckingham's technical background, aerospace contacts and track record in the sector constituted key building-blocks, while the management and contractual strengths of Barnes brought critical balance to the business side of the new venture.

When Payne, Buckingham and Barnes were seeking a site for Routes AstroEngineering, Kanata, on the outskirts of Ottawa, provided a fitting location for the start-up given its close juxtaposition to the CRC and the presence of other federal research centres and aerospace companies in the capital. An added advantage was the breadth and depth of the labour pool around the capital. The breadth of technologies represented in the local labour market was one of its strengths. Given the nature of Routes' work, it often needed people with different skills from those employed in-house; for example, it was able to secure the services for a few days of an engineer with expertise in fibre optics.

direction". He considered that he could pursue the area himself and brought together a four-person team to exploit the technology, supported by an extensive group of financial

Table 4. Incubator organisation of founder, by type of firm

Previous employer of founders	Technology product firms		Technology service firms		Total	
	n	Percentage	n	Percentage	n	Percentage
Small firm	85	36.8	47	28.8	132	33.5
Large firm	49	21.2	50	30.7	99	25.1
Medium firm	44	19.0	24	14.7	68	17.3
Federal government	24	10.4	23	14.1	47	11.9
Provincial government	3	1.3	0	—	3	0.8
University	21	9.1	12	7.4	33	8.4
Other	5	2.2	7	4.3	12	3.0
Total number of founders	231	100.0	163	100.0	394	100.0

Source: fax survey.

Table 5. Location of founder's previous workplace

Number of founders who worked	Technology product firms		Technology service firms		Total	
	n	Percentage	*n*	Percentage	*n*	Percentage
Less than 30 miles from location	204	87.9	130	86.1	334	87.2
30–199 miles from location	9	3.9	4	2.6	13	3.4
200 miles or more from location	19	8.2	17	11.3	36	9.4
Total	232	100.0	151	100.0	383	100.0

Source: fax survey.

backers. Another entrepreneur commented that "the skills which I had used to build up that particular business [in the incubator organisation] I could apply to something on the outside which would further my career". Four firms—all established by a single founder—were established as a result of the founder wishing to work for himself. Just one firm, engaged in leading-edge research in the life science field, was technology-driven. It was focused on more fundamental research which had been started within the incubator organisation.

In contrast, only two firms had been started for negative reasons. In one case, the founder had fallen out with senior staff in the firm he worked for. In the other case the founding team found themselves being steered away from the area of technology which was of most interest to them (Routes: Astro Engineering, Figure 1).

> Two of the three of us at least were getting a little bit away from the kind of area that we had our best expertise in ... so we returned to our roots, to do basic space work which is what two of us had been doing for some years in the past ... The two of us were becoming somewhat redirected into things like military electronics and things like that which were not really our areas of expertise or interest ... Our fourth partner joined us a little bit later, he left for essentially the same reasons. He was into an engineering function where he was doing an awful lot of paper handling and specification brochures. He is a designer and that is what he wanted to do.

Entrepreneurial Geographies

Over four-fifths of firms which responded to the fax survey had at least one founder who had been working within 30 miles (i.e. commuting distance) of the location of their new venture immediately prior to its formation. Taking *all* members of the founding team into account, 88 per cent (204) of all product technology founders for whom information was available and 86 per cent (130) of the founders of technology service firms worked within 30 miles of the start-up location immediately prior to starting their own business (Table 5). Just 12 firms in each group had no local founders (16 per cent of technology service firms and 11 per cent of technology product firms). The propensity of the *principal* founder to have been working locally immediately prior to start-up is slightly less: 85 per cent in technology product firms and 78 per cent in technology service firms. In the interview sample, all but one of the founders had been working in the Ottawa region when they set up their firm.

At first sight these findings—which indicate that most of the founders worked locally immediately prior to start-up—would appear to suggest that much of the entrepreneurial activity in Ottawa is a *local* phenomenon. However, this disguises the fact that many founders had been born and brought up elsewhere and had moved to the region during their adult lives for employment reasons. Indeed, only 6 members of the founding teams (17 per cent) of the companies in the interview survey were born in Ottawa and 2 others were born elsewhere but moved during childhood to Ottawa (Table 6). Thus, the majority of entrepreneurs had spent their formative years elsewhere and in most cases had come to Ottawa because of employment

Table 6. Place of birth of firm founders

Place of birth	Total	
	n	Percentage
Ottawa	6	16.7
Rest of Ontario	5	13.9
Maritime Provinces	5	13.9
Quebec	3	8.3
Western Canada	3	8.3
USA	1	2.8
UK	9	25.0
Rest of world	4	11.1
Total	36	100.0

Note: Information was not available for seven founders who were no longer with their firm.
Source: interview survey.

Figure 2. SiGem Inc. *Sources*: interview; Anderson (2000); Hill (2000); *Ottawa Business Journal* (2000).

SiGem Inc. was established in Kanata in 1997. The firm specialises in the development of wireless tracking devices and exploited technology held within SiGe Microsystems, a sister company. SiGe, a spin-off from the National Research Council (NRC), designs high-speed chips for the wireless Internet market. SiGe Microsystems held a licence from the University of Bristol and it was this licence and technology which had been developed around it which was transferred to SiGem. Three individuals played key roles in the establishment of SiGem Inc. When SiGe was established in 1995, John Roberts worked with Derek Houghton of the NRC and David Edwards whom Roberts knew when the two of them worked with the Strategic Microelectronics Consortium, an industry association based in Ottawa. Roberts and Edwards continued their working relationship through the formation of SiGem. The third member of the team was Geoff Haynes. It was he who found the technology in the University of Bristol and brought it to the attention of SiGe.

SiGem's three founders were all born in the UK. Roberts was born and brought up in Wales and studied for a Bachelors degree in electrical engineering at the University of Wales, Cardiff, before taking his Masters at the University of Swansea. He worked as an engineer for GE Semiconductor in Essex and for Siliconix in Wales before emigrating to Ottawa in 1972. In Ottawa he joined Microsystems International, which proved to be a breeding-ground for entrepreneurs in the early 1970s. Roberts left the firm in 1975 and the next phase of his career saw him working with another large firm in Ottawa, Control Data, before he joined Mosaid as its second employee in 1978. It was while working with Mosaid that Roberts developed a taste for working in small companies. Roberts then became involved in the start-up of a string of technology ventures including Calmos in 1983 which went on to become Tundra. As time went on, he played an increasingly important role in the financing of new ventures, acting as a gather-in of funds from those with modest amounts to invest, in addition to putting in funds of his own. Specialising in company start-ups and raising funding for new ventures, he has played a key role in the creation of over 20 technology firms in Ottawa.

Edwards was born in England and had moved to Canada with his family at the age of 9. After studying mechanical engineering, he changed direction, becoming professionally qualified as an accountant. Roberts and Edwards met soon after Roberts arrived in Ottawa when both worked for Microsystems International, where Edwards was a financial analyst. During his career Edwards worked for organisations as diverse as the EDC and the UN. Extensive periods of time travelling abroad meant that he did not have extensive local networks which were of benefit to SiGem; however, as financial officer at the SMC, working with Roberts, he had extensive experience of working with technology companies.

Haynes, who was brought up in the south of England, studied engineering in Brighton before working as an engineer with GE Semiconductor, which is where he met Roberts. Later, Haynes moved into sales and worked for a string of US multinationals, developing an extensive personal and professional network. Based primarily in the UK, Haynes became a co-founder of SiGem, partly as a result of his identification of the company's core technology and due to his great ability to raise money, through his extensive networks and from his friends. Indeed, he raised over $750 000 for SiGem from his personal networks around the world. Haynes headed up SiGem's European sales operation.

SiGem, which went public in September 1999, was established with venture capital and angel funding to the tune of over $3 million. Investments included financing from the BDC and the Canadian Science and Technology Growth Fund and from a number of business angels from as far afield as Silicon Valley, Germany and England, as well as Ottawa. It attracted a number of notable Ottawa technology players to its board including Leo Lax (who ran the Affiliates Program within Newbridge Networks) and Doug Smeaton who was formerly Senior Vice President and General Manager of Mitel Semiconductor and served on the boards of a number of Ottawa start-ups, including Philstar Semiconductor and SiGe Microsystems. From an initial staff of eight, the firm grew organically and by acquisition with early deals including the global positioning system company, GMSI Inc.

opportunities. In the case of SiGem Inc—for example, all 3 of its founders were born in the UK (Figure 2). The founders in the other case studies (Figure 1, Figure 3) had also

Figure 3. Bridgewater Systems Corporation. *Source*: interview; www.bridgewatersystems.com.

In 1997, Bridgewater Systems Corporation was founded by Doug Somers and Russ Freen who spun out of Newbridge Networks Corporation in Kanata. At the time of the start-up, Somers was Vice President (VP) and General Manager of the Network and Services Management business unit. Running the software business unit inside Newbridge, Somers recognised that he could apply the skills which he had used in building up that particular business to something on the outside which would further develop his own career. Freen, who headed up a very large R&D group which developed products for Somers' unit, was intrigued by the idea of starting a company and by the area of the Internet.

Somers moved from Halifax, Nova Scotia, to Ottawa with his family at the age of 12. He attended Queens University, where he studied applied sciences in electrical engineering, before joining Northern Telecom to work in Belleville, Ontario. Somers moved back to Ottawa when his job and most of the unit within which he was working relocated to the city. Somers worked on voice/data integrated services network products at the company in mainly technical roles as a result of which he acquired knowledge and skills which proved valuable in the start-up of Bridgewater. Key areas included a strong technical understanding about communications networks, an appreciation for manufacturing, particularly hardware manufacturing, and what it meant to make changes and perform upgrades to product out in the field. On an organisational level, he gained a sound understanding of dealing with larger organisations, a group from which many of Bridgewater's customers were drawn. After around 9 years working with Northern Telecom, Somers joined Newbridge Networks in Kanata. His roles at Newbridge provided him with further key knowledge and skills, such as opportunity recognition, market analysis, sales and distribution, and business plan development. After 7 years, Somers and Freen left Newbridge to found Bridgewater in Kanata. Freen took on the role of VP R&D and Operations in Bridgewater, while Somers became President and CEO. Freen was brought up in Toronto but moved to Ottawa to join Bell Northern Research's computer science group in 1977. Four years later, he moved to Mitel where he worked until he joined Newbridge in 1987. Freen worked at Newbridge until 1993 when he left Ottawa to work, first of all, for Motorola in Vancouver and then for Raynet in San Francisco, in fibre optics. Freen returned to Ottawa in 1996 when he was recruited into Newbridge once again, to work in its network management group.

When setting up Bridgewater, Somers and Freen recognised that Newbridge could act as a channel for its products. Newbridge was also interested in putting money into the company. A deal was brokered whereby Somers and Freen agreed not to hire out of Newbridge and steal its employees. Given their experience in technology firms in Ottawa they were able to use their networks and contacts developed throughout their careers to hire from the labour pool or through their contacts in the Ottawa area and in some cases hire-in staff with whom they had worked previously. The experience of working for some of the largest and best-known technology firms in Ottawa provided Somers and Freen with not only valuable experience gained through the varied positions which they had held, but also extensive personal networks which proved extremely useful in other ways as they established and grew their business. For example, their pedigree assisted in establishing a high level of credibility within the business and financial community prior to start-up.

moved to Ottawa. Moreover, the move to Ottawa has occurred at various career stages and by no means always at the start of their career. Indeed, only 11 principal founders (out of 20) had held their first permanent job in Ottawa. Analysis of interview data related to the length of time spent working in the Ottawa region prior to establishing the survey firm revealed that the majority of founders had worked in Ottawa for quite some time before establishing their enterprise. Over 68 per cent of founders had worked in the region for more than 10 years before setting up the firm; only 6 per cent had fewer than 5 years of local work experience. During this period, social and professional networks had developed to create rich and varied relationships on which the founders were able to draw as they established their venture.

In-migrants from elsewhere in Canada and from overseas have therefore been an important source of entrepreneurs for Ottawa's technology cluster. Ottawa has been an attractive destination in which to find work for a number of reasons. First, a number of organisations in the Ottawa region were perceived to offer exciting job opportunities because they were at the forefront of their fields. In particular, BNR/Nortel has played, and continues to play, an important role in attracting skilled individuals to the region. Future entrepreneurs were also drawn to Ottawa by employment and research opportunities within the federal government (including the NRC and the CRC) and the University of Ottawa. Secondly, the growth of the cluster has created an increasing range of alternative employment opportunities within the ICT sector in particular. Places with 'thick' labour markets are attractive to highly qualified workers because they allow the possibility of pursuing a 'horizontal' career and also offer the prospects of moving to another organisation in the event that the initial job did not work out (Florida, 2002a). The latter scenario rarely proved to be the case and job moves tended to be stimulated by the desire to broaden experience. For the same reason, places with 'thick' labour markets are also attractive to professional dual-career households.

Entrepreneurial Histories

The previous section has established that the founders of technology ventures in Ottawa have typically worked for more than one organisation before starting their own venture. Only a minority of founders left their *initial* employer in Ottawa to start their own firm. The expertise and knowledge-base which entrepreneurs drew upon to start their own business have therefore been the product of periods of work in several organisations, both in the Ottawa region and elsewhere in Canada and abroad. Although entrepreneurs did bring direct experience to their business from the incubator organisation, this was not the only source of learning and experience that they drew upon to start their own businesses.

In the case of the interview companies, the majority of founders had worked for more than one organisation prior to starting their own business. Considering just the principal founders, in only six cases had they worked for a single organisation prior to starting their venture. However, during that period of employment, which in one case totalled 35 years, four of the six founders had worked in different roles and, in some cases, different geographical locations within the one organisation. The result of this interorganisational (and in some cases intraorganisational) mobility was that most entrepreneurs had accumulated a considerable range of experience and expertise prior to starting their own firm. The typical career path was a relatively narrowly focused technical or research position early in their career (influenced in part by their education) and a subsequent move into managerial roles such as technical sales or business development prior to becoming entrepreneurs themselves. These different job positions enabled the founders to develop different skill-sets and at least some of the skills and knowledge required to start their own venture. One founder commented that in his first permanent job, a technical role within a large private-sector firm, he developed skills which had proved very valuable in the context of his own venture. Specifically, he gained

> a greater appreciation for dealing with larger organisations ... and certainly some technical information about communications networks, because that is where I first got my teeth stuck into them ... and an appreciation for manufacturing, particularly hardware manufacturing, and what it means to have lots of product out in the field and how you make changes and upgrades to it.

In contrast, the skills which he acquired while working for his last employer included

> the analysis of the market, understanding of how you determine various opportuni-

Table 7. Sources of technology on which firm was based

Technology source	Technology product firms		Technology service firms		Total	
	n	Percentage	*n*	Percentage	*n*	Percentage
Personal R&D	48	42.9	39	50.6	87	46.0
Previous employer	45	40.2	41	53.2	86	45.5
University	34	30.4	28	36.4	62	32.8
Government	24	21.4	24	31.2	48	25.4
Another company	23	20.5	13	16.9	36	19.0
Other	4	3.6	1	1.3	5	2.6

Note: multiple responses possible.
Source: fax survey.

> ties with clients and knowledge of some of the sales and distribution issues ... some of the marketing issues, in terms of launching the company ... as well as some of the issues in determining the size of the available market ... skills in terms of being able to put together business plans, that was very, very well developed and I had quite a good system for that.

Thus, previous employment provided founders with at least part of the skill-set required to start their own venture. This process, evident in the wider interview sample, is illustrated by the case of Doug Somers, co-founder of Bridgewater Systems Corporation (Figure 3).

The consequence of this career mobility is that the entrepreneurs had typically accumulated an extensive range of experience. Amongst the founders of the businesses that were interviewed, 22 had primarily technical experience when they set their firm up, of which 14 were the principal entrepreneurs, 12 had a combination of technical and business skills, while the remaining 9 had predominantly business expertise.

Sources of Technology

Evidence from the fax survey on the sources of technology that the founders drew upon when establishing their venture are shown in Table 7. This indicates that the majority of firms drew on just one source of knowledge. Personal R&D and previous employer were the most important sources, accounting for a little under half of the firms in each case. The more detailed information from the interview survey revealed that in all instances where there was reliance on an external source, at least one of the founders had worked for the organisation from which the technology was acquired immediately prior to start-up. However, 11 of these firms (55 per cent) were based on technologies from multiple sources: universities (11),[9] private-sector firms (10), federal/provincial government (7), own R&D (8) and not disclosed (1). As the career histories evidence indicates, during their university education and previous employment the founders of these technology firms developed a portfolio of technological know-how which they subsequently drew upon to establish their new venture. So here again our evidence serves to weaken the significance of the incubator organisation. The incubator organisation was certainly the source of technology for some firms. However, it is necessary to recognise the contribution of other organisations which the founders had worked for earlier in their careers in contributing to the ideas that the new venture was based on.

Resourcing the Start-up Business

Both the incubator organisation and the other organisations that the entrepreneurs worked for were also important sources of start-up resources, notably employees and finance.

Most new firms relied upon the expertise of the founder or founders to launch the business. However, sooner or later it became necessary to recruit additional senior staff and strengthen the board of directors. One of the ways of reducing risk and uncertainty is to recruit employees who are known to the founders and so can be trusted. A number of the new firms therefore used their previous organisations as sources from which they recruited some key employees.[10] In some instances, the entrepreneurs had worked with these individuals earlier in their careers and, when setting up their own firm, had approached them with an offer of employment. The connections which the founders had made whilst with these earlier employers were also an important means of accessing networks through which they identified suitable employees.

Finance is another important resource required for establishing the business. Personal finance was the most important source of start-up capital (75 per cent). This became less significant as the businesses developed, so that by the time that the interviews took place (mid 2000) only 45 per cent were still using the personal financial resources of their founders. Meanwhile, the reliance upon venture capital (both business angels and venture capital funds) increased from 15 per cent at start-up to 30 per cent by 2000. The career history of entrepreneurs is an important influence on their ability to raise finance in three respects. First, venture capital fund managers rely on referral networks as a first filter in their investment appraisal (Florida and Kenney, 1988). Whether or not an entrepreneur has access to such referral networks is determined by their professional networks: these, in turn, are shaped by the nature of their work and their employer. Secondly, our interviews with venture capitalists (Mason *et al.*, 2002) noted that the companies that entrepreneurs previously worked for send out a signal about the quality of the technology. One commented as follows: "people at Nortel that are working on leading-edge technologies … that really wakes us up". Indeed, several venture capitalists specifically mentioned their interest in investing in businesses "involving ex-Nortel guys". Another venture capitalist talked about his attempts to stimulate deal flow by encouraging Nortel employees to start their own businesses

> I meet these guys socially and through organisations. [I say] "What do you do?" [They say] "I'm designing this." [I say] "That's really exciting. Why don't you leave here, start your own company. We'll back you."

Thirdly, venture capitalists use the previous employers of the entrepreneurs to do their due diligence. As one venture capitalist explained

> It's real easy [to do this] in Ottawa. This is a community where most of the people are spin-outs of spin-outs. Two phone calls and I can find out everything. … For the most part, you are dealing with teams and at least some of the team members come out of the Ottawa community and have worked with other people. Because I have six or seven investments in semiconductors, there are not many people in the Ottawa area in the semiconductor industry that I don't already know or know someone who knows them, who has worked with them in the past and so on.

Another noted that "Ottawa is a small town, so typically the individual worked at Nortel at some stage in his career and you can find someone who worked alongside him at one point". A third commented: "I look at where they worked … If they've worked at half a dozen places there's got to be one of those places where I know somebody".

Conclusion

Contemporary research on clusters has been characterised by a number of problems (Malmberg and Maskell, 2002). First, most research studies have been static in design, focusing on interfirm relations and patterns of interactions at a specific point in time and attempting to draw out conclusions about

Table 8. Inputs and influences in the entrepreneurial process

	Incubator organisation	Previous employers	Other sources
Trigger event	***	*	*
Opportunity identification	***	*	*
Partners	**	**	**
Key employees	**	**	**
Technical know-how	***	**	*
Customer leads	***	**	*
Board members	**	**	**

Key: *** = usually; ** = sometimes; * = rarely

dynamic processes from cross-sectional data (Staber, 1997). Secondly, most empirical research is based on case studies, the selection of which is biased to high-technology industries and regional success stories (Mytelka and Farinelli, 2000), leading to accusations of a research tradition characterised more by stories than by statistics (Head *et al.*, 1995). Thirdly, and a consequence of the combined influence of the first two criticisms, where extensive empirical research has been undertaken into the cluster phenomenon, the empirical results do not provide clear-cut support for the theoretical arguments on cluster development and functioning (Martin and Sunley, 2003; Swann, 1998; Malmberg and Maskell, 2002; Hudson, 1999). Fourthly, much of the extant research has examined issues of cluster development and functioning—in terms of localisation and agglomeration economies (Parr, 2002) and regional collective learning processes (Keeble and Wilkinson, 2000a, 2000b)—at the expense of a consideration of the entrepreneurial dynamics underlying cluster origins and development trajectories (Feldman, 2001). Finally, where the entrepreneurial dynamics issue has been addressed, it is in the context of the debate over the incubator organisation, understood as the immediate past employer, and the embeddedness of the entrepreneurial process in the local milieu.

In this paper, we have responded to these criticisms and have focused specifically on the entrepreneurial dynamics of cluster formation as a dynamic process. Despite Head *et al.*'s (1995) comments, however, we have maintained a case-study emphasis, both because this ideographic approach (to use Malmberg and Maskell's (2002) phrase) is the most appropriate focus for process-oriented research, and to reflect the argument that all clusters are different and have their own distinctive development paths. Based on a detailed case study of Ottawa, we draw three major conclusions on entrepreneurial dynamics in technology clusters.

First, the idea that there is a single incubator organisation (of a key founding individual) for technology-based firms is oversimplistic and provides too restrictive a view of the influences on new venture creation. Specifically, as the majority of firms are multiple founder start-ups, it is important to consider the influence of the incubator organisations of all members of the founder team. Certainly, the incubator organisations from which the founders came generally provided the setting for opportunity identification and hence the trigger event. They also provided an environment in which entrepreneurs met prospective future business partners. The incubator organisations also played a role in the founding process by providing key employees, technological know-how and important leads to customers and suppliers. However, the incubator organisation in the traditional sense was not the only important influence on the new firm development process. Some organisations in which entrepreneurs previously worked played a role in the provision of key resources (Table 8). While the incubator organisation was important at the start-up stage of a number of businesses, as time went on the influence of other organisations, and connections that the entrepreneurs had developed through their association with them, became more prominent. Some senior executives and directors were identified through earlier employers or via people whom the entrepreneurs had met while with these organisations. Generally, the incubator organisation was less influential in venture growth than it had been in the venture creation process. In other words, entrepreneurial intellectual capital is developed throughout the career of would-be entrepreneurs and a number of organisations other than the incubator organisation(s), sometimes in different locations, contribute to the process. To focus solely on the last employer therefore severely restricts our understanding of entrepreneurial career development and ignores both the temporal and geographical dimensions of entrepreneurial learning and knowledge accumulation. The expertise and knowledge-base which entrepreneurs develop are the product of employment in a number of organisational settings and locations.

Secondly, entrepreneurs are spatially mobile during their pre-start-up life and career and the in-migration of entrepreneurial capital is central to cluster development (Saxenian, 2001). Mobile entrepreneurs bring new ideas, different perspectives and different networks, which they utilise in developing their business.[11] Learning can occur both outside and within the region where it is applied during the entrepreneurial process. Like Britton (2003), our findings therefore challenge the emphasis in much of the literature on technology clusters which emphasises their local embeddedness, by highlighting the way in which they are enriched by interregional and international relationships. Our evidence on the geographical mobility of technology entrepreneurs argues for a reorientation of thinking, away from the traditional focus on the location of economic activity (of which the clusters research industry is only the latest manifestation), towards the economic geography of talent (Florida, 2002a, 2002b). Increasingly, it is being recognised that the clustering of talented human capital is the driving-force behind the growth and development of cities and regions and underpins regional innovation (Glaeser, 2000; Glaeser *et al.*, 2001; Mathur, 1999; Mahroum, 1999, 2000). As Florida observes

> talent and high-technology industry work independently and together to generate higher regional incomes. In short, talent is a key intermediate variable in attracting high-technology industries and generating higher regional incomes. (Florida, 2002b, p. 744)

Thirdly, and following on from this, the discussion of the Ottawa case points to the obvious, but often overlooked, point that it is organisations that attract talent to places. 'Magnet organisations'—that is, high-reputation technically oriented organisations, offering positions in a range of functional areas—play a crucial role in the development of technology clusters by attracting highly educated and skilled scientists and engineers into a region. Such organisations provide jobs for both indigenous employees and those attracted into the region by the quality of the employment opportunities and the quality of life on offer (a key issue in the economic geography of talent: Florida, 2002a, 2002b). They help to increase the technology skill-base within the region and provide an environment in which employees may enhance their skills. In addition, a period of employment within a 'magnet organisation' may provide access to key social and professional networks, which may prove invaluable during the entrepreneurial process, and may lend credibility when approaching external investors for start-up or development funding. Some entrepreneurs may spin-out directly from the magnet organisation, but the majority move to other employers within the cluster, working for several organisations, including smaller businesses and, in the longer term, may spin-out to create their own ventures. To some extent, the identification of the important role played by these magnet organisations (such as BNR/Nortel and federal government research labs in the Ottawa case) is an example of the role of institutional fit in the development of a cluster (Malmberg and Maskell, 2002). In an argument based on the establishment of critical mass, the development of a dominant institutional pattern makes possible the attraction of entrepreneurs with ambitions to set up in the particular industry

> This is ... why many of the most talented 'wannabes' within the film industry tend to end up in Hollywood and many of the world's best specialists in information and communication technology are attracted to Silicon Valley (Malmberg and Maskell, 2002, p. 441).

However, the Ottawa case makes clear the importance of magnet organisations in establishing the basis for the development of a cluster: in this case at least, it could be argued that the cluster would not have developed as it did in the 1990s without the core institutional pattern which was initially established in the 1940s. Future clusters research which is process-oriented and longitudinally based can usefully build on this initial investigation of the entrepreneurial dynamics of cluster development by examining the role of magnet organisations, the related economic geography of talent and human capital and the pattern of entrepreneurial histories and geographies which underlie this process.

Notes

1. These statistics are derived from an annual survey of technology firms in Ottawa that is undertaken by OCRI (the Ottawa Centre for Research and Innovation). Statistics Canada, which uses a narrower definition of a technology firm, calculates that there were 56 000 high-tech jobs in Ottawa in 2003 (compared with 69 000 in 2001) (*Ottawa Business Journal*, 23 June 2003).
2. The context for this was a judicial decision in the US in the 1950s to force the Western Electric Company to divest itself of its subsidiary, the Northern Electrical Manufacturing Company (now Nortel). Cut-off from its source of research, Northern Electrical searched for a location to establish its own R&D facility. Ottawa was chosen largely because of its proximity to the NRC labs and the CRC (Wolfe, 2003).
3. Both Doyletech, an Ottawa-based consulting firm, and PricewaterhouseCoopers, have produced company genealogies for the Ottawa technology cluster to show the spin-off process. PwC have subsequently produced similar 'maps' for other Canadian technology clusters (Smith, 2002).
4. See the following for accounts of Ottawa's technology cluster in the 1970s and 1980s: Mittelstaedt, 1980; Sweetman, 1982; Steed and DeGenova, 1983; McDougall, 1986; Steed, 1987.
5. The majority were face-to-face interviews. However, two were conducted over the telephone.
6. One interview included all of the members of the founding team and in another two cases other members of the founding team were brought in to the interview to answer the questions about themselves. In most of the remaining cases, the lead entrepreneur was very familiar with the career histories of their team members and so was able to provide the required information.
7. For example, biophotonics, which is the use of photonics in biomedical processes and products (Chamberlin and de la Mothe, 2003).
8. For example, in one case a team member came from the same town in India as one of the other team members.
9. Some of those who cited university sources had been doctoral students engaged in leading-edge, technology research projects earlier in their careers which provided at least some of the technology used to start their business. Overall, just over three-quarters of the founders of the interview cases had university degrees (30 per cent of them had PhDs), mainly in science or engineering, and a further 18 per cent had college diplomas.
10. In one case, the new business received funding from the incubator and as part of the arrangement had agreed not to take any of its staff.
11. For example, Leng (2002) notes that Taiwanese entrepreneurs in Silicon Valley have used their ethnic networks to access venture capital from Taiwan-based venture capitalists.

References

Anderson, M. (2000) Insights raise $15.6M in first-round financing, *The Ottawa Citizen*, 19 May.

Bahrami, H. and Evans, S. (1995) Flexible recycling and high-technology entrepreneurship, *California Management Review,* 37(3), pp. 62–89.

BEAUDRY, C. and SWANN, G. M. P. (2001) *Growth in industrial clusters: a bird's eye view of the UK.* SIEPR Working Paper 00-38, Stanford University.

BRITTON, J. N. H. (2003) Network structure of an industrial cluster: electronics in Toronto, *Environment and Planning A,* 35, pp. 983–1006.

BYGRAVE, W. D. (1988) *The entrepreneurship paradigm (I): a philosophical look at its research methodologies.* Paper presented at the *Entrepreneurship Doctoral Consortium, Babson Research Conference*, University of Calgary.

CHAMBERLIN, T. and MOTHE, J. DE LA (2003) Northern light: Ottawa's technology cluster, in: D. A. WOLFE (Ed.) *Clusters Old and New: The Transition to a Knowledge Economy in Canada's Regions*, pp. 213–234. Montreal: McGill-Queen's University Press.

COOPER, A. C. (1970) The Palo Alto experience, *Industrial Research,* May, pp. 58–84.

COOPER, A. C. (1973) Technical entrepreneurship: what do we know?, *R&D Management,* 3(2), pp. 59–64.

COOPER, A. C. (1986) Entrepreneurship and high technology, in: D.L. SEXTON and R.W. SMILOR (Eds) *The Art and Science of Entrepreneurship*, pp. 153–168. Cambridge, MA: Ballinger.

COOPER, A. C. and FOLTA, T. (2000) Entrepreneurship and high technology clusters, in: D. L. SEXTON and H. LANDSTRÖM (Eds) *The Blackwell Handbook of Entrepreneurship*, pp. 348–367. Oxford: Blackwell.

COOPER S. Y. (1996) *Small high technology firms: a theoretical and empirical study of location issues.* Unpublished PhD thesis, Heriot-Watt University, Edinburgh.

CROSS, M. (1981) *New Firm Formation and Regional Development.* Farnborough: Gower.

DOERINGER, P. B. and TERKLA, D. G. (1996) Why do industries cluster?, in: U. STABER, N. SCHAEFER and B. SHARMA (Eds) *Business Networks: Prospects for Regional Development*, pp. 175–189. Berlin: Walter de Gruyter.

FELDMAN, M. P. (2001) The entrepreneurial event revisited: firm formation in a regional context, *Industrial and Corporate Change,* 10, pp. 861–891.

FLORIDA, R. (2002a) *The Rise of the Creative Class.* New York: Basic Books.

FLORIDA, R. (2002b) The economic geography of talent, *Annals of the Association of American Geographers,* 92, pp. 743–755.

FLORIDA, R. L. and KENNEY, M. (1988) Venture capital, high technology and regional development, *Regional Studies,* 22, pp. 33–48.

FOTHERGILL, S. and GUDGIN, G. (1982) *Unequal Growth: Urban and Regional Employment Change in the UK.* London: Heinemann.

FUJITA, M., KRUGMAN, P. and VENABLES, A. (2000) *The Spatial Economy: Cities, Regions and International Trade.* Cambridge, MA: MIT Press.

GARNSEY, E. (1998) The genesis of the high technology milieu: a study in complexity, *International Journal of Urban and Regional Research,* 22, pp. 361–377.

GARNSEY, E. and CANNON-BROOKES, A. (1993) The Cambridge phenomenon revisited: aggregate change among Cambridge high-technology companies since 1985, *Entrepreneurship and Regional Development,* 5, pp. 179–207.

GLAESER, E. L. (2000) The new economics of urban and regional growth, in: G. CLARK, M. GERTLER and M. FELDMAN, (Eds) *The Oxford Handbook of Economic Geography*, pp. 83–98. Oxford: Oxford University Press.

GLAESER, E. L., KOLKO, J. and SAIZ, A. (2001) Consumer city, *Journal of Economic Geography,* 1, pp. 27–50.

HEAD, K. RIES, J. and SWENSON, D. (1995) Agglomeration benefits and location choice: evidence from Japanese manufacturing investments in the United States, *Journal of International Economics,* 38, pp. 223–247.

HENDRY, C. and BROWN, J. (2001) Local skills and knowledge as critical contributions to growth of industry clusters in biotechnology, in: W. DURING, R. OAKEY and S. KAUSER (Eds) *New Technology-Based Firms in the New Millenium*, pp. 127–140. Oxford: Pergammon.

HILL, B. (2000) NRC hatchling leaves the nest: SiGe sets up shop in the west end, *The Ottawa Citizen*, 7 April.

HUDSON, R. (1999) The learning economy, the learning firm and the learning region: a sympathetic critique of the limits to learning, *European Urban and Regional Studies,* 6, pp. 59–72.

KEEBLE, D. (2000) Collective learning processes in European high-technology milieux, in: D. KEEBLE and F. WILKINSON (Eds) (2000) *High-technology Clusters, Networking and Collective Learning in Europe*, pp. 199–229. Aldershot: Ashgate Publishing Ltd.

KEEBLE, D. and WILKINSON, F. (Eds) (1999) Special issue: Regional networking, collective learning and innovation in high technology SMEs in Europe, *Regional Studies,* 33(4), pp. 295–400.

KEEBLE, D. and WILKINSON, F. (2000a) High technology SMEs, regional clustering and collective learning: an overview, in: D. KEEBLE and F. WILKINSON (Eds) *High-technology Clusters, Networking and Collective Learning in Europe*, pp. 1–19. Aldershot: Ashgate Publishing Ltd.

KEEBLE, D. and WILKINSON, F. (Eds) (2000b)

High-technology Clusters, Networking and Collective Learning in Europe. Aldershot: Ashgate Publishing Ltd.

KRUGMAN, P. (1991) *Geography and Trade*. Cambridge, MA: MIT Press.

LAWTON-SMITH, H. (1991) The role of incubators in local industrial development: the cryogenics industry in Oxfordshire, *Entrepreneurship and Regional Development,* 3, pp. 175–194.

LAWTON-SMITH, H. (1996) National laboratories and regional development: case studies from within the UK, France and Belgium, *Entrepreneurship and Regional Development,* 8, pp. 1–17.

LENG, T.-K. (2002) Economic globalization and IT talent flows: the Taipei/ Shanghai/ Silicon Valley triangle, *Asian Survey,* 42, pp. 230–250.

LINDHOLM-DAHLSTRAND, A. (2000) Large firm acquisitions, spin-offs and links in the development of regional clusters of technology intensive SMEs, in: D. KEEBLE and F. WILKINSON (Eds) *High-technology Clusters, Networking and Collective Learning in Europe*, pp. 156–181. Aldershot: Ashgate Publishing Ltd.

MACKINNON, D., CUMBERS, A. and CHAPMAN, K. (2002) Learning, innovation and regional development: a critical appraisal of recent debates, *Progress in Human Geography,* 26, pp. 293–311.

MAHROUM, S. (1999) Global magnets: science and technology disciplines and departments in the United Kingdom, *Minerva,* 37, pp. 379–390.

MAHROUM, S. (2000) Highly skilled globetrotters: mapping the international migration of human capital, *R&D Management,* 30, pp. 23–31.

MALMBERG, A. and MASKELL, P. (2002) The elusive concept of localization economies: towards a knowledge-based theory of spatial clustering, *Environment and Planning A,* 34, pp. 429–449.

MARSHALL, A. (1890) *Principles of Economics*. London: Macmillan.

MARTIN, R. and SUNLEY, P. (2003) Deconstructing clusters: chaotic concept or policy panacea?, *Journal of Economic Geography,* 3, pp. 5–35.

MASON, C. M. and HARRISON, R. T. (2003) *After the exit: entrepreneurial recycling and regional economic development*. Paper to the *23rd Babson Kauffman Foundation Entrepreneurship Research Conference*, Babson College, MA, June.

MASON, C. M., COOPER, S. Y. and HARRISON, R. T. (2002) Venture capital in high technology clusters: the case of Ottawa, in: R. OAKEY, W. DURING and S. KAUSER (Eds) *New Technology-Based Firms in the New Millenium*, pp. 261–278. Oxford: Pergammon.

MATHUR, V. K. (1999) Human capital-based strategy for regional economic development, *Economic Development Quarterly,* 13, pp. 203–216.

MCDOUGALL, B. (1986) Digital dreamers: scientific entrepreneurs have placed Ottawa on the high-tech map, *Small Business*, December, pp. 20–24.

MILLER, R. and CÔTÉ, M. (1987) *Growing the Next Silicon Valley*. Lexington, MA: Lexington Books.

MITTELSTAEDT, M. (1980) Ottawa: the new high-tech haven, *Canadian Business*, June, pp. 43–46, 82–85, 87, 105.

MYTELKA, L. K. (1999) Competition, innovation and competitiveness: a framework for analysis, in: L. K. MYTELKA (Ed.) *Competition, Innovation and Competitiveness in Developing Countries*, pp. 15–27. Paris: OECD,

MYTELKA, L. K. and FARINELLI, F. (2000) *Local clusters, innovation systems and sustained competitiveness*. UNU/INTECH Discussion Papers No. 2005, Maastricht, Netherlands.

NECK, H. M., COHEN, B. D. and CORBETT, A. C. (1999) A genealogy and taxonomy of high technology new venture creation within an entrepreneurial system, in: P. D. REYNOLDS, W. D. BYGRAVE, S. MANIGART *ET AL.* (Eds) *Frontiers of Entrepreneurship 1999*, pp. 541–555. Babson Park, MA: Babson College.

OAKEY, R. (1995) *High Technology New Firms: Variable Barriers to Growth*. London: Paul Chapman Publishing Ltd.

OAKEY, R. P., FAULKNER, W., COOPER, S. Y. and WALSH, V. (1990) *New Firms in the Biotechnology Industry*. London: Pinter Publishers.

OECD (Organisation for Economic Co-operation and Development) (1998) *Fostering Entrepreneurship*. Paris: OECD.

OECD (1999) *Boosting Innovation: The Cluster Approach*. Paris: OECD.

OECD (2001) *Innovative Clusters: Drivers of National Innovation Systems*. Paris: OECD.

OECD–DATAR (2001) *World Congress on Local Clusters*. Paris: OECD.

O'FARRELL, P. (1986) *Entrepreneurs and Industrial Change*. Dublin: Irish Management Institute.

Ottawa Business Journal (2000) YOUtopia.com brings gift certificates online, 20 June.

PARR, J. B. (2002) Agglomeration economies: ambiguities and confusions, *Environment and Planning A,* 34, pp. 717–731.

PINCH, S., HENRY, N., JENKINS, M. and TALLMAN, S. (2003) From 'industrial districts' to 'knowledge clusters': a model of knowledge dissemination and competitive advantage in industrial agglomerations, *Journal of Economic Geography,* 3, pp. 373–388.

PORTER, M. E. (1990) *The Competitive Advantage of Nations*. London: Macmillan.

PORTER, M. E. (1998) *On Competition.* Cambridge, MA: Harvard Business School Press.

PORTER, M. E. (2000) Location, competition and economic development: local clusters in the global economy, *Economic Development Quarterly*, 14, pp. 15–31.

PORTER, M. E. and ACKERMAN, F. D. (2001) *Regional Clusters of Innovation*. Washington, DC: Council on Competitiveness (www.compete.org).

ROBERTS, E. B. (1991) *Entrepreneurs in High Technology*. Oxford: Oxford University Press.

SAXENIAN, A. (2001) The role of immigrant entrepreneurs in new venture creation, in: C. BIRD-SCHOONHOVEN and E. ROMANELLI (Eds) *The Entrepreneurship Dynamic*, pp. 68–108. Stanford, CA: The Stanford University Press.

SCOTT, A. J. (1998) *Regions and the World Economy*. Oxford: Oxford University Press.

SCOTT, A. J. (Ed.) (2001) *Global City Regions: Trends, Theory and Policy*. Oxford: Oxford University Press.

SEGAL QUINCE AND PARTNERS (1985) *The Cambridge Phenomenon*. Cambridge: Segal Quince and Partners.

SHAPERO, A. (1984) The entrepreneurial event, in: C. A. KENT (Ed.) *The Environment for Entrepreneurship*, pp. 21–40. Lexington, MA: Lexington Books.

SMITH, R. K. (2002) Techmaps: a tool for understanding social capital for technology innovation at a regional level, in: J. J. CHRISMAN, J. A. D. HOLBROOK and J. H. CHUA (Eds) *Innovation and Entrepreneurship in Western Canada*, pp. 59–75. Calgary: University of Calgary Press.

STABER, U. (1997) An ecological perspective on entrepreneurship in industrial districts, *Entrepreneurship and Regional Development*, 9, pp. 45–64.

STEED, G. P. F. (1987) Policy and high technology complexes: Ottawa's 'Silicon Valley North', in: F. E. I. HAMILTON (Ed.) *Industrial Change in Advanced Economics*, pp. 261–269. Beckenham: Croom Helm.

STEED, G. P. F. and DEGENOVA, D. (1983) Ottawa's technology-oriented complex, *Canadian Geographer*, 27, pp. 263–278.

STORPER, M. (1993) Regional 'worlds' of production: learning and innovation in the technology districts of France, Italy and the USA, *Regional Studies*, 27, pp. 433–455.

SWANN, G. M. P. (1998) Towards a model of clustering in high technology industries, in: SWANN, G. M. P., PREVEZER, M. and STOUT, D. (Eds) *The Dynamics of Industrial Clustering: International Comparisons in Computing and Biotechnology*, pp. 52–76. Oxford: Oxford University Press.

SWANN, P. and PREVEZER, M. (1996) A comparison of the dynamics of industrial clustering in computing and biotechnology, *Research Policy*, 25, pp. 1139–1157.

SWANN, G. M. P., PREVEZER, M. and STOUT, D. (Eds) (1998) *The Dynamics of Industrial Clustering: International Comparisons in Computing and Biotechnology*. Oxford: Oxford University Press.

SWEETMAN, K. (1982) Ottawa is also our high-tech capital, *Canadian Geographic*, February/March, pp. 20–30.

US SMALL BUSINESS ADMINISTRATION (2000) *Developing high technology communities in San Diego*. Office of Advocacy, US Small Business Administration, Washington, DC.

WOLFE, D. A. (2002a) Social capital and cluster development in learning regions, in: J. A. HOLBROOK and D. A. WOLFE (Eds) *Knowledge, Clusters and Learning Regions*, pp. 11–38. Kingston: McGill-Queen's University Press.

WOLFE, D. A. (2002b) *Knowledge, learning and social capital in Ontario's ICT clusters*. Paper presented to *Canadian Political Science Association Annual Meeting*, University of Toronto, May (http://www.utoronto.ca/progris/recentpub.htm).

WOLFE, D. A. (2003) *Clusters from the inside and out: lessons from the Canadian study of cluster development*. Paper presented to the *DRUID Summer Conference*, Copenhagen, June (http://www.utoronto.ca/progris/recentpub.htm).

WORLD BANK (2000) *Electronic Conference on Clusters*. Washington, DC: World Bank.

Clusters from the Inside and Out: Local Dynamics and Global Linkages

David A. Wolfe and Meric S. Gertler

[Paper first received, March 2003; in final form, December 2003]

1. Introduction

Interest in cluster development has exploded in recent years across North America, Europe and newly industrialised countries. This interest has been prompted, in part, by fascination with the success of Silicon Valley at reinventing itself through successive waves of new technology; and, in part, by the efforts of other regions to emulate the Silicon Valley model. A growing number of clusters around the globe, from Scotland to Bangalore and from Singapore to Israel, claim direct lineage to the original model in northern California (Miller and Coté 1987; Bresnahan *et al.*, 2001; Rosenberg, 2002). The perceived success of Silicon Valley, and the claims by other regions to have replicated its formula for success, have stimulated a widespread interest by policy analysts and consultants eager to assist national, regional and local governments in growing their own clusters. This fascination with using the leading success stories as a model for the development of new clusters has vastly outstripped our current understanding of the key factors or elements that support the growth of clusters. It is not even clear whether there is a unique paradigm for cluster development that cuts across the diverse array of regions and industrial sectors currently attempting to apply the concept as the key to their economic development strategy.

The relevant body of literature has applied the cluster concept in two different, and sometimes contradictory, ways: first, as a functionally defined group of firms and supporting institutions that produce and market goods and services from a group of related industries that are concentrated in a specific geographical locale; secondly, as an overarching framework to guide policy-makers in the design of initiatives to promote that development. The underlying rationale for the first concept is to generate analytical results that can provide insights into the forces that contribute effectively to cluster development and thus provide guidance to local and regional policy-makers in crafting their development strategies. The more applied practitioners who work with the second approach often draw upon the results of the first in drafting policy guidelines, but in a rather limited way. Too often their interpretation of the more analytical cluster studies amounts to little more than the elaboration of lists of the 'critical factors' for cluster development derived from individual studies of the most successful cases. These lists provide relatively little in the way of effective guidance for policy-makers trying to apply the lessons learned to their local economy—which may be based on different economic sectors and facing radically different economic prospects. Frequently, the two strands of research, the empirical and the prescriptive, tend to work at cross-purposes, with the policy goals sometimes predetermining the analysis, rather than the other way around. A key challenge for those interested in applying the concept of clusters from either perspective is to respond to the concerns raised by Martin and Sunley (2003) that academic analysts are being seduced by the lure of the 'cluster brand' at the expense of serious analysis of whether the presence or absence of clusters actually contributes to sustained economic development in local and regional economies.

This paper reports on the initial findings of a broad comparative study of cluster development across a wide range of industrial sectors and virtually all regions of the Canadian economy, conducted by members of the Innovation Systems Research Network.[1] It presents an overview of some of the key conceptual issues in cluster analysis that are emerging from both the analytical and prescriptive literature noted above and uses that overview as the context for exploring initial findings emerging from the ISRN study. The national study is comprised of 26 cases, which aim to identify the presence of significant concentrations of firms in the local economy and understand the process by which these regional-industrial concentrations of economic activity are managing the transition to more knowledge-intensive forms of production. The study's design is unique in terms of the large number and breadth of cases being studied using a common methodological framework and approach. One of the challenges for cluster analysis is to accommodate the diverse array of industrial sectors and geographical locales in which clusters are found. The danger is that generalising from a limited number of case studies in specific sectors, such as information technology, or specific regions, such as high-growth areas of the leading industrial economies, may lead to inappropriate policy conclusions for the broad cross-section of regions and sectors to which they are applied. The key questions posed in each of our cases are

(1) What role do local institutions and actors play in fostering this transition to more innovative, knowledge-intensive production?
(2) How important is interaction with non-local actors in this process?
(3) How dependent are local firms on unique local knowledge assets, and what is the relative importance of local versus non-local knowledge flows between economic actors?
(4) How did each local industrial concentration evolve over time to reach its present state, and what key events and decisions shaped its path?
(5) And, finally, to what extent do these processes, relationships and local capabilities constitute a true cluster? What

are the key relationships, linkages and processes that ground the cluster in its existing location?

The initial results are surprising in that they contradict some of the most commonly accepted arguments in the literature. It is also clear that the national and regional contexts in which these cases have evolved are of great importance in shaping their specific evolutionary trajectories. In particular, the open nature and smaller size of the Canadian economy relative to that of the US appear to explain the apparently distinctive and divergent characteristics of Canadian clusters (or putative clusters). These findings provide a strong note of caution for policy-makers seeking a generic or 'cookie cutter' approach to clusters as the prescription for the economic development challenges they face.

2. Emerging Themes in the Cluster Literature

While the cluster literature is expanding rapidly and becoming ever more diverse, a number of broad themes pre-occupy cluster analysts. In particular, three stand out. The first is the issue of path dependence: how do cluster dynamics become established and can they be seeded, particularly through the action of public-sector agencies? Despite the ever-increasing base of empirical case studies available, there remains a striking lack of consensus over how clusters are started and to what extent their emergence can be set in motion by conscious design or policy interventions. One approach in the literature adopts an historical perspective to unearth the origins and evolution of specific clusters. According to Malmberg and Maskell (2002), these historiographic studies attribute the emergence of the cluster to some natural or social factor endemic to a particular location that triggers or stimulates a certain kind of activity by a local entrepreneur. Once the initial activity is launched, its expansion is sustained by the emulation effect as other firms spin off from the anchor firm or engage in related activities. Equally important is the attraction, or embeddedness, of firms to the region in which they originate and the infrequency of relocation—in other words, the force of inertia.

As straightforward as this analysis may appear, many such accounts have difficulty dating the precise origins of individual clusters and identifying the critical or initial founding event. In the case of the most celebrated cluster, Silicon Valley, no such consensus on its origins exists. The common launch event for many is the decision by William Shockley to move to California and establish his semiconductor company in 1956 and the subsequent decision by seven of his key employees to leave to establish Fairchild Semiconductor, which became the source of most of the major semiconductor firms in the Valley. Other accounts date the origins of the Valley from the decision by David Packard and William Hewlett to found their company in a garage in Palo Alto in 1939. Yet Timothy Sturgeon (cited in Kenney, 2000b, pp. 3–4) argues that the real roots of the cluster should be dated as far back as the formation of the Federal Telegraph Company in 1909 with the ensuing spin-offs laying the basis for the Valley's early electronics industry. The critical issue is how to draw policy lessons on the formation of clusters when their precise origins are so difficult to ascertain. And where, in particular, does policy fit into a seemingly random or serendipitous process?

The second key theme concerns the nature of knowledge and learning in clusters. Within economic geography, clusters have generally been perceived in one of two ways. The first approach, dating back to the work of Alfred Marshall, views clusters as the product of traditional agglomeration economies, where firms co–located in the cluster benefit from the easier access to, and reduced costs of, certain collective resources, such as a specialised infrastructure or access to a local labour market for specialised skills (Porter 1998). The second view emphasises the role of knowledge and learning processes in sustaining clusters, often on the basis of local flows of spatially sticky tacit knowledge. This second approach also emphasises that knowledge flows in clusters are not necessarily restricted to the local level—dynamic clusters usually develop strong connections to other clusters through the international sharing of knowledge (Bathelt *et al.*, 2002). This draws attention to the need to understand how local clusters are situated within an international hierarchy, in those cases where the local knowledge-base provides one element in a more complex set of knowledge flows.

The final theme concerns the scales of analysis. While much of the cluster literature focuses predominantly on the influence of local factors on cluster development, there is growing recognition that clusters are embedded in a broader institutional matrix at the regional, national and even supranational levels. The central question involves the nature of the relationship between the local cluster and other analytical frames of reference, such as national or regional innovation systems. If we accept that clusters should be defined primarily in local terms, then the issue of how they fit into broader institutional frameworks must be addressed. In the eyes of some, clusters can be defined in relatively self-contained terms, with little attention paid to the role that higher levels of spatial analysis contribute to the success of local clusters. Given the parallel interest in the concept of innovation systems—at the national, regional and sectoral levels—it is not surprising that some analysts have attempted to specify the nature of the linkages and the relative contributions made by the different spatial levels to economic competitiveness. There is an emerging interest in the need to understand how clusters are inserted into these broader scales of analysis and how the latter both support and constrain the trajectories for growth and development within the cluster.

2.1 Path Dependency and the Creation of Clusters

According to Michael Porter, clusters are seeded in a variety of ways; however, their growth can only be facilitated by building upon existing resources. They cannot be started just anywhere from scratch. The key assets that determine the viability of a cluster are firm-based. Of particular importance is the emergence of a lead or anchor firm for the cluster. Whole clusters can develop out of the formation of one or two critical firms that feed the growth of numerous smaller ones. It is the emergence of these core or anchor firms that is so difficult to predict, yet so central to the history of many leading clusters. Examples of the role played by this kind of anchor firm are found in the case of Medtronic in Minneapolis, MCI and AOL in Washington, DC, (or, as we shall discuss below, NovAtel in Calgary). In other instances, the presence of major anchor firms in a local cluster can act as a magnet, attracting both allies and rivals to the region to monitor the activities of the dominant firm. This is the case with San Diego, where Nokia, Ericsson and Motorola all established their CDMA wireless research efforts to benefit from Qualcomm's leadership in the field (Porter *et al.*, 2001). Other analysts emphasise the role played by highly skilled labour, or a unique mix of skill assets, in seeding the cluster. Either way, the process also requires a long time to take root.

This does not mean that the public sector has no role to play in catalysing cluster development. The public sector—broadly defined—encompasses federal, state/provincial/regional and local governments, as well as public research institutes like Canada's National Research Council and institutions of higher education. The impact of public-sector interventions on cluster development can be positive or negative, as well as intentional or inadvertent in character. One emerging hypothesis suggests that the public interventions that seem to have the most effect in seeding the growth of a cluster are ones that contribute to the development of the asset-base of skilled knowledge workers.

The catalytic role of the federal laboratories in the origin of knowledge-intensive clusters is central in Feldman's (2001) account of the emergence of the current telecommunications clusters in the Washington–Baltimore corridor. Feldman's analysis emphasises the importance of entrepreneurship in driving the development of that cluster. She traces the roots of this entrepreneurial drive to the massive wave of downsizing and outsourcing that occurred in the US federal government in the late 1970s and 1980s. As a result of this process, employment conditions in the federal public service became less secure and future prospects deteriorated. In the same period, public-sector pay scales began to lag significantly behind those for executives in the private sector. An increased emphasis on outsourcing by the federal government provided a further inducement for prospective entrepreneurs to leave the government and start firms to supply goods and services back to their former employer. Other policy initiatives launched in the early 1980s facilitated the licensing and transfer of technology from federal laboratories and provided further support for innovation in small businesses.

> Enterprising scientists licensed technology out of their own university or government research labs to start new companies and chose to locate the new companies near their existing homes (Feldman, 2001, p. 878).

Although cluster creation was clearly not the principal objective in the policy decisions she cites, the inadvertent role played by public policy in the formation of the cluster cannot be overlooked. The lesson here is that the evolutionary paths for cluster creation are highly variable. Public-sector decisions can affect cluster trajectories in a variety of ways, although the impacts are often unpredictable and often unintended. While this growing consensus in the literature on the origins of clusters and the nature of evolutionary paths explains the presence of significant agglomerations of firms in specific locales, it does not fully account for the benefits that firms derive from cluster membership nor whether firms located in the cluster are more innovative or economically competitive than those found in more dispersed locations.

2.2 Knowledge and Spillovers in Clusters

The benefits of clustering and the potential advantages that firms derive from cluster membership are addressed in a second theme found in the literature—the role of knowledge and spillovers in clusters. One stream of literature stresses that the key advantages are derived from the agglomeration economies afforded by the cluster. These agglomeration economies arise primarily from the ready access to a collective set of resources available to firms co-locating in the same region or locale. This perspective is adopted in the work of Michael Porter, although he embellishes the benefits attributed to traditional agglomeration economies by setting out the competitive advantages derived from the effects of his 'diamond'. Porter stresses that the location of a firm within the cluster contributes to enhanced productivity, higher wages and greater innovativeness by providing access more easily and/or cheaply to specialised inputs, including components, machinery, business services and personnel, as opposed to the alternative, which may involve vertical integration or obtaining the needed inputs from more remote locations. Sourcing the required inputs from within the cluster also facilitates communication with key suppliers in the sense that repeated interactions with local supply firms in the value chain creates the kind of trust conditions and the potential for conducting repeated transactions on the basis of tacit, as well as more codified, forms of knowledge. Clusters offer distinct advantages to firms in terms of the availability of specialised and experienced personnel. The cluster itself can act as a magnet drawing skilled labour to it. Conversely, the location of specialised training and educational institutions in the region provides a steady supply of highly qualified labour to the firms in the cluster (Porter, 1998).

While not diminishing the importance of these agglomeration economies, another stream of literature suggests that the underlying dimension that confers competitive advantages on the firms located in the cluster is shared access to a distinctive local knowledge-base. The central argument in this stream is that the joint production and transmission of new knowledge occur most effectively among economic actors located close to each other. Proximity to critical sources of knowledge, whether they are found in public or private research institutions or grounded in the core competencies of lead or anchor firms, facilitates the process of acquiring new technical knowledge, especially when the relevant knowledge is located at the research frontier or involves a largely tacit dimension. Knowledge of this nature is transmitted most effectively through interpersonal contacts and the interfirm mobility of skilled workers. However, Breschi and Malerba (2001) argue that this approach overestimates the benefits of physical proximity alone. They argue that sheer proximity is not sufficient to account for local knowledge spillovers. In their view, the body of research on local knowledge spillovers overlooks the broader set of factors and conditions that support the effective transfer of knowledge in clusters.

> A key feature of successful high-technology clusters is related to the high level of embeddedness of local firms in a very thick network of knowledge sharing, which is supported by close social interactions and by institutions building trust and encouraging informal relations among actors (Breschi and Malerba, 2001, p. 819).

In other words, the degree to which firms can tap into a common knowledge-base at the local level depends on more than just spatial proximity, cultural affinity or corporate culture. In this sense, there is a strong interdependence between the economic structure and social institutions that comprise the cluster. The institutional context of the cluster defines how things are done within it and how learning transpires. As Gertler has argued, it is a function of institutional proximity—the common norms, conventions, values and routines that arise from commonly experienced frameworks of institutions existing within a regional setting (Gertler, 2003, p. 91).

It is also critical to differentiate between different kinds of knowledge spillovers. Much of the literature on knowledge spillovers and, in particular, the role of tacit knowledge, presumes that the knowledge being shared is highly technical in nature and results largely from the transfer of research results between regionally embedded research institutes and private firms. However, technical research results are only one element of the kinds of knowledge flows that contribute to the competitive dynamics of a successful cluster. One of the most important sources of knowledge flows is the knowledge embodied in highly qualified personnel which flows directly from research institutes to private firms in the form of graduates and also moves between firms in the form of mobile labour. There is a strong suggestion in the literature that the recombining of talent in new constellations through labour mobility and the spinning-off of new start-up firms is one of the most important sources of innovation in dynamic clusters (Saxenian, 1994; Brown and Duguid, 2000). A third form of knowledge flows involves entrepreneurial skills. This is often one of the least well documented, but most critical, elements of successful clusters. Closely related to this is knowledge about external market conditions. For small and medium-sized enterprises, an essential piece of knowledge they must acquire to grow and expand concerns the competitive conditions in external markets and which ones constitute the most suitable targets for expansion. Entrepreneurial skill and market information can be transmitted through the cluster via a variety of mechanisms—some formal and some informal—but one of the most important is frequent peer-to-peer mentoring and knowledge sharing that is organised through local civic associations. The dynamic role played by civic associations in facilitating this form of knowledge flow underlines the importance of the local and regional institutional structures once again. The final dimension of knowledge sharing crucial for the success of the cluster is the kind of infrastructural knowledge resources found in the specialised legal, accounting and financial firms that are essential to the success of individual firms in the cluster. These kinds of services often provide vital support to the individual firms in the cluster.

In an attempt to elaborate further the role that knowledge plays in sustaining clusters, Maskell (2001) has proposed a knowledge-based theory of the cluster. He suggests that the primary reason for the emergence of clusters is the enhanced knowledge creation that occurs along two complementary dimensions: horizontal and vertical. Along the horizontal dimension, clusters reduce the cost of co-ordinating dispersed sources of knowledge and overcoming the problems of asymmetrical access to information for different firms producing similar goods and competing with one another. The advantages of proximity arise from continuous observation, comparison and monitoring what local rival firms are doing, which act as a spur to innovation as firms race to keep up with or get ahead of their rivals. The vertical dimension of the cluster consists of those firms that are complementary and interlinked through a network of supplier, service and customer relations. Once a specialised cluster develops, local firms increase their demand for specialised services and supplies. Furthermore, once the cluster has emerged, it acts as a magnet drawing in additional firms whose activities require access to the existing knowledge-base or complement it in some significant respect (Maskell, 2001, p. 937). In critical respects, this knowledge-based conception of the cluster takes for granted key aspects of the Porter diamond, in its assumption that firms co-located in the cluster tend to be rivals in the same product markets or part of a locally based supply chain, and that close monitoring of competitors or tight buyer–supplier interaction are key elements that tie the firm to the cluster. While these conditions may hold for the most developed clusters in their respective industrial or product segments, there is growing evidence (see following sections of this paper) to suggest that they do not apply universally to all clusters—especially those in more specialised niches or at an earlier stage of cluster development.

If this is the case, then it opens up a new line of inquiry about the relationship between the global and the local, and complicates considerably the question we posed at the outset: just what is it that ties the group of firms to a specific location? A knowledge-based theory of the cluster must recognise that relatively few clusters are completely self-sufficient in terms of the knowledge-base they draw upon. As the innovation process changes to involve the development of ever more complex technologies, the production of these technologies requires the support of sophisticated organisational networks that provide key elements or components of the overall technology (Kash and Rycroft, 2000). While some elements of these complex technologies may be co-located in an individual cluster, increasingly the components of these networks are situated across a wide array of locations. This suggests that the knowledge flows that feed innovation in a cluster are often both local and global. Bathelt *et al.* (2002) maintain that successful clusters are those that are effective at building and managing a variety of channels for accessing relevant knowledge from around the globe. However, the skills required when dealing with the local environment are substantially different from the ones needed to generate the inflow and make the best use of codified knowledge produced elsewhere, and these different tasks must be managed by the cluster. They maintain that an accurate model of the knowledge-based cluster must account for both dimensions of these knowledge flows.

Bathelt *et al.* refer to these two kinds of knowledge flows as 'local buzz' and 'global pipelines' respectively. Following Storper and Venables (2003), 'buzz' arises from the fact of physical co-presence. It incorporates both the broad general conditions that exist when it is possible to glean knowledge from intentional face-to-face contacts, as well as the more diffuse forms of knowledge acquisition that arise from chance or accidental meetings and the mere fact of being in the same location. Buzz is the force that facilitates the circulation of information in a local economy or community and it is also the mechanism that supports the functioning of networks in the community. In this context, it is almost impossible to avoid acquiring information about other firms in the cluster and their activities through the myriad number of contact points that exist. Pipelines, on the other hand, refer to channels of communication used in distant interaction, between firms in clusters and sources of knowledge located at a distance. Important knowledge flows are generated through network pipelines. The effectiveness of these pipelines depends on the quality of trust that exists between the firms in the different nodes involved. The advantages of global pipelines derive from the integration of firms located in multiple selection environments, each of which is open to different technical potentialities. Access by firms to these global pipelines can feed local interpretations and the usage of knowledge that contibuted to the emergence of successful firms and clusters elsewhere. Firms need access to both local buzz and the knowledge acquired through international pipelines. The ability of firms to access such global pipelines and to identify both the location of external knowledge and its potential value depends very much on the internal organisation of the firm, in other words, its 'absorptive capacity'. The same can be said of local and regional clusters (Bathelt *et al.*, 2002).

However, the precise mix of the global and local knowledge flows present in individual clusters must of necessity be indeterminate. There is increasing evidence to suggest that, even in the most advanced clusters, a growing proportion of the knowledge-base is not exclusively local. The most recent work on Silicon Valley suggests that the production involved in local clusters is part of a complex production chain that is connected into global production networks. The most dynamic of multinational corporations and a larger proportion of emerging small and medium-sized enterprises are embedded in a variety of specialised clusters around the globe. Both types of firms use their presence in the local clusters to access specialised bodies of knowledge created by the local research institutions or tap into a specialised skill-set or unique technical knowledge developed by cluster-based firms. However, rarely are the local knowledge-bases of these clusters, or the production activities of the firms embedded in them, completely self-contained. Rather, according to Sturgeon (2003, p. 200), "what gets worked out in the clusters is exactly the codification schemes that are required to create and manage spatially dispersed but tightly integrated production systems". A greater proportion of the production of complex technologies in sectors ranging from information technology to automotive assembly occurs in these modular production networks with activities dispersed across a wide range of global locations. What take place in the clusters of the more industrialised economies are the core interactions between lead firms and key suppliers that resist easy codification, such as design, development of prototypes and determining the validity of manufacturing processes. The production of high-value-added or low-volume products also takes place in these locations. He implies that there is a geographical hierarchy of clusters within specific industrial sectors, with Silicon Valley acting as the key location for standard-setting activity in information technology (Sturgeon, 2003, p. 220).

A marked pattern of stronger global (vs local) relations emerges even more clearly in a recent study of opto-electronics clusters in six locations. This study found that extraregional commercial linkages are more important than localised ones due to the highly diversified nature of the end-user markets and the complexity of the technologies involved in assembling an end-product for the market. The individual clusters in each of the six case-study regions are dominated by a dominant local actor: either a strong research centre or a lead firm that serves as a catalyst to bring together the firms in the cluster. However, due to the nature of the technologies involved and the intrafirm and interfirm dynamics, there is little local co-operation and few traded relationships among firms within the individual clusters. What the firms in the clusters do share is their common linkage to the leading institution or firm and their common interest in stimulating and maintaining the critical supply of highly skilled labour (Hendry *et al.*, 2000, pp. 140–141).

2.3 Placing Clusters in a Broader Context

The complex role of both local and non–local knowledge sources in the dynamics of even the most advanced clusters draws our attention to the relationship between the local dimensions of the cluster and the other levels of governance within which they are embedded. If, as we have argued above, institutions are the hidden glue that holds clusters together, the implicit question is whether the institutional structures relevant to cluster dynamics are exclusively those found at the local level. A number of studies have recently focused on the relationship between the concept of the cluster and others used to analyse the innovative capacity of regional and national economies, principally the innovation systems approach. Bunnell and Coe argue for a shift in focus away from forms of analysis that privilege one particular spatial scale as the basis for analysing and understanding the nature of innovation towards those that emphasise the relationships that exist between and across different spatial scales. They adopt the concept of 'nested scales' from Swyngedouw, but suggest that this should not be conceived in a hierarchical or deterministic sense, but rather as involving effects that can move in multiple directions across the scales (Bunnell and Coe, 2001, p. 570).

Thus clusters can be seen as nested within, and impacted by, other spatial scales of analysis, including regional and national innovation systems, as well as the kind of global relationships and forces implied by the 'pipelines' discussed above, each of which adds an important dimension to the process of knowledge creation and diffusion that occurs within the cluster. Various elements of each of these spatial levels of analysis may have significance for the innovation process. For instance, the national innovation system, as analysed by Nelson (1993) or Lundvall (1992), may play a preponderant role in establishing the broad framework for research and innovation policies, in providing a national system of research organisations, in establishing the rules of corporate governance that influence firm behaviour, in setting the rules of operation for the financial system that determine the availability of different sources of financing and time-horizons for new and established firms and, finally, for setting the broad framework for the industrial relations, employment and training systems that influence job paths, interfirm mobility and skill levels for the labour force. Levels of regional specialisation as encompassed in the concept of regional innovation systems developed by Cooke and others play an important role in affecting cluster performance through the provision of the regional/state/provincial research infrastructure, specialised training systems, the broad education system, policies for physical infrastructure and the investment attraction dimensions (Cooke *et al.*, 1997; Cooke, 1998). At the local level, varying degrees of civic associationalism, particularly the business–higher education link, influence cluster development. The local level can also play an important role in the provision of infrastructure such as roads and communication links, as well as in the governance of the primary and secondary education system.

The case of Silicon Valley clearly illustrates the way in which these differing scales of governance impact on the performance of local clusters. The cluster exists within the distinctive features of the US system of innovation—with its unique system of laws, regulations and conventions governing the operation of capital markets, forms of corporate governance, research and development and other relevant factors. A number of these features are central to the story of Silicon Valley's growth and development, including the highly decentralised nature of the post-secondary education system with complementary and interlocking roles for both the federal and state governments. Changes introduced in the 1970s and 1980s in capital gains tax rates and the tax treatment of stock options, as well as the rules governing investments in venture capital by pension funds, stimulated the growth of the venture capital industry, a critical factor for the development of the ICT cluster. The federal government played a central role as the initial customer for many of the early products of the cluster. It was also the primary funder for much of the critical research and development that has underpinned the growth of these clusters (Rowen 2000). Thus the concept of 'nested scales' of analysis deepens our understanding of the multiple factors that influence the development trajectory of a cluster and, ultimately, its economic performance. From a policy perspective, it also draws attention to the role that higher levels of government play in creating the conditions that support cluster development (Porter *et al.*, 2001).

3. Methodological Approaches to Cluster Studies

One of the key challenges in attempting to draw a consistent set of conclusions from the rapidly expanding opus of cluster studies is the diverse array of methodological approaches used in the studies. The first approach employs a diverse set of statistical-analytical tools, of differing sophistication, to measure the degree of clustering found in local and regional economies. A second approach involves the conduct of case studies of individual clusters or several clusters on a comparative basis. These case studies can involve a wide range of clusters all located within one country or a select group of similar clusters located across different countries. The intent is to use a standard framework to compare the individual cases or benchmark them against the presumed leader or role model for the clusters. Another approach focuses on the analysis of public policies and strategies explicitly designed to promote the establishment and/or growth of individual clusters or sets of clusters. This latter approach is frequently undertaken for a regional or municipal development authority with the goal of benchmarking the relative performance of the region's clusters and providing policy prescriptions for improving their competitive success. This last category usually includes some combination of both quantitative and case-study methodologies.

3.1 Statistical Approaches to Cluster Analysis

One of the most common techniques employed by analysts to identify the presence of clusters within a specific geographical locale is the use of the employment location quotient, which is a ratio of employment shares for a particular industry: the regional industry's share of total regional employment over the national industry's share of total national employment. A quotient greater than one identifies those industries that may constitute the components of local clusters, since it indicates a higher degree of specialisation in the industry regionally than exists at the national level. This is usually interpreted to reflect the degree of competitive advantage enjoyed by the industry locally, relative to its status elsewhere in the country.

A more sophisticated version of this method of analysis is found in the growth–share matrix used by some analysts to provide a maximum amount of information about the relative strength of a local cluster. The growth–share matrix combines three specific measures of local industrial strength in one diagram: the absolute size of the sector in the region, measured in terms of employment; the average annual regional growth rate in employment for the sector, and its location quotient. The representation of the growth–share matrix in graphical form provides a powerful visual medium for depicting the relative economic strengths of a regional or local economy. The use of the growth–share matrix also provides an easy way to benchmark local and regional economies against other localities where the analysis has previously been done and is useful for highlighting the relative strengths and competitive challenges facing a region (Information Design Associates and ICF Kaiser International, 1997, pp. 41–45). One critique of this methodology is that location quotients are largely an industry–based technique derived from traditional statistical categories such as Standard Industrial Classifications (SIC) and, consequently, offer little insight into the interdependencies between sectors that ought to characterise dynamic local clusters. Ultimately, they are only useful if employed in association with other methods that provide some degree of information on industrial interdependence (Bergman and Feser, 1999, ch. 3).

A more sophisticated version of this technique is represented in the ambitious undertaking by Michael Porter through the Institute for Strategy and Competitiveness at the Harvard Business School. The Institute's Cluster Mapping Project uses statistical techniques to profile the performance of regional economies in the US over time, with a special focus on clusters. Economic profiles of the 50 US states and the District of Columbia were prepared for the National Governors Association Initiative "State Leadership in the Global Economy" using this approach. The detailed profiles of each state provide analyses of major concentrations of employment for both traded and untraded clusters. The Cluster Mapping Project uses information drawn from the County Business Patterns data on employment, establishments and wages by four-digit SIC codes, plus patent data on location of inventor, to identify the core clusters in a region using the correlation of industry employment within geographical areas. The dominant clusters in a region are identified using a location quotient analysis to identify those that are relatively more concentrated based on the region's total employment. Applying this methodology, the Cluster Mapping Project has identified 41 types of clusters in US economy, differentiated between traded, resource-driven and locally oriented clusters (Porter *et al.*, 2001, pp. 18–28; Porter, 2003).

Despite the apparent sophistication of these techniques, they are not without their critics. First, the empirical approaches to cluster identification tend to overlook the nature of cluster life cycles. Clusters frequently go through specific stages of development and the identification of the stage of development for an individual cluster is very important to an analysis of the cluster dynamics. Empirical methodologies that focus exclusively on a statistical snapshot of the cluster at a specific point tend to ignore an analysis of its trajectory of development (Breschi and Malerba, 2001). Empirical analyses that incorporate the rate of growth of employment in the cluster can partially compensate for this shortcoming, but failure to account for this factor means that two clusters on a radically different development path may appear to be quite similar in a statistical snapshot at one point. More generally, their value is limited by the fact that they fail to capture the critical contribution made by soft factors, such as trust and social capital, as well as the organisational dynamics of the cluster. Thus, they only hint at the role played by non-market-based processes, or untraded interdependencies (Storper, 1997).

3.2 Case Studies

Many analysts reject the argument that clusters can only be adequately studied by using statistically oriented methods. They argue instead that the growth and innovation dynamics of clusters can only be properly captured by using qualitative research techniques, especially in-depth interviews with a broad cross-section of cluster participants or ethnographic accounts of the cluster's evolution from leading members. The most common approach in this category is the intensive case study of an individual cluster—the most studied being Silicon Valley. The original model was Saxenian's (1994) case study of Silicon Valley undertaken in the early 1990s and the comparison she provided with Route 128 in Massachusetts. Saxenian drew upon the growing body of literature on the dynamics of regional network-based industrial systems to highlight the similarities and the differences between the two regions. Firms in network systems compete in global markets and collaborate with distant customers and suppliers, but their most strategic relationships are often local because of the critical importance of face-to-face communication for rapid product development. The variable that determines the relative performance of firms in different regionally based networks is the nature of its industrial system, which includes three important dimensions—the indigenous mix of institutions and culture in the region; the structure of the industrial system; and the internal organisation or industrial culture that prevails in firms in the region (Saxenian 1994, pp. 5–7).

Saxenian's study of Silicon Valley and the insights it affords have been complemented by two recent volumes edited by Kenney (2000a) and Lee *et al.* (2000). Both provide a series of studies that enrich our understanding of the historical trajectory of Silicon Valley's development, its institutional underpinnings and its operating dynamics. The papers in these volumes trace some of the critical junctures in the history of the Valley and, especially, the central role played by key anchor firms in stimulating the growth of related firms at different stages in the Valley's evolution. The influence of forces at different spatial scales is also highlighted, in particular the key support mechanisms provided by the federal government, including defence procurement and critical funding for pre-commercial research. The nature of entrepreneurship, interfirm relationships and the role of knowledge flows in the Valley are also covered.

Although these analyses offer competing explanations of the underlying dynamics that have sustained the growth of the Valley's firms through successive waves of technological innovation, their authors agree that its dynamism can be attributed to the nature of its 'ecosystem' which involves the continuous creation of a multitude of diverse, specialised firms and support organisations that constantly interact with each another to accelerate the innovation process. Saxenian and a number of colleagues have also completed a broad comparative case study of a number of emerging regions attempting to emulate Silicon Valley. The regions covered in this study include Ireland, India, Cambridge in the UK, Israel, Scandinavia, Taiwan and northern Virginia within the US. The key factor driving the growth of these clusters is the ready supply of skilled human capital that attracts managerial talent and entrepreneurs into the cluster. Public policy can support this tendency in a number of ways, but these authors are highly critical of attempts to jumpstart clusters or make top–down or directive efforts to promote them (Bresnahan *et al.*, 2001).

Other notable projects employing a case-study approach include the five detailed studies undertaken by Michael Porter for the US Council on Competitiveness. The Council's Clusters of Competitiveness Initiative examined five regions in the US: Atlanta, Pittsburgh, the Research Triangle, San Diego and Wichita, selected to provide a diverse sample based on size, geography, economic maturity and relative degree of economic success. The case studies used a variety of research methodologies to obtain data on the five regions, including data from the Cluster Mapping Project described above, a set of regional surveys designed and conducted specifically for the Initiative and in-depth interviews with business and government leaders in each region. The study identified a set of factors that contribute to the evolution of regional economies. Successful regions leverage their unique mix of assets to build specialised clusters. They do not try to pick winners, but build on their existing assets to create unique economic strengths that offer competitive advantages to firms based in the region. Building strong regional economies is not an overnight phenomenon. It takes decades of effort to develop existing assets, create new ones, link firms to this regional asset-base and attract inward investment to the cluster. Finally, they conclude that collaborative institutions play a critical role in building regional economies by facilitating the flow of information, ideas and resources among firms and supporting institutions (Porter *et al.*, 2001, pp. x–xiii).

It is apparent from the preceding review that the case-study approach can yield important insights into the nature and dynamics of regional industry clusters and the sources of their success. The most effective case studies transcend the limitations of the purely statistical approach to shed new light on the underlying social and institutional dynamics that create the extrafirm dimensions of the cluster's strength. The limitation of these studies, however, is that it may be difficult to compare findings across individual cases if they have not been derived within a common study framework. While the best of them illuminate the relative strengths of a particular cluster, the lack of comparability limits our appreciation of why certain clusters succeed to a greater extent than others. The comparative study by Bresnahan *et al.* and Michael Porter's work for the Council on Competitiveness, which introduced a degree of comparability into the case studies, take an important step in overcoming this limitation. They provide a useful model for other studies in the design of their own research methodologies.

4. Cluster Evolution in Canada: What Have We Learned So Far?

The ISRN's national study of cluster development employs a range of empirical methods to document and understand the emergence and evolution of local clusters in different regions of Canada. It has been designed to allow us to examine—whenever possible—the same type of industry in two or more different regions in Canada. At the same time, we are also studying multiple industrial cases in the same region. Each case is being examined using a common research methodology, based primarily on in-depth interviews with key cluster participants, although supplemented by statistical analysis at the regional and national levels (Gertler and Levitte, 2003; Amara *et al.*, 2003). Each case study addresses a common set of features including

(1) the size and composition of the actual or potential cluster;
(2) the history of the cluster's evolution, including key events (intentional and accidental);
(3) the nature of relationships between firms, and between firms and the research infrastructure;
(4) the geographical structure of these relationships;
(5) the role of finance capital (especially angel investors and venture capitalists);
(6) the role of local associative behaviour; and
(7) other forces contributing to (or inhibiting) the growth of the cluster.

In this way, we hope to discern intrasectoral commonalities, as well as differences in experience that may have arisen due to regional influences and histories.

The selection of industries covered reflects the breadth and structure of the Canadian economy, resisting the temptation to focus solely on a narrow list of 'new economy' cases. The cases range from highly knowledge-intensive activities such as biotechnology, photonics and wireless equipment, telecommunication equipment and aerospace, to more traditional sectors such as steel, automotive parts, specialty food and beverages, and wood products. The cases are distributed across both metropolitan and non-metropolitan regions, reflecting the unique geography of Canada's national economy.[2] We are employing a common research framework and interview guide to analyse all 26 cases. Each case study is based on in-depth, semi-structured interviews with a range of stakeholders drawn from 5 different groups, with the total number of interviews conducted ranging from a minimum of 50 to more than 100, depending on the size and complexity of each case. Specific interview guides have been developed for each of these stakeholder groups:

(1) 'Lead' (large, technologically dynamic, export-oriented), smaller and mid-sized firms, including suppliers.
(2) Industry associations, chambers of commerce, local political leaders and 'civic entrepreneurs'.
(3) Government agencies (federal, provincial, local).
(4) Universities, colleges and other institutions for research and training (including offices of technology transfer/commercialisation as well as relevant departments and individual researchers).
(5) Financial sector (venture capitalists, banks, other).

The first wave of case studies commenced in mid 2001, with most research projects slated to last three years. Two sets of preliminary results have been presented at annual meetings of the ISRN held in May 2002 and 2003, and the first set are available in published form (Wolfe, 2003). What follows is a description of some key indicators of cluster dynamics and properties, and a discussion of common themes emerging across many of the case studies.

4.1 Key Cluster Indicators: How Do We Know a Cluster When We See One?

In contrast to many pre-existing studies of clusters, we have been careful to treat the existence of a local cluster as a hypothesis to be verified through investigation, instead of an *a priori* assumption. Given this orientation, we need some systematic methodology for discriminating between the *bona fide* cases and the imposters. The research completed thus far, and the theoretical and conceptual literature from which we draw our inspiration, have led us to emphasise *flows* and *dynamics* over stocks and static measures of innovativeness. They also point quite clearly to the centrality of knowledge and learning processes, both embodied and otherwise. At this stage, the analysis focuses on four categories of indicators: inflows, outflows, local social dynamics and historical path dynamics.

Inflows. One clear way to confirm the existence of unique, distinctive local knowledge-based assets is by tracking three different forms of inflow. Capital inflows, in the form of venture capital investments, foreign direct investments, or mergers and acquisitions, indicate that investors have identified the local presence of local knowledge assets and capabilities. This seems to have been the case in Ottawa's information technology sector, where Cisco (US) and Alcatel (France) both acquired local firms during the 1990s to partake of the optical and telecommunications expertise embedded in the region through the presence of Nortel and JDS Fitel (now JDS Uniphase) (Chamberlin and de la Mothe 2003). More recently, non-local venture capitalists have continued to invest aggressively in Ottawa firms with high growth potential throughout the post-2000 downturn in both the telecom and photonics sectors. The same phenomenon is evident in the case of Calgary's wireless industry, where Intel has invested directly in new R&D capacity (Langford *et al.*, 2003).

Inflows of people are, in our view, an especially robust indicator of local dynamism. It is now increasingly well established that highly educated, talented labour flows to those places that have a 'buzz' about them—the places where the most interesting work in the field is currently being done. One way to track this is through the inflow of 'star scientists' or by tracking the in-migration of tomorrow's potential stars (post-docs). The recent analysis of this geography in the context of the Canadian biosciences demonstrates that centres such as Vancouver, Montreal and Toronto have exerted a powerful attractive force. Moreover, those firms that have developed working relationships with such stars have experienced significantly higher employment growth between 1997 and 2002 (Queenton and Niosi, 2003).

Another approach, promoted by Florida and colleagues (Florida, 2002; Gertler *et al.*, 2002), utilises a more broadly defined measure of 'talent' and has documented its strong geographical attraction to the presence of other creative people and activities locally. Of course, in-bound talented labour represents knowledge in its embodied form flowing into the region. Hence, such flows act to reinforce and further accentuate the knowledge assets already assembled in a particular region. In Canada, cities such as Toronto, Vancouver, Montreal and Calgary stand out as leading centres for the attraction and retention of highly educated and creative workers. One should also be able to track knowledge inflows directly, in their disembodied form. This would be monitored through licensing of intellectual property produced elsewhere, or through local citation of externally generated patents, as is suggested in the case study of the Saskatoon biotech cluster (Ryan and Phillips, 2003).

Outflows. Dynamic, innovative clusters of economic activity should also be discernible by the things that flow outward to the rest of the world. Of course, Porter's own methodology for identifying clusters starts with this point, by attempting to document locally produced goods and services that are traded on world markets. But a more complete analysis would need to go beyond these relatively tangible flows, to consider some important but intangible outflows. Foremost among these would be outflows of knowledge, as monitored through various formal modes of intellectual property transfer (such as licensing or patent citations). We would argue that this kind of activity provides perhaps the best indicator of wider recognition of the unique capabilities and knowledge assets of a region. As noted below, emerging evidence from our biotechnology case studies confirms that dynamic firms in Canadian clusters are indeed the origin point for knowledge outflows to commercial partners in the US, Europe and Asia (Gertler and Levitte, 2003).

Local social dynamics. This is the starting-point for most of the literature in economic geography and related fields over the past 15–20 years. This literature has tended to focus on local social dynamics almost to the complete exclusion of all else, including the important non-local flows discussed above. Relevant here, of course, is evidence of co-operation and network-based behaviour, particularly those forms that promote the circulation of knowledge locally. But, as Malmberg and Maskell (2002) point out, competition is as much a part of the story as is collaboration. The dense local clustering of competing firms provides a vitally important opportunity for mutual monitoring and observation, itself a crucial form of knowledge flow. Our case studies are beginning to document the circulation of labour and entrepreneurs between local firms (or other organisations such as research institutes) through the collection of information on career histories, spin-off activity and the process of new firm formation. As noted below, the case-study evidence to date suggests that informal monitoring of other firms' activities as well as learning through the circulation of labour among firms are relatively more important sources of knowledge flows than formal collaborations among the local firms or dense networks of buyer–supplier relationships. Other key markers of local social dynamics include the presence of community-level institutions for associative governance (public, private and hybrid). Such institutions have the potential to promote social interaction and reflexive behaviour leading to successful adaptation and resilience in the face of competitive challenges from abroad. And as Maskell and Malmberg (1999) have argued, because of the path-dependent nature of such local institutions, they are usually quite difficult to replicate, making them a key component of the region's distinctive and unique asset-base.

Historical path dependencies. Following on from the previous point, perhaps the most discerning test of 'true' cluster dynamics is one that assesses the alleged cluster's resilience and robustness over time, in the face of severe shocks and dislocations. How has the region fared under such circumstances? How effectively have its firms and institutions adapted and evolved in response to such pressures for change? To what extent can firms take advantage of opportunities to learn from success (manifest in the form of successful spin-offs and demonstration effects from successful competitors and/or role models)? In an important respect, the post-2000 meltdown in the telecom and information technology sectors is providing an important laboratory for studying how individual clusters in city-regions such as Ottawa, Waterloo and Calgary respond to these 'external' shocks and the degree to which the 'extrafirm' institutional supports afforded by the location within a cluster serve to cushion the shock and facilitate both the adjustment strategy on the part of individual firms, as well as a broader process of firm collapse and regeneration within the cluster at large.

Related to this idea is another question: how is failure handled? In the most dynamic regions, failure is recognised as a learning opportunity, such that potential investors may see entrepreneurs who have experienced past failure as lower-risk prospects if they have learned valuable lessons in the process (Saxenian 1994; Best 2001). Similarly, the failure or downsizing of large, once-successful firms represents a potential opportunity for regional renewal, since highly educated and experienced knowledge assets are released back into the local economy. Our assertion is that successful clusters capitalise on such events by absorbing these valuable assets back into productive activity—for example, by facilitating and supporting the process of new firm formation. Less dynamic places will tend to squander such opportunities by permitting or encouraging out-migration. One case that we are following closely is that of Ottawa's telecom and photonics cluster. Local surveys indicate that close to 20 000 jobs have been shed by large firms such as Nortel and JDS Uniphase since the onset of the downturn. Nevertheless, the number of firms generated by the cluster has increased by 300. No one in the local economy expects all of these to survive and grow, but the rate of new firm formation as well as the continued inflow of venture capital during the downturn are compelling indicators of the cluster's vitality.

4.2 Case Studies in Canada: Common Themes and Emerging Findings

The interim findings of those cases in progress reveal both commonly shared experiences and unique local circumstances concerning the forces shaping each region's evolution over time. Our observations are structured around five dominant themes.

Learning. Learning has been found to be the key economic process unfolding in each of the cases. Learning is instrumental in enabling old industries to adapt to changing competitive conditions in the global economy, as well as new ones to become more successful innovators. The learning processes have been identified as present both within individual firms and across firm boundaries in the form of learning from other firms, research institutions, industrial associations and related institutional elements of the cluster. Moreover, we have uncovered instances of both local and non-local learning relationships across our range of case studies. However, one of the most notable findings to date has been that non-local learning relationships appear to be more significant than the existing literature would have us believe. Not surprisingly, given the openness and strong export orientation of much of the Canadian economy, many of the firms interviewed in our case studies indicate that their markets and competitors are overwhelmingly outside the region and the country. Thus far, this tendency appears to be especially marked in sectors such as ICT, biotechnology and aerospace. This suggests that at least two corners of Porter's famous diamond—sophisticated and demanding local customers and strong rivalry between local competitors—are not consistently present in the Canadian context. Also notable is the fact that there seems to be relatively little of the diverse specialisation that characterises the larger ICT clusters, such as Silicon Valley. However, location within the cluster does serve as a spur to learning and innovation, as the local buzz within the clusters ensures that firms are well informed about what others are doing. As we shall discuss below, learning also seems to occur at the cluster-wide level through community-based organisations and both formal and informal processes of mentoring.

Labour. One of the most consistent findings thus far concerns the centrality of skilled labour as the single most important local asset. The local endowment of 'talent' in the labour force is emerging as a crucial determinant of regional-industrial success. This endowment is created and maintained by the retention and attraction of highly educated, potentially mobile workers who are drawn to thick, deep, opportunity-rich local labour markets. The emergence of a strong, concentrated talent pool in local and regional economies also serves as a key factor in launching individual clusters along the path to sustained growth and development. Critical mass appears to be important here: until this is achieved, local employers will fight a losing battle in attempting to retain or attract the skilled talent they need, particularly in the context of a highly competitive North American labour market for highly educated workers. Once this status is achieved, this sets in motion a positive, self-reinforcing circle through which regions with a critical mass of highly skilled workers in a particular sector are able to attract still more workers of this kind. The initial source of the local talent pool can be highly varied, with both government laboratories and local anchor firms playing a key role in developing the early talent base. Post-secondary educational institutions also play a central role in many of the health-based biotech clusters, but seem to be less critical for the initial launch of many of the other clusters. In many of the cases we are studying it appears that post-secondary institutions are followers, not leaders in key areas of technology. However, once industry has demonstrated leadership in the area and the cluster begins to grow, post-secondary institutions seem particularly adept at expanding their programmes and offering in the areas of strength required by the cluster. Their capacity to expand the local talent pool thus becomes critical in accelerating the pace of cluster development.

A fascinating case that demonstrates this effect most clearly is the information technology cluster in Waterloo, Ontario. The roots of this cluster are linked to the decision of a group of local business leaders to create a new university in the region in the late 1950s. Even more influential were the subsequent decisions to focus the core strengths of the university in the sciences, math and engineering and to establish what has become one of the most successful co-operative education programmes in North America. The founders of many of the firms that populate this cluster—including well-known success stories such as Research in Motion (RIM)—are graduates of the university and many started their firms with core technologies developed while they were at the university or through their practical experience in their co-operative terms (Wolfe, 2002).

Leadership. While one of the hallmarks of cluster-based development is its highly decentralised, socially organised network of relationships between local economic actors, the research thus far has highlighted the role that leadership can play in differentiating one firm (or one region) from another. Moreover, this is exercised at two different but equally important scales. First and foremost, the quality and nature of leadership within the firm are crucial in explaining the different strategic approaches taken by firms in the same industry and region, as well as their ultimate competitive success. Perhaps the most vivid example of this comes from the steel industry case study (Warrian and Mulhern, 2003), in which the very different paths taken by leading firms such as Stelco and Dofasco—both integrated steel producers operating from the City of Hamilton—have been strongly shaped by radically divergent attitudes towards co-operation with local research organisations. Dofasco has been far more aggressive than Stelco in pursuing relationships with local institutions of research and higher learning. Similarly, Bombardier, Canada's leading aerospace producer, has differentiated itself from the competition (and its home-base in Montreal from other aerospace-producing regions around the world) by its corporate strategy of buying assets (both tangible bricks and mortar as well as intangibles such as knowledge) and managing them skillfully, rather than by building them from scratch.

Leadership is also expressed at a social scale: at the level of the community. Here, our early findings point to the key role of 'civic entrepreneurs' in catalysing the development of new and emerging industries such as telecom equipment in Ottawa (Chamberlin and de la Mothe 2003), wireless equipment in Calgary (Langford *et al.*, 2003) or the emerging multimedia sector in Nova Scotia's Cape Breton Island (Johnstone and Haddow, 2003). These community leaders—who are more often than not from the private sector—help to animate local processes of strategic visioning, galvanise socially organised activities to upgrade the innovative capabilities of local firms and represent the common, collective interest of firms in the industry when required.

Legislation and labs: the role of public institutions and organisations. Our case studies also reveal the subtle but pervasive influence of institutional forces, exerted in a number of different ways and at a number of spatial scales. While private-sector initiative and ingenuity are of obvious importance, provincial and national institutional frameworks play a key role in shaping the trajectory of regional-industrial evolution by making certain kinds of strategic choices by firms easier, and others more difficult. They have also played a leading role in building the knowledge infrastructure in different regions of the country: universities, colleges, government labs and other research and technology-transfer organisations. Through the direct creation of crown corporations or government labs at both the federal and provincial levels, they help to produce critical knowledge-based assets for the region. Examples such as Alberta Government Telephone and its role in fostering the Calgary wireless industry through firms such as Novatel demonstrate vividly the potential influence of publicly funded entities in triggering the emergence of new industries and firms (Langford *et al.*, 2003). Similarly, the National Research Council labs in Ottawa, Montreal and Saskatoon have served as important attractors of private firm investment—in telecom, health-based biotechnology and agricultural biotechnology—as well as a generator of significant numbers of spin-off firms started up by former employees (Niosi and Bas, 2000, 2003). Finally, publicly funded agencies have been found to play crucial roles as '*animateurs*', working side-by-side with private and not-for-profit organisations at the local level to organise reflexive learning processes at the level of industries and communities.

However, not all of this public-sector influence is exerted through conscious decision-making. An illustration of the inadvertent role that public policy can play is provided by the case of the telecommunications equipment cluster in Ottawa, which traces its origins partly to the judicial decision in the US that forced the Western Electric Company to divest itself of its subsidiary, the Northern Electrical Manufacturing Company (now Nortel) in the late 1950s. Cut off from its sources of innovation and research, Northern Electric searched for a location to establish its own research facility. It eventually bought a substantial tract of land on the outskirts of Ottawa to be the home of Bell Northern Research, largely because it viewed the presence of the federal government's National Research Council labouratories and the Communications Research Centre as a substantial draw for the highly skilled research scientists and engineers it expected to populate its own facility. Many of the leading entrepreneurs in the Ottawa telecommunications and photonics cluster began their careers as researchers for BNR (Chamberlin and de la Mothe 2003). This case should caution us to avoid looking only for the direct effects of government policy on cluster development.

Location. While our work began from the premise that 'geography matters', we recognise the perils of presupposing the importance of place, rather than examining this proposition through systematic study. What is emerging from our cases is a more nuanced understanding of the importance of proximity to the creation and maintenance of learning dynamics for firms and industries. As already noted, the cases document a consistent tension between local and non-local relationships and knowledge flows—in other words, the dynamic tension that exists between local buzz and global pipelines. Moreover, they are leading us to appreciate the specificity of particular case-study circumstances, in which regional, national, sectoral and historical variation are significant. For example, the studies of Montreal's aerospace industry, Saskatoon's agri-biotech sector (Ryan and Phillips, 2003), Calgary's wireless industry (Langford, Wood *et al.*, 2003) and Hamilton's steel industry (Warrian and Mulhern, 2003) reveal that much of the knowledge-base required for innovation and production is acquired through relatively straightforward market transactions, often from non-local (even global) sources.

Perhaps the most vivid examples come from the life sciences, where firms in Canada's leading biotech clusters (such as Montreal, Toronto, Vancouver and Saskatoon) have strong non-local backward and forward linkages. Recent analysis of Statistics Canada's national survey of biotechnology firms (Gertler and Levitte, 2003) reveals the complex, dual geography of relationships in which successful firms are embedded. On the one hand, they tap into global knowledge markets by hiring highly qualified personnel from abroad. They also take advantage of other global flows of knowledge, through the use of scientific publications and databases, by licensing their intellectual property to foreign partners, or by licensing the intellectual property of foreign firms for their own use. When they develop collaborative relations with other firms, for both research and marketing purposes, these are both local and global in nature. On the other hand, they rely heavily on local sources of investment capital from private sources (angel investors, family and friends) and are highly likely to have spun off from another local company or research institution at some point in their past.

Nevertheless, there is still an important role to be played by local institutions and actors that enable local firms to exploit this knowledge effectively and combine it with other local assets and capabilities for success. While global knowledge flows are certainly important to the competitive success of local firms, the local knowledge/science-base represents a major generator of new, unique knowledge assets. Local universities and research institutes constitute an important part of this base as 'anchors' that generate highly skilled graduates, spin-off start-ups and new, publicly available knowledge (often developed interactively with other local partners outside the sphere of the university). In many cases, it appears that one or a few 'anchor' firms or 'lead' institutions play a critical role in these processes. Examples from our ongoing work include biotechnology in Montreal, telecom and photonics in Ottawa (Chamberlin and de la Mothe, 2003), steel in Hamilton, particularly as produced by Dofasco (Warrian and Mulhern, 2003) and the evolving information technology cluster in New Brunswick.[3]

5. Conclusion

It should be clear from the above discussion that the large and varied international literature on cluster emergence, evolution and policy offers much in the way of rich detail. At the same time, it suffers from an inconsistency of definitions and methodological approaches that compromises the value of the findings flowing from this work. It should be equally clear that the approach adopted in the ISRN project differs from most of the work performed under the rubric of 'cluster studies' in several important ways. First, much of the earlier work presumes the importance of 'the local' and then sets out to find indicators that confirm this. In contrast, our approach is to treat the possible existence of cluster dynamics as an hypothesis to be investigated and either verified or rejected. For this reason, we continue to ask ourselves: when, or under what circumstances, does spatial proximity matter, and why? Secondly, our relatively large number of case studies across a broad spectrum of both mature and emerging industries, in large metropolitan regions as well as smaller urban centres, provides a solid basis for comparison and for the development of a more robust theory of cluster development. Thirdly, in stark contrast to the vast majority of work on clusters, the indicators we have fashioned for this project emphasise dynamic processes and change over comparative statics. Furthermore, drawing inspiration from recent conceptual work on knowledge-based theories of the firm, innovation processes and the cluster, we have favoured knowledge-based indicators of cluster dynamism and success. Fourthly, rather than adhering to a purely quantitative style of analysis, our view is that quantitative and qualitative analytics are mutually complementary and can render a far more complete story of local innovative dynamics than can quantitative measures alone. Finally, our overarching interest in innovation systems—both regional and national—has encouraged us to situate our analysis of cluster evolution within the broader institutional framework that shapes the behaviour and practices of firms. At the same time, our conceptual approach also emphasises the importance of firm-based and community-level agency (leadership), as well as the potential significance of serendipity, local historical accidents and path dependence. As a result, we are better positioned to highlight and understand the distinctive paths followed by individual cases.

The picture already emerging from our study departs substantially from the received wisdom—most notably concerning the alleged importance of a strong local customer-base and strong local competition in spurring the emergence and evolution of dynamic, knowledge-based clusters. Nor is it evident from our findings that direct, non-market interaction and knowledge sharing between local firms in the same industry are rampant. Our evidence suggests that, where such interaction occurs, it is indirect and mediated through civic associations and other local organisations. While this form of local learning is considerably more prevalent between firms and their local suppliers, not all inputs are locally sourced. In particular, it appears that a large component of the knowledge inputs to local production—at least in certain sectors—is drawn from well outside the region.

The findings reported in this paper represent the results to date from a substantial number of the cases in our study. The next stage of the analysis involves a systematic comparison of the results across similar cases in different regions and different cases in the same region. The goal of this analysis is to enrich not only our conceptual understanding of the process of cluster evolution, but also our insight into the role that public policy—at a range of geographical scales—might play in either promoting or discouraging this process. This nuanced understanding will hopefully provide a more effective guide for policy-makers across a range of geographical scales, as well as an appreciation of the intersecting impacts of both local and global factors on the process of cluster development.

Notes

1. The Innovation Systems Research Network is a cross-disciplinary, national network of researchers funded by the Social Sciences and Humanities Research Council of Canada, with additional support from other federal and provincial departments and agencies. In 2001 the ISRN launched a five-year study of industrial clusters across Canada. The authors are co-ordinators of this study. More details on the network, its members and the current cluster study can be found at the website: http://www.utoronto.ca/isrn.
2. The framework and milestones document which provides more detail on the research project, as well as the presentations from the 2003 meeting, can be found at: http://www.utoronto.ca/isrn/clusters.htm. The project is scheduled to conclude in 2005.

3. The New Brunswick cluster is of particular interest because of efforts by the provincial government to use the local telecommunications firm (NBTel) as the 'anchor' for an emerging ICT cluster and the recently adopted strategy by the federal government's National Research Council to accelerate the cluster's growth by locating a branch of its Institute for Information Technology in the provincial capital, Fredericton (Davis and Schaefer, 2003).

References

AMARA, N., LANDRY, R. and OUIMET, M. (2003) *Milieux innovateurs: determinants and policy implications*. Paper presented at the. *DRUID Summer Conference. Elsinore*, Denmark, June. (http://www.druid.dk/conferences/summer2003/Papers/Amara_Landry_Ouimet.pdf).

BATHELT, H., MALMBERG, A. and MASKELL, P. (2002) *Clusters and knowledge: local buzz, global pipelines and the process of knowledge creation*. DRUID Working Paper No. 02–12, Copenhagen (http://www.druid.dk/wp/pdf_files/02–12.pdf).

BERGMAN, E. M. and FESER, E. J. (1999) *Industrial and Regional Clusters: Concepts and Comparative Applications*. The Web Book of Regional Science. Regional Research Institute, West Virginia University (http://www.rri.wvu.edu/WebBook/Bergman-Feser/contents.htm).

BEST, M. H (2001) *The New Competitive Advantage: The Renewal of American Industry*. Oxford: Oxford University Press.

BRESCHI, S. and MALERBA, F. (2001) The geography of innovation and economic clustering: some introductory notes, *Industrial and Corporate Change,* 10(4), pp. 817–833.

BRESNAHAN, T., GAMBARDELLA, A. and SAXENIAN, A. (2001) 'Old economy' inputs for 'new economy' outcomes: cluster formation in the new Silicon Valleys, *Industrial and Corporate Change,* 10(4), pp. 835–860.

BROWN, J. S. and DUGUID, P. (2000) Mysteries of the region: knowledge dynamics in Silicon Valley, in: C.-M. LEE, W. F. MILLER, M. G. HANCOCK and H. S. ROWEN (Eds) *The Silicon Valley Edge*, pp. 16–39. Stanford, CA: Stanford University Press.

BUNNELL, T. G. and COE, N. M. (2001) Spaces and scales of innovation, *Progress in Human Geography,* 24(4), pp. 569–589.

CHAMBERLIN, T. and MOTHE, J. DE LA (2003) nothern light: ottawa's technology cluster, in: D. A. WOLFE (Ed.) *Clusters Old and New: The Transition to a Knowledge Economy in Canada's Regions*, pp. 213–234. Kingston: School of Policy Studies, Queen's University.

COOKE, P. (1998) Introduction: origins of the concept, in: H. BRACZYK, P. COOKE and M. HEIDENREICH (Eds) *Regional Innovation Systems: The Role of Governances in a Globalized World*, pp. 2–25. London: UCL Press.

COOKE, P., URANGA, M. G. and ETXEBARRIA, G. (1997) Regional innovation systems: institutional and organizational dimensions, *Research Policy,* 26, pp. 475–491.

DAVIS, C. H. and SCHAEFER, N. V. (2003) Development dynamics of a start-up innovation cluster: the ICT sector in New Brunswick, in: D. A. WOLFE (Ed.) *Clusters Old and New: The Transition to a Knowledge Economy in Canada's Regions*, pp. 121–160. Kingston: School of Policy Studies, Queen's University.

FELDMAN, M. (2001) The entrepreneurial event revisited: firm formation in a regional context, *Industrial and Corporate Change,* 10(4), pp. 861–891.

FLORIDA, R. (2002) *The Rise of the Creative Class: And How It's Transforming Work, Leisure, Community and Everyday Life*. New York: Basic Books.

GERTLER, M. S (2003) Tacit knowledge and the economic geography of context, *Journal of Economic Geography,* 3, pp. 75–99.

GERTLER, M. S. and LEVITTE, Y. (2003) *Local nodes in global networks: the geography of knowledge flows in biotechnology innovation.* Paper presented at the *DRUID Summer Conference*, Elsinore, Denmark, June (http://www.druid.dk/conferences/summer2003/Papers/Gertler_Levitte.pdf).

GERTLER, M. S., FLORIDA, R., GATES, G. and VINODRAI, T. (2002) *Competing on creativity: placing Ontario's cities in North American context.* A report prepared for the Ontario Ministry of Enterprise, Opportunity and Innovation and the Institute for Competitiveness and Prosperity, Toronto (http://www.utoronto.ca/progris/Competing%20on%20Creativity%20in%20Ontario%20Report%20(Nov%2022).pdf).

HENDRY, C., BROWN, J. and DEFILLIPPI, R. (2000) Regional clustering of high-technology-based firms: opto-electronics in three countries, *Regional Studies,* 34(2), pp. 129–144.

INFORMATION DESIGN ASSOCIATES and ICF KAISER INTERNATIONAL (1997) *Cluster-based economic development: a key to regional competitiveness*. Economic Development Administration, US Department of Commerce, Washington, DC.

JOHNSTONE, H. and HADDOW, R. (2003) Industrial decline and high technology renewal in Cape Breton: exploring the limits of the possible, in: D. A. WOLFE (Ed.) *Clusters Old and New: The Transition to a Knowledge Economy in Canada's Regions*, pp. 187–212. Kingston: School of Policy Studies, Queen's University.

KASH, D. E. and RYCROFT, R. W. (2000) Patterns of innovating complex technologies: a framework for adaptive network strategies, *Research Policy,* 29, pp. 819–831.

KENNEY, M. (Ed.) (2000a) *Understanding Silicon Valley: The Anatomy of an Entrepreneurial Region.* Stanford, CA: Stanford University Press.

KENNEY, M. (2000b) Introduction, in: M. KENNEY (Ed.) *Understanding Silicon Valley: The Anatomy of an Entrepreneurial Region*, pp. 1–12. Stanford, CA: Stanford University Press.

LANGFORD, C. H., WOOD, J. R. and ROSS, T. (2003) The origins of the Calgary wireless cluster, in: D. A. WOLFE (Ed.) *Clusters Old and New: The Transition to a Knowledge Economy in Canada's Regions*, pp. 161–185. Kingston: School of Policy Studies, Queen's University.

LEE, C.-M., MILLER, W. F., HANCOCK, M. G. and ROWEN, H. S. (Eds) (2000) *The Silicon Valley Edge: A Habitat for Innovation and Entrepreneurship.* Stanford, CA: Stanford University Press.

LUNDVALL, B.-Å. (Ed.) (1992) *National Systems of Innovation: Towards a Theory of Innovation and Interactive Learning.* London: Pinter Publishers.

MALMBERG, A. and MASKELL, P. (2002) The elusive concept of localization economies: towards a knowledge-based theory of spatial clustering, *Environment and Planning A,* 34, pp. 429–449.

MARTIN, R. and SUNLEY, P. (2003) Deconstructing clusters: chaotic concept or policy panacea?, *Journal of Economic Geography,* 3(1), pp. 5–35.

MASKELL, P. (2001) Towards a knowledge-based theory of the geographic cluster, *Industrial and Corporate Change,* 10(4), pp. 921–943.

MASKELL, P. and MALMBERG, A. (1999) Localised learning and industrial competitiveness, *Cambridge Jounal of Economics,* 23, pp. 167–185.

MILLER, R. and COTÈ, M. (1987) *Growing the Next Silicon Valley: A Guide for Successful Regional Planning.* Lexington, MA: Lexington Books.

NELSON, R. R. (Ed.) (1993) *National Innovation Systems: A Comparative Analysis.* New York: Oxford University Press.

NIOSI, J. and BAS, T. G. (2000) The competencies of regions and the role of the national research council, in: J. A. HOLBROOK and D. A. WOLFE (Eds) *Innovation, Institutions and Territory: Regional Innovation Systems in Canada,* pp. 45–65. Kingston: Queen's School of Policy Studies.

NIOSI, J. and BAS, T. G. (2003) Biotechnology megacentres: Montreal and Toronto regional systems of innovation, *European Planning Studies,* 11(7), pp. 789–804.

PORTER, M. E (1998) Clusters and competition: new agendas for companies, governments, and institutions, in: M. E. PORTER (Ed.) *On Competition*, pp. 197–287. Cambridge, MA: Harvard Business Review Books.

PORTER, M. E (2003) The economic performance of regions, *Regional Studies,* 37(6&7), pp. 549–578.

PORTER, M. E., MONITOR GROUP, ON THE FRONTIER and COUNCIL ON COMPETITIVENESS (2001) *Clusters of Innovation: Regional Foundations of US Competitiveness.* Washington, DC: Council on Competitiveness.

QUEENTON, J. and NIOSI, J. (2003) *Bioscientists and biotechnology: a Canadian study.* Paper presented at the *Fifth Annual Meeting of the Innovation Systems Research Network,* Ottawa, May (http://www.utoronto.ca/isrn/documents/Queenton_bioscientists.pdf).

ROSENBERG, D. (2002) *Cloning Silicon Valley: The Next Generation of High-tech Hotspots.* London: Pearson Education Ltd.

ROWEN, H. (2000) Serendipity or strategy: how technology and markets came to favour Silicon Valley, in: C.-M. LEE, W. F. MILLER, M. G. HANCOCK and H. S. ROWEN (Eds) *The Silicon Valley Edge: A Habitat for Innovation and Entrepreneurship,* pp. 184–199. Stanford, CA: Stanford University Press.

RYAN, C. D. and PHILLIPS, P. W. B (2003) Intellectual property management, in: D. A. WOLFE (Ed.) *Clusters Old and New: The Transition to a Knowledge Economy in Canada's Regions,* pp. 95–120. Kingston: School of Policy Studies, Queen's University.

SAXENIAN, A. (1994) *Regional Advantage: Culture and Competition in Silicon Valley and Route 128.* Cambridge, MA: Harvard University Press.

STORPER, M. (1997) *The Regional World: Territorial Development in a Global Economy.* New York: The Guilford Press.

STORPER, M. and VENABLES, A. J. (2003) *Buzz: face-to-face contact and the urban economy.* Paper presented at the *DRUID Summer Conference on Creating, Sharing and Transferring Knowledge: The Role of Geography, Institutions and Organizations,* Copenhagen/Elsinore, June.

STURGEON, T. J. (2003) What really goes on in Silicon Valley? Spatial clustering and dispersal in modular production networks, *Journal of Economic Geography,* 3, pp. 199–225.

WARRIAN, P. and MULHERN, C. (2003) Learning in steel: agents and deficits, in: D. A. WOLFE (Ed.) *Clusters Old and New: The Transition to a Knowledge Economy in Canada's Regions,* pp. 37–62. Kingston: School of Policy Studies, Queen's University.

WOLFE, D. A. (2002) *Knowledge, learning and social capital in Ontario's ICT clusters.* Paper presented at the *Annual Meeting of the Cana-*

dian Political Science Association, Toronto, May (http://www.utoronto.ca/progris/pdf_files/Ontario%27s%20ICT%20Clusters.pdf).

WOLFE, D. A. (Ed.) (2003) *Clusters Old and New: The Transition to a Knowledge Economy in Canada's Regions*. Kingston: School of Policy Studies, Queen's University.

Innovation and Clustering in the Globalised International Economy

James Simmie

[Paper first received, February 2003; in final form, November 2003]

Introduction

The meteoric rise of Michael Porter's cluster concept has taken policy-makers if not academics by storm in recent years. Government and regional policy-makers in many parts of the world have been beguiled by the possibilities promised for improving the competitiveness of their national and regional economies. Much of this is a clear case of policy-making outrunning the evidence to support the policy in question. Despite the popularity of the cluster idea, there is, as yet, not much empirical evidence to suggest what the dynamics of clustering achieve in different circumstances and locations. In many cases, it is simply identified with concentrations of co-located firms. Policy-makers in particular have been anxious to identify almost any group of firms located within their administrative jurisdictions as a cluster.

The study of benefits accruing to firms grouped together in a given location is not new. Agglomeration economies and the advantages of externalities arising from co-location have been analysed since at least the time of Marshall (1920). Part 2 of this paper, therefore, summarises briefly the central arguments of this well-known literature and seeks to establish whether or not Porter's notion of clusters provides any advance over the concept of agglomeration economies.

The argument developed is that, although Porter accepts that localisation economies form part of the advantages derived from local clustering, he underestimates the continuing significance of urbanisation economies. A number of studies of innovative and high-technology firms have shown the importance of location in large metropolitan agglomerations for the irregular use of other firms, services and institutions on a pick-and-mix basis rather than as part of regular networking.

A further problem with the cluster concept that is identified in this section is the fact that it does not provide an *a priori* way of identifying the geographical scale or boundaries of a cluster. These appear to depend on functional interlinkages that may extend all the way from intercountry connections down to a minority confined within cities. This means that clusters cannot be defined as geographical objects of study. Instead, it is necessary to start with the kinds of linkages that competitive firms use and then to assess how far these are confined within particular localities.

Within this context, part 3 goes on to examine the claims made by Michael Porter for the distinctive contributions that clustering dynamics make to innovation. Porter's arguments are summarised briefly. These show his initial concern with the drivers of competitiveness. He argues that competitiveness rests on productivity and that this in turn is driven by innovation. Innovation allows firms to compete successfully in world markets by introducing new and differentiated products and services.

Porter provides six hypotheses on why clustering delivers innovation. These are as follows. Clustering delivers higher rates of innovation because it allows rapid perception of new buyer needs. It concentrates knowledge and information and the knowledge-based economy is most successful when knowledge resources are localised. It facilitates on-going relationships with other institutions including universities. It allows the rapid assimilation of new technological possibilities. Finally, it provides richer insights into new management practices.

Two kinds of evidence are used to evaluate these claims. These are, first, some of the secondary literature that analyses the nature and geographical extent of linkages used by competitive, high-technology and innovative firms. These show that such linkages are notable by their absence in detailed empirical studies. They were found to be weak across Europe in the opto-electronics sector, among small high-technology firms in the UK, in the ITEC sector in the South East and among innovative companies in the Greater South East. What is particularly striking here is the distinct lack of detailed empirical analyses of networking and interlinkages between competitive firms. Many studies content themselves with demonstrating the simple co-location of firms or by connecting institutional and organisational boxes with networks for which they have no real evidence.

Secondly, data from the third Community Innovation Survey are used to test the first four of Porter's hypotheses. In all cases, it is shown that local or regional networks and linkages are not regarded as important as national and international ones by the leading innovative companies. These data do not therefore support Porter's hypotheses concerning the ways in which local clustering dynamics deliver innovation.

In contrast, where Porter has emphasised the significance of traded clusters irrespective of whether or not they are confined to a particular locality, a somewhat different perspective emerges with respect to the significance of geographical proximity. In this instance, trading linkages between different sectors, institutions, suppliers and customers are important for innovation. In these circumstances, strong linkages at most geographical levels seem to be a particular characteristic of the most innovative firms. The subject matter of this paper therefore contrasts with the excellent evaluation by Martin and Sunley (2003) of Porter's arguments that the processes driving the development of new economic knowledge and its application and commercialisation in innovation are facilitated by localisation.

The concluding section suggests that innovative firms are part of an internationally distributed system of innovation. They use localities as places primarily to operate from, rather than within. Firms and clusters that do not have this outward orientation are liable to suffer from too much intellectual inbreeding and lock-in. From this perspective, local clustering is just as likely to deliver economic decline, low productivity and a lack of innovation as the reverse. The economic landscape is littered with sites that used to contain co-located firms that were once market leaders but have since slipped into terminal decline. In the UK, such industries include—for example, motorcycles, consumer electronics and yacht construction.

Is Clustering Different from Agglomeration?

Economists from Marshall (1920) onwards have generally agreed that agglomeration economies arising from firm concentrations in particular places confer economic advantages. Thus the concept of clustering must add a distinctive and new dimension to this old debate if it is to be regarded as making additional contributions to competitiveness, productivity or innovation. Any distinctive advantages derived from clustering need to be in addition to those provided by simple agglomeration.

Turning first to the base line of agglomeration advantages, Marshall (1920) argued that spatial concentration could confer external economies on firms as they concentrated in particular cities. These economies mainly take the form of increasing returns to scale as firms are able to take advantage of large pools of skilled labour, local markets and the easy transmission of new ideas. Hoover (1937, 1948) developed this idea. He grouped the sources of agglomeration advantages into internal returns to scale, localisation and urbanisation economies. Localisation economies arise as a particular industry concentrates in a given location. This leads to the development of local expertise, special skills and advantages that are peculiar to the industry in question. Urbanisation economies on the other hand arise from the more general characteristics of a city. They include the multiplicity of specialised business services, infrastructure and cultural and leisure facilities that may be used by any firm in the locality rather than being of use mainly to a single economic sector.

Arthur (1994) also addressed the question of why particular industries often congregate in a limited number of locations. His main argument is that industries tend to concentrate in a limited number of cities not necessarily because those places have any intrinsic advantages but because some historical accident placed certain firms there initially and this minor concentration attracts a high proportion of later entrants. Although this overstates the role of serendipity, there is clearly a number of less than rational reasons why firms locate in one city rather than another.

Gordon and McCann (2000) identify three 'ideal types' of local firm concentrations that should be distinguished from clustering. These are pure agglomeration, the industrial complex and social networks. In the model of pure agglomeration, Gordon and McCann (2000) correctly point out that there is no presumption of co-operation between firms beyond what is in their own individual interests in an atomised and competitive environment. Agglomeration allows for intermittent and changing local interactions as a result of pure chance, the presence of large numbers of other firms and ecological processes of natural selection among firms.

> Within this approach the only reason why we might observe spatial industrial clustering is that the individual firms, in aiming to minimise their observable spatial transaction costs, have implicitly or explicitly determined that this is best achieved by locating close to other firms within the particular input–output production and consumption hierarchy of which they are a part (Gordon and McCann, 2000, p. 518).

In other words, industrial concentration may be the result of entirely unconnected individual-firm decisions where those firms reach similar decisions but without any necessary interactions or linkages between them. In the absence of any evidence to the contrary, it should be assumed that most of the concentrations of co-located firms mapped by the DTI (1999, 2001) and Bennett *et al.* (1999) are of this type. There is no evidence to show that they involve any linkages beyond those of basic economic activity.

The second type identified by Gordon and McCann (2000) is the industrial complex. Here, firms co-locate in particular areas in order to minimise their transport costs. Traditional locational arrangements between coalmines and steel works or power stations might be an example of such an industrial complex. Here, the location of natural resources and their uses are the driving force of concentration.

The third type of economic concentration is that based on social networks. This has been inspired by the work of Granovetter (for example, 1985). There are three special characteristics of the social network model. These are that, because of the trust relationships that are said to exist, firms are willing to undertake risky co-operative joint ventures, to reorganise their relationships and, finally, to act as a group in support of mutual goals. Although there is nothing inherently spatial about social network relationships, it is often argued that they are easier to maintain when participants are located within 'reasonable' proximity to one another.

A fourth type of economic relationship that is becoming more significant as knowledge becomes an increasingly important factor of production, is the concept of knowledge spillovers (for example, Audrestch and Feldman, 1996). In this instance, locally produced knowledge is said to become a semi-public good because it is impossible to prevent it spilling over from the public science base or other firms in the area. Again, these types of relationship are said to place a premium on proximity as knowledge may decay with distance.

Porter tends to slide over these distinctions between different reasons for firms to group together. He argues—for example, that on the one hand localisation economies are part of the more general phenomena of business clusters. On the other hand, he goes on to argue that these are not formed as a result of urbanisation economies. In fact, he argues that urbanisation economies are becoming less important as a source of competitive advantage because they are so widely available in cities across the global economy. This view is not supported by evidence on the Greater London region provided by Simmie and Sennett (1999) and Gordon and McCann (2000). In both these studies, using different sources of data, the innovative performance of firms in the region is shown to be enhanced by their ability to access multiple sources of knowledge, suppliers, labour and international markets on a 'pick-and-mix' or 'promiscuous' basis. Their innovative performance is enhanced as a direct result of the urbanisation economies provided by a large metropolitan region like London.

Thus it is argued here that, at the local level, there are several plausible explanations of why innovative firms group together that do not rely on the concept of clusters at all. For example, unorganised urbanisation economies are very significant in the fast-moving world of innovation, as are knowledge spillovers. These are at least partially unintentional rather than part of collaborative firm strategies. The network paradigm may be closer to Porter's local cluster hypothesis, but this raises the question of how general a phenomenon this actually is.

Thus the first major question raised in this paper is: how may clusters be distinguished from other forms of local firm groupings? Porter's well-known definition of clusters is that they are

> geographical concentrations of interconnected companies, specialised suppliers, service providers, firms in related industries and associated institutions (e.g. universities, standards agencies, trade associations) in a particular field that compete but also co-operate. Clusters, or critical masses of unusual competitive success in particular business areas, are a striking

> feature of virtually every national, regional, state, and even metropolitan economy, especially in more advanced nations (Porter, 2000, p. 15).

In this definition, there are three key features that distinguish, at least in part, the dynamics of clustering from those of agglomeration economies. The first is that clusters are subject to vigorous national or regional technological competition that drives product and process innovation. Secondly, commonalities and complementarities interlink participating firms. These crucial linkages consist both of vertical buying and selling chains and horizontal complementary products and services, specialised inputs and the use of related technologies. The third main characteristic of clustering interactions is that they take place between geographically proximate groups of interlinked companies. It is argued that co-location encourages value-creating networking between firms.

A major prerequisite needed to investigate how different these characteristics are from other forms of firm grouping, is to be able to define the geographical scale of a cluster in order to identify how local is the competition that drives them and what contributions linkages and networks make to competitiveness, productivity and innovation. Porter does not provide much assistance in dealing with this problem. He says that

> Drawing cluster boundaries often is a matter of degree and involves a creative process informed by understanding the linkages and complementarities across industries and institutions that are most important to competition in a particular field. The strength of these 'spillovers' and their importance to productivity and innovation are the ultimate boundary-determining factors (Porter, 2000, p. 17).

So, while there is a strong implication in the cluster concept that intensity of interactions places limits on the geographical extent of clusters, in practice Porter is prepared to countenance definitions based on linkages extending over almost any distance. As a result, clusters can extend anywhere from "large and small economies in rural and urban areas, and at several geographical levels (for example nations, states, metropolitan regions, and cities" (Porter, 1998, p. 204) or even as far as "a network of neighbouring countries" (Porter, 1998, p. 199).

Martin and Sunley (2003) correctly argue that both the geographical and the industrial definitions of clusters in Porter's work are ridiculously elastic. They go on to say that there is nothing inherent in the concept itself that provides a way of defining the spatial range of a cluster or which are the key dynamic processes at different geographical scales. This imprecision helps to account for the enormous popularity of the concept in policy circles. It can mean all things to all policy-makers. It does mean, however, that it is very difficult to investigate the dynamics of clustering starting from a spatial definition.

For this reason, this paper does not attempt to define what 'local' in the context of innovation and clusters means. Instead, the focus is on innovation *per se* and how this maps onto space. First, however, Porter's analysis of the relationships between competitiveness, productivity, innovation and clustering is outlined.

Innovation and Clusters

Porter arrived at the notion of clusters as a result of following his arguments on the nature and characteristics of competition. Accordingly, he argues that

> Competition is dynamic and rests on innovation and the search for strategic differences. Close linkages with buyers and suppliers and other institutions are important, not only to efficiency, but also to the rate of improvement and innovation. Location affects competitive advantage through its influence on productivity and especially on productivity growth (Porter, 2000, p. 19).

In this respect, the key to successful competition is based first on the ability to produce continuous streams of innovation and, secondly, to position a company strategically in the market-place in such a way as to produce products that are both different from and superior to those of rivals.

For Porter, innovative capacity is the key to productivity and competitiveness can be equated with productivity. He argues that

> A region's or nation's standard of living (wealth) is determined by the productivity with which it uses its human, capital and natural resources. The appropriate definition of competitiveness is productivity (Porter, 2002, p. 1).

In this line of reasoning, innovation leads to productivity and hence to competitiveness. The latter forms the basis of national prosperity and the standard of living enjoyed by citizens. In the context of globalisation. First World economies need to concentrate on high-value-added products and services and to be innovative in doing so (Porter, 2003a). In these economies, "Productivity and innovation—not low wages, low taxes, or a devalued currency—are the definition of competitiveness" (Porter, 2000, p. 30).

Porter's analysis has been influenced by the work of Piore and Sabel (1984). Like them, he argues that "Businesses have been deverticalising and developing systems of flexible production" (Porter, 2003b). In the examples cited in Emilia–Romagna and Baden–Württemberg, this has entailed developing networks of firms that both collaborate and compete. In these instances, space matters because physical proximity reduces the transaction costs of such a system of production. This closely parallels the network paradigm of Granovetter (1985) noted above.

But, to the extent that the sharpness and pervasiveness of the 'new industrial divide' is exaggerated, then the significance of local space attributes in the use and upgrading of factors of production is also overstated. Multinational, multilocational companies play a major role in the globalised economy. Their complex spatial divisions of labour use different factors of production in different locations, all of which contribute to eventual outputs. In such cases, space matters but not so much because of local uses of factors of production so much as the interconnections between different places involved in internationally distributed value chains (Castells, 1987; Amin and Thrift, 1992; Simmie, 2002a, 2002b).

Despite the fact that globalised economic interactions are increasing in importance, Porter argues that such linkages mitigate disadvantages rather than create advantages. He asserts that "distant sourcing is a second-best solution compared with accessing a competitive local cluster in terms of productivity and innovation" (Porter, 2000, p. 32). As a result, he emphasises the significance of microeconomic conditions and the ability to improve them in order to improve the competitiveness of the macro economy in general. This leads him to the famous representation of the key dynamics of a micro economy as a diamond.

As is well known, the Porter diamond is a model of the effects of localities on competition in terms of four interrelated influences. These are factor inputs ranging from tangible assets such as physical infrastructure to information, the legal system and university research institutes that all firms draw on in competition. The context for firm strategy and rivalry refers to the rules, incentives and norms governing the type and intensity of local rivalry. Demand conditions in a locality have much to do with whether firms can and will move from imitative, low-quality products to competing on differentiation. Advancement means the development of more demanding markets. Finally, related and supporting industries and the connections across industries are regarded as fundamental to competition, to productivity and to the direction and pace of new business formation and innovation (Porter, 2000, pp. 18–20).

Originally, Porter argued that there is a broad range of geographical levels of competitiveness. These range all the way from the world economy down to cities and metropolitan areas and include broad economic areas, groups of neighbouring nations, nations, states and provinces. Latterly, he has become more insistent that the intensity and quality of interaction within the diamond are enhanced when the firms involved are geographically localised, despite the fact that he also now emphasises the significance of trading as opposed to non-trading clusters. Thus, on the one hand, he argues that the microeconomic conditions favouring competitiveness, productivity and innovation are increasingly associated with local geographical clusters. On the other hand, he also argues that the key to economic growth lies with trading clusters that by definition must have international linkages.

The advantages claimed for firms operating within clusters include increased competitiveness, higher productivity, new firm formation, growth, profitability, job growth and innovation. This beguiling siren call is based on three main lines of argument. First, with respect to competition, Porter argues that, in the past, companies thinking about competition have been too dominated by what happens inside the organisation. According to him, clustering suggests that there is a good deal of competitive advantage that lies outside the firm and even outside the industry in which they work. Among other things, this means that competition within clusters can be a positive-sum game in which productivity improvements and trade can expand the total market in such a way that many locations can prosper provided that they become more productive and innovative (Porter, 2000, p. 28).

The dynamics of clustering are said to deliver competitiveness in four main ways. They stimulate the fast diffusion of new product and process technologies. They help to upgrade suppliers through competition and intense co-operation with customers on R&D. Collective action places pressure on political institutions to support the creation of specialised factors such as specific training and research centres. Finally, they stimulate firms to fund and develop links with local training and research centres (CBI, 1999, p. 2). All these features could also contribute to innovation and extend over international as well as local value chains.

In an important and increasingly stressed part of Porter's argument (for example, 2003), and following Keynes, Porter argues that it is primarily export-oriented clusters that drive regional prosperity. Exporting clusters tend to pay higher wages than those serving purely local markets and so they help to pull up other wages in the regional economy. Export clusters are, however, much more likely to have national and international linkages than to be based on purely local connections. The critical importance of these extended linkages in the context of a globalised international economy calls into question the relative significance of the kinds of limited and local connections so often stressed by local policy-makers supposedly following Porter's analysis.

The significance of innovation to trading clusters can hardly be exaggerated. The ability to produce products and processes that are new to the market is critical to gaining market share and exports. In order to access adequate markets, innovative companies internationalise very rapidly (Keeble *et al.*, 1998).

Porter claims that there are a number of advantages to be gained with respect to the key activity of innovation by operating in a cluster. These advantages include the ability to perceive and react to new buyer needs more quickly due to the proximity of demanding and sophisticated customers. In addition, firms can see the evolution of new technologies and understand their implications and possibilities more quickly. Local relationships, including those with universities, are said to facilitate this process (Porter, 2000, p. 23).

From this perspective, the cluster concept has become increasingly associated with the 'new' or 'knowledge' economy. The argument here is that the processes that drive the development of new economic knowledge and its application and commercialisation in innovation are facilitated by localisation (Martin and Sunley, 2003). Norton (2001), who argues that the success of the US in the 'new' economy derives directly from the growth of large and dynamic clusters of innovation and entrepreneurialism, supports this idea. Baptista has also argued that "geographical concentration is of foremost importance for organisational improvement and technological innovation" (Baptista, 1996, p. 60).

In summary, Porter argues that localised clusters deliver innovation because

—They allow rapid perception of new buyer needs.
—They concentrate knowledge and information.
—They allow the rapid assimilation of new technological possibilities.
—They provide richer insights into new management practices.
—They facilitate on-going relationships with other institutions including universities.
—The knowledge-based economy is most successful when knowledge resources are localised.

Some of these arguments will be tested later below using data from the third Community Innovation Survey (CIS 3).

First, however, it should be reiterated that locally delimited clusters could and have had exactly the opposite effects to those listed above. As Martin and Sunley point out

> Economic landscapes are littered with local areas of industrial specialisation that were once prosperous and dynamic but have since gone into relative or even absolute decline (Martin and Sunley, 2003, p. 26).

Too much intellectual inbreeding can lead to 'lock-in'. This is a process that arises if too much reliance is placed upon local knowledge and face-to-face relationships. This may work within certain parameters, but makes the cluster very vulnerable to major shifts in technology such as the development of electronic as compared with mechanical watches and digital publishing compared with hot lead type-setting.

Cluster advocates offer no explanation of how the dynamics of clusters can go into reverse in this way. They are also quiet on the issue of how cluster policies should respond to these dangers.

What Is the Evidence on Innovation and Clusters?

In the absence of ways of *a priori* identifying clusters spatially, the approach adopted in this paper is not to consider geography first, but to start with firms that are thought or known to be competitive and innovative. Once such firms have been identified, the next stage is to examine the nature and geographical extent of the linkages that they have with other firms and organisations and to evaluate how significant they are for innovation.

Within this general framework, several empirical studies have been undertaken of high-technology and/or innovative firms. At the international level, Cantwell and Iammarino (2000) studied the networks of multinational corporations (MNCs). They found that networks for innovation conformed to a hierarchy of regional centres. The degree of specialisation of foreign-owned plants depended on the position of the region in the hierarchy. This implies that innovation flourishes as a result of urbanisation economies rather than cluster dynamics that are supposed to be powerful in many different types of region.

Also at the international level, Hendry *et al.* (2000) studied the opto-electronics industry in six European regions. They examined the extent and significance of localised inter-company trading and network relationships. What they found was that national and international relationships were much stronger than local ones (Hendry *et al.*, 2000, p. 229).

These findings echo those of an earlier study of small high-technology firms by Oakey (1984). He reached two clear conclusions. On the one hand, linkages between firms identified as being within a cluster and restricted to less than 10km were extremely meagre (Oakey, 1984, p. 411). Secondly, he found that many of the inputs and outputs of high-technology small firms are international in both their origins and destinations (Oakey, 1984, p. 403).

Keeble *et al.* (1998) studied relatively new technology-intensive small firms in Cambridgeshire and Oxfordshire. They found that such firms internationalise early in their existence. This is mainly because their regional or even national markets were not large enough to provide sufficient demand for their highly specialised niche products and processes.

The greater South East region is widely considered to be home to many of the UK's most competitive firms. As a result, there have been several studies of high-technology and/or innovative firms in the region. The CBI analysed the ITEC sector in the UK. Much of it is concentrated in the Thames Valley. Many of the members of the working group established by the CBI to study ITEC worked in this sector in the Thames Valley. Despite setting out with the intention of confirming the existence of clustering dynamics, they were forced to conclude that "cluster interactions as we have defined them are quite weak" (CBI, 1999, p. 3). They went on to say that the levels of both formal and informal interactions between firms in the sector were much lower than in the US.

Gordon and McCann (2000) examined the evidence for industrial complexes and strong social networks in the London region. They concluded that the picture that emerged was much more akin to the model of pure agglomeration than to clustering dynamics. As a result of their findings, they argued that the social network model, which is a key element in clustering, was not a pre-condition for innovation success.

Simmie and colleagues (1999, 2002) have also shown that local linkages are weak or relatively unimportant among innovative firms located in Hertfordshire and the Greater South East in general. Their evidence showed that innovative firms benefited from urbanisation economies arising from the infrastructure, labour pools and sheer numbers of other firms in the region. In addition, linkages with national suppliers and international customers were regarded as significantly more important than those within the locality.

Finally, in studies of one of the UK's most likely cluster candidates by Henry and Pinch (2001), the lack of institutional infrastructure in 'Motor Sport Valley' emerged as a key finding. The firms in this competitive industry are notable for their numerous global investment linkages. At the local level, individual firms and the labour market are the key institutions.

These studies suggest that innovative firms use national and international linkages as much if not more than local networks. The most successful among them appear to use the advantages of multiple but often *ad hoc* local linkages often of a type related to urbanisation economies in the development of their innovations. But, at the same time even the smaller firms among them develop international linkages early in new product life cycles. They are frequently characterised by comparatively high rates of exports. Such firms form the backbone of trading nodes in the international economy. In so far as they are involved in clustering dynamics, these tend to be at the national and even international levels rather than predominantly local.

It is clear that, as Oinas and Malecki (2002) have pointed out, innovation is an internationally distributed system rather than an activity primarily confined within a given local cluster. All the evidence cited here suggests that national and international linkages are as significant for innovation as are more local networks. This provides at least a *prima facie* case suggesting that innovation must be understood in terms of trading nodes in an international system that encompasses both local and international knowledge spillovers and multilayered economic linkages extending over several different spatial scales.

Empirical Evidence

These propositions can be tested further with respect to innovation by using the results of the Community Innovation Survey 3 (CIS 3). This is becoming the most comprehensive Europe-wide survey of innovation. Local agents for each of the 15 member-states conduct it on a 4-yearly cycle. The methodology

Table 1. Relative competitiveness of non-innovators and leading innovators, 1998–2000

Growth	Mean scores (percentages)	
	Non-innovators	Leading innovators
Turnover	6.4	19.5
Exports	29.1	100.2
Capital expenditure	29.5	46.3
Employees	7.2	9.8

Source: Community Innovation Survey 3.

is based on the recommendations of the Oslo manual (OECD/Eurostat, 1997).

In the UK, the Office of National Statistics (ONS) conducted the latest survey in 2001 for the Department of Trade and Industry (DTI). It involved a 2-stage sample of all firms in the UK. In the first stage, 13 315 firms were sent a postal questionnaire in April 2001. A top-up survey of 6287 was conducted in November of the same year. This produced a total sample of 8172 firms. The results were weighted to represent all firms in the production and construction industries, wholesale trade (excluding motor vehicles), financial intermediation and business services. The weighted results constitute the best estimates of the innovation activities of firms across the entire UK for the period 1998–2000. The results are available from the DTI as an SPSS file to approved users.

The results show that, during the 3 years leading up to the millennium, some 10.3 per cent of firms in the UK introduced a product innovation that was new to the firm. A further 7.8 per cent introduced one that was new to their particular market. As a result, some 18.1 per cent of firms had introduced a product innovation during the period, while 81.9 per cent had not. For the purposes of the analysis presented in this paper, the characteristics of the leading 7.8 per cent of innovative firms will be compared with the 81.9 per cent of non-innovators. Process innovators will not be considered in this particular analysis.

The relative competitiveness of the leading innovators is shown in Table 1. It shows that the mean percentage growth in exports among these firms over the 3-year period was 100 per cent. This compares with 29 per cent for non-innovators. Partly as a result of this competitive performance in international markets, the turnover of leading innovators increased by three times more than that of non-innovators. These results provide a strong indication that it is fair to describe the leading innovators in this sample as competitive firms.

The data also suggest that leading innovators are productive firms. Thus, on the one hand, they increased both their turnover and exports by around three times that of non-innovators. On the other hand, they achieved these results by increasing their capital expenditure by only 16 per cent and their labour force by around 2 per cent more than non-innovators. In other words, they achieved much higher increases in performance than the increases in inputs of capital and labour that they needed to generate this relative success.

Taken together, the data suggest strongly that the leading firms with respect to innovation are also both competitive and productive. According to the cluster thesis, local or regional linkages should play significant roles in assisting firms to be competitive, productive and innovative. This paper focuses especially on Porter's six arguments listed above on why cluster dynamics are supposed to contribute to innovation. The CIS 3 does not always provide exactly the data needed to test these hypotheses and so not all of them can be explored using this

Table 2. Locations of main markets (percentages)

	Non-innovators (n = 101 422)	Leading innovators (n = 9731)
Local	34.7	10.8
Regional	20.3	11.6
National	35.6	54.3
International	7.5	22.5

Source: Community Innovation Survey 3.

source. Nevertheless, where it does contain appropriate information this is used in the analysis that follows.

The first argument is that local clustering allows rapid perception of new buyer needs. Table 2 examines the locations of the main markets of non-innovators and leading innovators. The data show more or less the reverse of the local cluster hypothesis. It may be seen that the locations of the main markets and therefore new buyer needs of the sampled firms are more likely to be local or regional for non-innovators than for the leading innovators. In contrast, the main markets for innovative firms are much more likely to be national or international. The most telling figure in Table 2 is that some 22 per cent of the main markets for leading innovators are international. In contrast, only 10.8 per cent are local and 11.6 per cent national. These figures suggest that in the UK local clustering does not contribute as much to the rapid perception of new buyer needs as do national and international linkages and networks. This finding supports similar results in surveys of innovative firms in a number of European city-regions by Simmie (2002a, 2002b) and Simmie *et al.* (2002).

Two other clustering hypotheses are that the activity concentrates knowledge and information, and that the knowledge-based economy is most successful when knowledge resources are localised. Table 3 shows a range of sources of knowledge and information used by non-innovators and leading innovators. These sources are grouped into five main categories starting with internal through markets, institutions, other to specialised sources. The relative importance of these different sources is ranked from not used (0) to high (3). The mean of these scores has been calculated for both non-innovators and leading innovators.

The highest mean score for importance attached to a particular source of knowledge is 2.29 by leading innovators for sources within the enterprise. In other words, such firms regard the concentration of knowledge and information within themselves as more important than any external source that may be associated with clustering. Leading innovators also rank quite highly some sources of knowledge that are not associated with space. These include specialised standards such as technical, health and safety, and environmental standards and regulations. These are usually set by government and industry bodies and have little or no relationship to clustering. Other notable examples of non-spatial sources of knowledge and information include the technical trade press and databases, along with fairs and exhibitions.

Among leading innovators, clients or customers—who have already been shown to be national or international rather than regional or local—are ranked as reasonably significant (1.8) sources of knowledge or information. The suppliers of equipment and materials, who might be based more locally, are ranked 1.68 in importance. All the institutions, such as universities, GREs, public and private research institutes, that are supposed to concentrate knowledge and information in localities, are ranked as relatively unimportant sources.

These figures suggest that the leading innovators in the UK do not depend for their success on localised concentrations of

Table 3. Sources of knowledge or information used in innovation 1998–2000

	Mean scores	
	Non-innovators ($n = 74\ 794$)	Leading innovators ($n = 9\ 568$)
Internal		
Within the enterprise	0.80	2.29
Other enterprises within the group	0.36	1.03
Market		
Suppliers of equipment, materials, etc.	0.87	1.68
Clients or customers	0.77	1.8
Competitors	0.55	1.09
Consultants	0.41	0.86
Commercial laboratories	0.15	0.46
Institutional		
Universities or other HEIs	0.17	0.52
Government research organisations	0.12	0.27
Other public sector	0.21	0.46
Private research institutes	0.12	0.31
Other		
Professional conferences, meetings	0.44	0.95
Trade associations	0.53	0.85
Technical/trade press, databases	0.55	1.13
Fairs, exhibitions	0.49	1.24
Specialised		
Technical standards	0.60	1.40
Health and safety and regulations	0.76	1.24
Environmental standards and regulations	0.67	1.11

Note: Degree of importance 0 = not used to 3 = high.
Source: Community Innovation Survey 3.

knowledge that are external to the firms themselves. Much of the knowledge they use is internalised within the firm. It is scarce and valuable new technical and economic knowledge and so they tend to be reluctant to share it with other firms. On the other hand, most of their external sources of knowledge tend to be non-spatial in character. These include externally set standards and regulations, impersonal technical press and databases together with face-to-face meetings at fairs and exhibitions. It does not appear, therefore, that localised clustering contributes much to the knowledge and information used by leading innovators.

A further advantage claimed for clustering as a contribution to innovation is that it facilitates on-going relationships with other institutions including universities. These kinds of networking and linkages are indeed one of the presumed most significant characteristics of the dynamics of localised clusters. Networking and linkages may be both formal and informal. Unfortunately, the CIS 3 does not distinguish between these two types and defines innovation co-operation as "active participation in joint innovation projects (including R&D) with other organisations. It does not imply that either partner derives immediate commercial benefit from the venture" (DTI, 2001, Question 13). This could be interpreted as implying formal rather than informal co-operation.

Table 4 shows the percentage of firms that had used (formal) co-operation arrangements. It is noteworthy that even among the leading innovators some two-thirds had not been involved in external co-operation ar-

Table 4. External co-operation arrangements (percentages)

	Non-innovators ($n = 97\ 417$)	Leading innovators ($n = 9\ 675$)
No	90.5	66.1
Yes	3.7	32.4

Source: CIS 3.

rangements. Thus, a substantial majority of innovative firms did not use formal external co-operation arrangements that are supposedly a key feature of the contribution of clustering to innovation. Approximately one-third of leading innovators did use external co-operation arrangements. Table 5 shows who these collaborators were and where they were located.

Table 5 breaks the type of collaborators down into four main groups. These are internal to the group, market participants, institutions and specialised. The main locations of these different types of collaborator are then divided between local, national, European and the US. Table 5 shows that for the leading innovators the proportions of local collaborators used of all types is uniformly less than those that are located nationally. More collaborators located in Europe than locally were used in the cases of other enterprises within the group, the suppliers of equipment, materials, components and software, clients or customers, and competitors. There were also more collaboration arrangements with clients or customers in the US than within the firms' own localities.

These data show that, while it is true that leading innovators used some local collaborative linkages, national collaborations were used more than those associated with local clustering. The data also show that international linkages, especially in Europe, are often more significant than local ones. These results suggest strongly that in the UK innovation performance is not dependent on the kinds of dynamics associated with the concept of local clustering. Instead, leading innovators are more accurately seen as parts of an internationally distributed system of innovation with important linkages between actors located nationally and in other advanced economies. The CIS 3 data therefore provide little support for the proposition that localised clusters deliver innovation.

Summary and Conclusions

In this paper, it has been argued that the analysis of clustering—that is to say, the dynamic interactions that may take place within a geographically defined cluster—should focus on what they deliver in terms of competitiveness, productivity and innovation. Performance in these areas was the original focus of Porter's (1990) seminal contribution to the analysis of the competitiveness of national economies. Initially, he argued that competition rests on innovation and the consequential differentiation of products in the market-place. The importance of innovation was also emphasised because of its contribution to productivity. Often, Porter equates productivity with competitiveness. Thus, his main line of argument is that innovation leads to new products and product differentiation. It also leads to productivity and this is the basis of competitive advantage.

The dynamics of these economic interactions are portrayed in the famous diamond. In this, particular relationships between factor input conditions, related and supporting industries, the context of a firm's strategy and rivalry and demand conditions can lead to the competitiveness of firms. These are all reasonable propositions. They do not initially or necessarily have any significant relationship to space.

The significance of space and location in Porter's analysis arises from the recognition that factors of production are used in particular places and therefore productivity arises from the ways in which these factors are used and upgraded in those localities. Under the influence of the new industrial divide thesis, Porter goes on to argue that companies have been deverticalising and breaking up into local networks of firms characteristic

Table 5. Local co-operation arrangements by type of collaborator (percentages)

	Local		National		European		USA	
	Non-innovators	Leading innovators	Non-innovators	Leading innovators	Non-innovators	Leading innovators	Non-innovators	Leading innovators
Internal								
Other enterprises within group	0.5	5.1	0.7	5.9	0.4	5.2	0.5	5.0
Market								
Suppliers of equipment, materials, comps. software	0.5	3.8	1.2	13.4	0.4	7.1	0.2	3.2
Clients or customers	0.6	4.5	0.9	14.0	0.3	7.3	0.2	4.8
Competitors	0.3	1.1	0.3	4.0	0.1	2.5	0.0	1.5
Consultants	0.5	2.6	0.6	7.6	0.1	1.2	0.1	1.4
Commercial laboratories or R&D enterprises	0.1	1.3	0.4	4.0	0.1	1.3	0.1	1.1
Institutional								
Universities or other higher education institutes	0.4	4.5	0.5	7.7	0.1	3.1	0.1	0.5
Government research organisations	0.3	0.7	0.3	3.1	0.1	0.6	0.1	0.3
Specialised								
Private research institutes	0.1	0.7	0.3	3.0	0.0	0.8	0.1	0.4

Note: Non-innovators: $n = 3776$; Leading innovators: $n = 3154$.
Source: Community Innovation Survey 3.

of Emilia–Romagna and Baden–Württemberg. These require local interactions and linkages. The intensity of these within local clusters of networked firms is argued to deliver greater competitiveness, productivity and innovation from those firms and consequentially from the localities in which they operate. Even so, the point of this is to develop a successful competitive strategy within the context of the global economy. The relative competitiveness of clusters is indicated by the abilities of the firms located within them to compete in world markets.

Leaving aside the arguments about the significance and generality of the new industrial divide, it is well known that most firms are concentrated or co-located in urban areas. Explanations for this have been offered by agglomeration theory. Porter accepts that localisation economies where firms in a particular industrial sector congregate together form part of the dynamics of clusters. In contrast, he argues that urbanisation economies that arise primarily from the size and variety of activities in a city are becoming less important because large cities are common. Studies of the locational advantages valued by innovative firms have shown quite the opposite. Thus, innovative firms tend to congregate in large metropolitan areas like London and Paris precisely because of the large numbers of firms, workers, services and other facilities that are also concentrated there. In doing so, they are often making independent and unconnected decisions which are not based on the need to establish regular interactions with other firms or institutions in the city.

The significance of urbanisation economies is also illustrated by the fact that studies of interfirm networking have shown that it tends to cascade from higher to lower levels in urban hierarchies and between high-level metropolitan centres. In other words, the economic linkages between firms show the importance of core metropolitan centres and their size rather than the significance of multiple and freestanding and dispersed clusters of companies. Part of the reason for this, in contrast to the new industrial divide thesis, is the continued and possibly growing importance of multinational corporations (MNCs). Their spatial divisions of labour often reflect and reinforce national and international urban hierarchies. As a result, many places remain parts of the internationally distributed value chains of MNCs. Among other things, this is reflected in the strength and importance of economic linkages between places rather than within them.

It is argued here, therefore, that clustering dynamics must be shown to make additional contributions to those of agglomeration economies to competitiveness, productivity and innovation for clustering to be regarded as a critical economic concept. Somewhat surprisingly, there is a limited amount of comparative evidence on the performance of firms operating within clusters as opposed to those that do not.

Following Keynes, Porter argues that it is export-oriented clusters that drive national prosperity. Studies of geographically identified clusters in the US, however, have shown that some two-thirds of them are primarily concerned with serving local needs. In the sense that competitiveness is indicated by the share taken of international markets, this finding suggests that the majority of clusters in the US do not contribute directly to the competitiveness of the national economy. Conversely, it does seem likely that trading clusters will form significant nodes in the increasingly globalised economy.

A major problem in analysing the economic contributions of clustering dynamics is that there are no criteria inherent in the concept that permit a consistent geographical definition of the spatial boundaries of clusters. According to Porter, the functional linkages within clusters may extend as widely as groups of countries or as narrowly as a single city. This suggests that the analysis of clusters should not start with their geography, but with the linkages between firms and institutions. This is the approach adopted in this paper with respect to the relationships between clustering and innovation. The argument here is that, if local clustering makes a significant contribution to the output of innovations in a particular locality, then this should be indicated by the numbers of local linkages and the importance attached to them by innovative firms.

Previous studies of high-technology and innovative firms do not support the local cluster thesis that multiple and strong local linkages contribute significantly to the output of innovations. Studies of innovative firms in the UK have shown that they value national and international linkages more than those that are local or regional. Even small high-technology firms appear to make greater use of international inputs and outputs than those within their operating localities. In some of the UK's most dynamic sub-regions and sectors such as the Thames Valley and motor sport, the types of interactions associated with clustering are weak and institutional relationships are thin. Nevertheless, the innovative industries in these areas and sectors often show a combination of both regional and international linkages associated with highly successful export performance.

Porter argues that clustering delivers higher rates of innovation because it allows rapid perception of new buyer needs. It concentrates knowledge and information. It allows the rapid assimilation of new technological possibilities. It provides richer insights into new management practices. It facilitates on-going relationships with other institutions including universities. And the knowledge-based economy is most successful when knowledge resources are localised.

The CIS 3 survey of all UK firms permits the testing of four out of these six hypotheses. It is the most comprehensive source of data on both innovative and non-innovating firms. It therefore allows comparisons to be made between the characteristics of these two major categories. Using this source, it was shown that firms that had introduced a product that was new to their market between 1998 and 2000 were much more competitive and productive than firms that had not done so. So the main question that was addressed in the analysis was: "Does the innovation that drives competitiveness and productivity benefit from the hypothesised dynamics of local clustering?"

With respect to the first hypothesis that local clustering leads to higher rates of innovation because it allows rapid perception of new buyer needs, the evidence suggests that the opposite is the case. The main markets for the outputs of the leading innovative companies in the UK were national and international rather than local and regional. What they needed most was therefore rapid perception of distant foreign and national buyer needs rather than those of customers and clients located in their own region.

The hypotheses that clustering concentrates knowledge and information and that the knowledge-based economy is most successful when knowledge resources are localised were also not supported by the evidence. In the first place, a majority of the knowledge used by firms to produce market-leading innovations was concentrated within the firms themselves. A major reason for this is that new economic knowledge is a very valuable asset and firms that can generate such knowledge are likely to keep it to themselves in order to reap the first-mover monopoly profits that follow.

In the second place, the evidence shows that, where firms use external sources of knowledge and information, the most highly rated of them tend to be non-spatialised. These include standards and regulations. They also include the technical press and databases along with intermittent trade fairs and exhibitions. Most of these sources are either ubiquitously available on the Web or are not necessarily located within the region of the company using them.

Finally, the idea that clustering facilitates on-going relationships with other institutions including universities is also not supported by the data. Some two-thirds of the leading innovators did not use external collaborators at all. Among those who did use them, greater importance was attached to national and European collaborators than to local ones. Institutional collaborations in general and those with universities in particular were not rated as important by the leading innovative firms.

Taken together, most of the evidence presented in this paper suggests that innovation performance in the UK is not dependent on localised cluster dynamics. Local clustering does not appear to deliver innovation. Given the close links between innovation, productivity and competitiveness, this raises questions over whether clustering adds to the delivery of better performance in any of these phenomena.

Contrary to the local clustering hypotheses, market-leading innovative firms seem to be more a part of an internationally distributed system of innovation. Their clients and customers are located around the advanced economies particularly in Europe and the US. The knowledge and information they employ in innovation are concentrated within the firms themselves or gathered from non-spatial sources such as government and industry standards and regulations. Intermittent face-to-face meetings at trade fairs, exhibitions and professional gatherings are also important. Urbanisation economies are also significant because the size of an agglomeration influences the variety of inputs and contacts that may be made on a pick-and-mix basis during the development of an innovation.

Given that innovation is an internationally distributed system but that parts of this system are highly concentrated in a limited number of city-regions, the proponents of clustering dynamics need to be able to explain the relative significance of both concentration and international linkages. Porter himself seems to be moving in this direction with his latest insistence that a limited number of trading clusters are the key to national economic growth. This will prove something of a disappointment to local policy-makers who have been trying to identify locally confined clusters as the basis of economic growth where in fact few if any traded activities exist at all.

References

ACS, Z. J. and VARGA, A. (2002) Geography, endogenous growth, and innovation, *International Regional Science Review,* 25(1), pp. 132–148.

AMIN, A. and THRIFT, N. (1992) Neo-Marshallian nodes in global networks, *International Journal of Urban and Regional Research,* 16(4), pp. 571–587.

ARTHUR, W. B. (1994) *Increasing Returns and Path Dependence in the Economy.* Ann Arbor, MI: University of Michigan Press.

AUDRESTCH, D. B. and FELDMAN, M. P. (1996) R&D spillovers and the geography of innovation and production, *The American Economic Review,* 86, pp. 630–640.

BAPTISTA, R. (1996) Research round up: industrial clusters and technological innovation, *Business Strategy Review,* 7(2), pp. 59–64.

BENNETT, R. J., GRAHAM, D. J. and BRATTON, W. (1999) The location and concentration of business clusters, business services, market coverage and local economic development, *Transactions of the Institute of British Geographers NS,* 24, pp. 393–420.

BRESCHI, S. (2000) The geography of innovation: a cross-sector analysis, *Regional Studies,* 34(3), pp. 213–229.

CANTWELL, J. and IAMMARINO, S. (2000) Multinational corporations and the location of technological innovation in the UK regions, *Regional Studies,* 34(4), pp. 317–332.

CASTELLS, M. (1987) Technological change, economic restructuring and the spatial division of labour, in: H. MUEGGE and W. B. STOHR (Eds) *International Economic Restructuring and the Regional Economy.* Aldershot: Avebury.

CBI (CONFEDERATION OF BRITISH INDUSTRY) (1999) *Summary progress report: information age partnership working group on ITEC clusters.* London: CBI.

DTI (DEPARTMENT OF TRADE AND INDUSTRY) (1999) *Biotechnology clusters: report of a team led by Lord Sainsbury, Minister for Science.* London: DTI.

DTI (2001) UK *Innovation Survey Questionnaire.* Newport: Office for National Statistics.

EUROPEAN COMMUNITY DGXII (1999) *Industrial districts and localised technological knowledge: the dynamics of clustered SME networking.* Final Report, Contract No. SOE1-CT97-1058 (DGXII–SOLS).

GORDON, I. R. and MCCANN, P. (2000) Industrial clusters: complexes, agglomeration and/or social networks?, *Urban Studies,* 37(3), pp. 513–532.

GRANOVETTER, M. (1985) Economic action and social structure: the problem of embeddedness, *American Journal of Sociology,* 91, pp. 481–510.

HENDRY, C., BROWN, J. and DEFILLIPI, R. (2000) Regional clustering of high technology-based firms: opto-electronics in three countries, *Regional Studies,* 34(2), pp. 129–144.

HENRY, N. and PINCH, S. (2001) Neo-Marshallian nodes, institutional thickness and Britain's 'Motor Sport Valley': thick or thin?, *Environment and Planning A,* 33, pp. 1169–1183.

HOOVER, E. M. (1937) *Location Theory and the Shoe and Leather Industries*. Cambridge, MA: Harvard University Press.

HOOVER, E. M. (1948) *The Location of Economic Activity*. New York: McGraw-Hill.

KEEBLE, D., LAWSON, C., LAWTON-SMITH, H. *ET AL.* (1998) International processes, networking and local embeddedness in technology-intensive small firms, *Small Business Economics,* 11, pp. 327–342.

MARSHALL, A. (1920) *Principles of Economics*. London: Macmillan.

MARTIN, R. and SUNLEY, P. (2003) Deconstructing clusters: chaotic concept or policy panacea?, *Journal of Economic Geography,* 3, pp. 5–35.

MCCANN, P. and SHEPPARD, S. (2003) The rise, fall and rise again of industrial location theory, *Regional Studies,* 37(6 & 7), pp. 649–663.

MILLER, P. (2001) *Enhancing the productivity of the British economy*. Trends Business Research Ltd (www.tbr.co.uk).

NORTON, R. D. (2001) *Creating the New Economy: The Entrepreneur and US Resurgence*. Cheltenham: Edward Elgar.

OAKEY, R. (1984) *High Technology Small Firms*. New York: St. Martin's Press.

OAKEY, R. P., FAULKNER, W., COOPER, S. Y. and WALSH, V. (1990) *New Firms in the Biotechnology Industry: Their Contribution to Innovation and Growth*. London: Frances Pinter.

OAKEY, R., KIPLING, M. and WILDGUST, S. (2001) Clustering among firms in the non-broadcast visual communications (NBVC) sector, *Regional Studies,* 35(5), pp. 401–414.

OECD (ORGANISATION FOR ECONOMIC CO-OPERATION AND DEVELOPMENT)/EUROSTAT (1997) *Oslo Manual: Proposed Guidelines for Collecting and Interpreting Technological Innovation Data*. Paris: OECD.

OINAS, P. and MALECKI, E. J. (2002) The evolution of technologies in time and space: from national and regional to spatial innovation systems, *International Science Review,* 25(1), pp. 132–148.

PACI, R. and USAI, S. (2000) Technological enclaves and industrial districts: an analysis of the regional distribution of innovative activity in Europe, *Regional Studies,* 34(3), pp. 97–114.

PIORE, M. J. and SABEL, C. F. (1984) *The Second Industrial Divide: Possibilities for Prosperity*. New York: Basic Books.

PORTER, M. E. (1990) *The Competitive Advantage of Nations*. London: Macmillan.

PORTER, M. E. (1998) *On Competition*. Cambridge, MA: Harvard Business School Press.

PORTER, M. E. (2000) Location, competition, and economic development: local clusters in a global economy, *Economic Development Quarterly,* 14(1), pp. 15–34.

PORTER, M. E. (2002) *Regional foundations of competitiveness and implications for government policy*. Paper presented to *Department of Trade and Industry Workshop*, April.

PORTER, M. E. (2003a) *UK Competitiveness: moving to the next Stage*. Lecture delivered at the London School of Economics, 22 January.

PORTER, M. E. (2003b) The economic performance of regions, *Regional Studies,* 37(6 & 7, pp. 549–578.

SIMMIE, J. M. (2002a) Trading Places in the Global Economy, *European Planning Studies,* 10(2), pp. 201–214.

SIMMIE, J. M. (2002b) Innovation, international trade and knowledge spillovers, *Italian Journal of Regional Science,* 1/2002, pp. 73–91.

SIMMIE, J. M. (2002c) Knowledge spillovers and the reasons for the concentration of innovative SMEs, *Urban Studies,* 39(5/6), pp. 885–902.

SIMMIE, J. M. and SENNETT, J. (1999) Innovative clusters: global or local linkages?, *National Institute Economic Review,* 170, pp. 87–98.

SIMMIE, J. M., SENNETT, J., WOOD, P. and HART, D. (2002) Innovation in Europe: a tale of knowledge and trade in five cities, *Regional Studies,* 36(1), pp. 47–64.

TRENDS BUSINESS RESEARCH (2001) *Business Clusters in the UK—A First Assessment: A Report for the Department of Trade and Industry by a Consortium led by Trends Business Research*. London: DTI.

Life Sciences Clusters and Regional Science Policy

Philip Cooke

[Paper first received, February 2003; in final form, September 2003]

1. Introduction

In this paper, the aim is to explore the likely effects upon science policy of changes in R&D-based clusters due to the rise of 'knowledge economies' (Dunning, 2000; Cooke, 2002). To advertise the argument beforehand, it is that the decline in R&D power of large corporations is accompanied by the rise of specialist research firms. The latter include, for example, those referred to as 'discovery companies' in Life Sciences, along with university and other research labs in proximity to which knowledge-intensive firms increasingly cluster. This is particularly pronounced in biotechnology, but also occurs in other knowledge-intensive sectors like information and communication technologies (ICT), new media and advanced business services. Broadcasters and *bourses* are stronger cluster magnets than universities in the last two cases. That Life Sciences activities cluster geographically along the knowledge value chain from exploration (basic research) through examination (clinical trials) to exploitation (dedicated biotechnology firms—DBFs) is widely known and shown.[1]

Continuing the argument, it will be shown that over the 1990s many regional governance agencies developed interest and capability in formulating policies to network regional innovation systems. To some extent, multilevel governance hierarchies have evolved, as suggested in Lundvall and Borrás (1997) and Cooke *et al.* (2000) where national governments are mainly responsible for delivering science policy and basic research funding, while regional governance systems (involving public and private actors) deliver innovation programmes. These are usually near-market incentives to firms to build innovation networks, access co-funding and engage in joint marketing to enhance innovative potential and competitiveness. In

Europe, the European Union was less directly involved than member-states in basic research funding, more in research and technology development (RTD) and, while co-funding *innovation* initiatives, leaving these to regions to deliver along with their own and any national programmes on the ground.

In what follows, the first main section will show how and why knowledge economies create regionalised innovation networks and clusters in Life Sciences and biotechnology. In section 2, the implications of this are explored for the national innovation system model of 'big science' and—for example, 'big pharma' following an expensive 'chance discovery' methodology, also involving big departmental research laboratories in universities. Then, finally, evidence is mobilised to show the emergence of regional science policy and regional science policy *funding* mechanisms. A model that responds to regional *science* funding requirements within the national basic science funding remit is also discussed.

2. Knowledge Economies and Their Regionalisation

A common misconception among non-regional scientists is that when regional analysis is done it inevitably means somehow ignoring other spatial, economic or political scales. As will be shown, the contrary is the actual position, particularly where science, technology and innovation are in focus. Indeed, what has been shown elsewhere (Lagendijk, 2001; Cooke, 2001b) is that, by excluding the *regional* level from analysis, major innovation interactions between key knowledge generation and exploitation actors are likely to be overlooked. As Dicken (2001) sees it, from the TNC perspective, regionalisation enables faster delivery, more customisation and smaller inventories than globalisation. But this does not mean TNCs become less global; rather, they use whatever advantages may be available to them in seeking value chain efficiencies. So it is with regard to what might be termed the 'knowledge value chain' that this exploration is directed.[2] What is this and how might it be changing? We know of the changed emphases in knowledge production proposed by Gibbons *et al.* (1994). Key differences involved are the move from Mode 1 to Mode 2 conventions like disciplinary purity to *transdisciplinarity*, organisational hierarchy to *flexibility and diversity* and value freedom to *reflexivity*. Related to reflexivity, the authors argue, are *quality*-related questions of a new kind concerning the competitiveness of knowledge outcomes in the market, or cost effectiveness and social acceptability.

Critique of this view focused on its technocratic introspection and failure to show science having to engage with social forces. In a subsequent book (Nowotny *et al.*, 2001), some of these authors respond with an acceptance of Latour's (1998) identification of a struggle between 'science' and 'research'. The former is stable, the latter unstable and the rise of research expresses a 'contextualisation' of science by society, politics and economy despite its claims to objectivity. Thus more multidisciplinary concerns and network interactions typify modern Life Sciences, as Powell *et al.* (1996) and Orsenigo *et al.* (2001) among many others show. Indeed, the former go as far as to assert that 'knowledge is in the networks', a revision of the traditional primacy of codified over tacit knowledge. Orsenigo and colleagues explain this in terms of the heterogeneous nature of the cognitive skills demanded in bioscientific research. The economic and ethical demands of 'research' give rise to collaborative learning through transdisciplinary network relationships. Science's engagement with society initially occurred on *its* terms, as 'public understanding of science', now rightly criticised for its patronising disposition. It is, accordingly, difficult to see voluntary *reflexivity* in knowledge production in Life Sciences and others. Rather, ethical regulatory powers and company protocols to constrain 'value-free' excesses in the field have had to be imposed externally. Contrariwise, as we shall see, the inclination for stock market imperatives to interfere with peer review norms of scientific reporting is increasingly driving announcements of scientific discovery and raising further ethical issues.

Thus, in 2001, Millennium Pharmaceuticals and Human Genome Sciences both made press announcements when experiments reached Phase 1 of clinical trials, at least three years away from possible approval.[3] In the past, such announcements would be made on applying for approval. Also, dedicated biotechnology firms (DBFs) have recently made announcements on experiments still at basic research stage, such as Advanced Cell Technology's claim to have cloned a human embryo and PPL's to have cloned pigs, both in advance of peer review and publication of results. This is doubly problematic when, at approval stage, large numbers of candidate treatments are rejected, as 30 were by the US Food and Drug Administration during 2001. These included Chiron's anti-sepsis drug, Immunex's cardiac infarction treatment and Maxim's melanoma product. A US head of bioethics was quoted as saying that "these companies must raise enormous amounts of money and the only way to do that is to put a hard spin on any good news" (Griffith, 2002).

An effect of the rise of research is that the mode of knowledge production has shifted, as revealed in Life Sciences R&D databases such as that of Pammolli *et al.* (2000) at the University of Siena. For example, of the 50 dedicated biotechnology firms (DBFs) in the database pursuing 'designer molecule' rather 'discovery' methodologies, 74 per cent operated in global, hierarchical networks with big pharma developers. From 1992 onwards, the incidence of R&D projects involving combinatorial chemistry, target-based screening, genomics and genomic libraries doubled. All but one of the 26 per cent of specialists followed the leaders after 1992, mostly in bioinformatics. Strikingly, 54 per cent of the DBFs responsible for originating all these R&D projects were in 4 key US clusters: Cambridge, MA (18 per cent), San Francisco-San Jose, CA (16 per cent), San Diego, CA (12 per cent), and Maryland (8 per cent). Hence we see a highly globalised, hierarchical knowledge generation model in which leading-edge research is initiated by multidisciplinary DBFs in clusters linking with (often many) large pharmaceutical firms, research institutes and other DBFs as developers. It is plain that the clusters are increasingly the locus of knowledge generation.

The rise of research over science explains the rise of DBFs over big pharma in new knowledge generation. But DBFs still need large drugs firms to fund their discovery programmes. Thus despite most of the major companies experiencing drought in the product pipeline, they accommodate to new realities. For example, Swiss company Novartis announced in early 2002 that Glivec, its already successful chronic myeloid leukaemia (CML) drug also works for stomach tumours. The US Food and Drug Administration (FDA) hastened approval for Glivec to save the lives of leukaemia sufferers and the product was granted orphan drug status in the US and the UK for gastrointestinal stromal tumours.[4] Novartis' methodology for developing Glivec was a prototype for the rational drug design or 'silver bullet' rather than 'chance' mode of drug discovery. This entailed genomic research to target the precise molecule giving rise to the mutation causing CML.

Actually, Glivec is rather unusual in that DBFs were not directly involved in the progress towards production of the therapeutic treatment. Rather, as may be seen in Table 1, the elapsed time from initial discovery to final approval was 40 years. This is too long for DBF and venture capitalist survival of the normal 10 years or so from proof of concept to hoped-for initial public offering of the DBF to the stock market. Nowadays, combinatorial chemistry that allows vast numbers of compounds to be screened rapidly and systematically through high-throughput screening (HTS), applied also to methods for sequencing genes, speeds up the process (although for big 'pharma's' failures also in this field, see Cooke, 2004). Research DBFs have advantages in swiftly bringing together networks of distinctively skilled researchers and technologists to target specific molecules.

Table 1. Institutional and corporate history of Novartis CLM treatment 'Glivec'

Date	Institution	Name	Indication
1960	University of Pennsylvania	Nowell/Hungerford	Blood chromosome 22 'Philadelphia chromosome'
1973	University of Chicago	Rowley	C22 translocated to C9 discovery
1986–87	Whitehead Institute, Cambridge MA	Baltimore	Bcr-Abl protein: Tyrosine Kinase (cell regulator)
1992	Dana–Farber Cancer Institute, Boston	Druker	Bcr-Abl > CM leukaemia; mutant enzyme jams cell-signals discovery
1993	Oregon Health Sciences University	Druker/Ciba–Geigy	Reagent and inhibitor for Tyrosine Kinase activities
1993	Ciba–Geigy	Leyden/Matter	ST1571 inhibitor compound (Glivec) selected
1998–2000	Novartis	Druker	Clinical trials and FDA approval
1998		Nowell/Rowley	Lasker Medical Research Award

Source: Journal of the National Cancer Institute, 5 January 2000 (http://www.nci.nih.gov/clinical_trials).

Thus it was university and private research institute scientists that conducted the fundamental research that resulted in Novartis releasing the world's first approved drug directly to turn off the signal of a protein known to cause cancer. In other words, the 'rational drug design' approach pioneered in cancer treatment by Novartis was really the culmination of university and research institute processes of origination, and final development through the company. Although it started as 'chance' discovery of the Philadelphia chromosome, it evolved into a process in which precise molecular targeting became possible. This is expected to become an important, possibly the paradigmatic, methodology in the post-genomic era.

In the Glivec case, the importance of particular centres of research excellence in a strong biotechnology region like Greater Boston is evident. Random discoveries were made elsewhere early on but key milestones were reached in the 1980s and early 1990s, first at the Whitehead Institute (Cambridge, MA), subsequently co-leader of the human genome project, and the Dana–Farber Cancer Institute (Harvard), in identifying and then understanding the mechanism causing a mutant enzyme to jam the signal that normally prevents massive overproduction of white blood cells, hence CML. This built on the prize-winning research elsewhere that first identified and, secondly, found where the key piece of missing DNA had translocated, which was a valuable research by-product. Thereafter, the main development technology moved with Brian Druker, the holder of the reagent that matched Ciba–Geigy's inhibitor compounds for Tyrosine Kinase activities, from Harvard to Oregon and Basel, Switzerland. This meant Ciba only had to check its bioinformatics library. Such is the esteem in which Greater Boston is held by Novartis that in 2002 the company announced the location of its new $250 million Genomics Research Institute there. In 2003, Aventis announced a larger ($350 million) genomics research investment in Canada's leading cluster at Toronto. Comparable research clustering occurred in San Diego around the Scripps, Salk and Burnham Institutes and San Francisco in relation to the University of California Medical School and continues in—for example, Seattle in relation to the Fred Hutchinson Cancer Institute and Cambridge, UK, in relation to many but particularly the MRC Molecular Biology Research laboratory which has hosted 11 Nobel prizewinners in its time.

A key interim conclusion here is *not* that only a few places in the world (mostly in the US) can ever pretend to global excellence in

Life Sciences and their biotechnology exploitation capabilities. Healthcare, let alone agro-food and environmental applications, is enormous, constituting some 20 per cent of GDP (DTI, 1999) in advanced economies. That biotechnology is far more thorough-going than an esoteric toolkit is testified to by the political scramble to find ways at regional level to facilitate development of relevant capabilities world-wide (see Cooke, forthcoming). In the UK—for example, research performance assessments have resulted in the majority of Life Sciences research funding being allocated within the so-called Golden Triangle among London, Oxford and Cambridge.

3. Strategic Science Policy and Science Funding: Do Regions Matter?

It has been argued thus far that knowledge production is becoming rather strongly regionalised in particular clusters. This is because of the growing importance of university and research institute laboratories to clusters of DBFs that exploit and commercialise basic scientific knowledge, with the support of venture capitalists and other business or legal services. Simultaneously, possibly more distant multinational pharmaceuticals companies are investors in milestone payments that fund the research in exchange for future expectations of licences or acquisitions. Thus the innovation system in this sector is both highly regionalised, for research and early exploitation, and highly globalised for development and, later, distribution and marketing. Is there evidence of a growth in *regional* science policy and consequent allocations? Moreover, if there is, does it take a different form from the state or pooled public (EU) one described above? And if globally networked regional clusters predominate in biotechnology, as there is abundant evidence they do (see Swann *et al.*, 1998; and papers in *Small Business Economics*, Special Issue, 2001), is this leading to new policy thinking? Is 'ground–up' strategic science policy and funding within specialist 'knowledge economy' regions occurring for other sectors too? In the space available, this can only be explored for biopharmaceuticals, but the resulting evidence makes the hypothesis worthwhile exploring further. The key point, tackled later, is that while regional technology and innovation policy are not new, regional *science* policy is, in the sense of regional administrations pressing for augmented basic research budgets or for central governments where policy supremacy for science allocations lies to find ways to assist less research-favoured regions à la EPSCoR (see below). Despite Ashcroft *et al.*'s (1995) critique of, for example, UK innovation policy, there is no evidence of any UK regional science strategies such as that promulgated by the Scottish Executive in 2001 because there were none at that time.

Strategic research (as defined by Irvine and Martin, 1984) has become less military and more civilian since the end of the Cold War (despite the rise of funding against bioterrorism). It is arguable it has been forced to become more 'relevant' in the sense of more market-facing and ethically sensitive, as we have seen (Nowotny *et al.*, 2001). Importantly, it has allowed a repositioning of major science policy priorities towards the health rather than the defence of civilian populations. Even in the 1980s, the US spent nearly 50 per cent of its national academic and related research budget on Life Sciences. In 1987, US public healthcare R&D (funded by the National Institutes of Health) was $7.4 billion; by 1999, it was $14.8 billion and for 2003 it was over $27.3 billion, causing the following comment from the director of science policy at the American Association for the Advancement of Science.

> As a result, NIH is now the 800-pound gorilla of the research community, accounting for 42 per cent of all non-defense R&D, more than half of all federally-funded basic research, and nearly two-thirds of all federal support for R&D in colleges and universities (Teich, 1999, p. 12).

The 2003 appropriation was 51 per cent of federal basic research, a $3.7 billion increase over 2002 (Burnell, 2002). To this must be added portions of National Science Foundation, National Aeronautics and Space Administration and Department of Energy (human genome) research budgets. It is abundantly clear that health care is driving the US basic research-funding portfolio as never before. In the UK and Germany totals for biosciences research are of the order respectively of $2 billion and $1 billion annually. Senker and van Zwanenberg (2001) estimated annual public EU Life Sciences R&D expenditure at $10 billion, about half that in the US.

The very large sums of research funding now going to regional biosciences/biotechnology clusters in the US and their younger equivalents in European countries give these locations both the resources and expertise to develop as implicit if not explicit 'centres of excellence'. Because of the perceived relevance and political virtue of life sciences research, health research budgets have mushroomed. Correspondingly, financial pressure on hospitals to treat not conduct research on patients (see below) has undermined clinical scientific opportunity to take advantage of the molecular biology revolution. Hence, such centres of excellence attract further funding, from their regional governments, from bilateral industry research investments. Thus, although Novartis and University of California Berkeley announced a $25 million licensing agreement for Life Sciences patents, a year later in 2002 Novartis announced its $250 million Genomics Research Institute investment in Cambridge, Massachusetts, to tap into the knowledge and talent arising from human genome research. There are also great resources from endowed institutes and medical foundations which, in the shape of those such as the Howard Hughes Medical Institute ($13 billion endowment) and in the UK the Wellcome Trust with a $21 billion endowment and $1 billion annual expenditure, are of major signifi-cance. The latter explicitly operates a centres of excellence programme, which in the UK involves the Universities of Glasgow (parasitology), Edinburgh (cell biology), Manchester (cell matrix), Cambridge (cancer jointly with Cancer Research), Oxford (human genetics) and London (history of medicine). The Howard Hughes Medical Institute primarily funds researchers rather than centres, but has laboratories at the universities of Maryland, California–Los Angeles, Washington–St Louis, Rockefeller and internationally.

In the US, there are, of course, numerous other charitable and corporate medical foundations, the largest of which are shown in Table 2. Grants from these augment the large-scale NIH awards and add further to the resource base of centres of excellence. Indeed, the more a regional centre is designated as such the more likely it is to attract further funding. The UK is unique in having a single charitable health research trust that spends per year an amount equivalent to the sums the top 10 US charitable foundations spend together. This makes the Wellcome Trust a strategic science funder and policy-maker in its own right in the UK. It is as, if not more, important as the UK government in determining bioscientific and health research expenditure flows, as a glance at selected highlights of its funding portfolio during 2000–01 demonstrates (Table 3).

The key point to note from Table 3 and below is the trend towards regional centres of excellence and the manner in which in the UK the Wellcome Trust increasingly sees its role as regenerating parts of the national innovation system for health which has been damaged by a lack of policy or underfunding crises that have had negative effects on important parts of the national system, especially the National Health Service. Thus although, as Table 3 shows for the May 2001 entry, Wellcome grants under the Strategic Research Investment Fund scheme were diffused, a third of the £125 million, some £40 million was awarded to Leicester, Edinburgh, Leeds and Manchester (UMIST) universities, seven awards went to London, three to Oxford, two each to Sheffield, Cambridge and Cardiff, and a further two to Edinburgh. This reflects an emergent picture underlined in the UK government report on biotechnology clusters (DTI, 1999). That proposed regional centres in the above-named places plus regions with collaborating regional universities

Table 2. Top 10 US medical research foundations and corporate foundations

Foundations	Grants 1999 ($ million)	Endowment 2000 ($ billion)	Corporate foundations	Grants 2000 ($ million)
1 R. W. Johnson	372	8.8	Aventis Foundation	41.6
2 D&L Packard	114	9.8	Proctor and Gamble Foundation	30.4
3 California Endowment	91	3.5	Merck Foundation	28.8
4 Whitaker Foundation	50	NA	Pfizer Foundation	25.5
5 B&M Gates Foundation	48	21.2	Eli Lilly Foundation	17.1
6 Burroughs Wellcome	37	0.8	Bristol-Myers Squibb	15.8
7 Rockefeller	36	3.6	Monsanto Foundation	14.0
8 D. Reynolds	35	1.3	Medtronic Foundation	12.0
9 Starr	34	4.5	Abbott Foundation	10.0
10 WM Keck	32	1.5	Glaxo Wellcome	7.0

Source: Lawrence (2001).

Table 3. Wellcome Trust grant announcements 2000–01

Date	Headline	Funding (£ million)	Recipients
April 2000	Joint infrastructure	129	Ulster, Dundee Birmingham JIF
July 2000	Joint infastructure	225	New (SRIF) Programme
July 2000	Genome bioinformatics	8	Cambridge (Sanger)
October 2000	Genome sequencing JV	8	Cambridge (Sanger)
October 2000	C. for molecular medicine	7	Cambridge (Addenbrookes)
April 2001	Science centres/infrastructure	76	ScottishUniversities
May 2001	Scientific research facilities	125	34 SRIF grants
July 2001	Synchrotron	110	Oxford
October 2001	Post-genomic research	300	Cambridge (Sanger)
November 2001	Clinical research facility	3.8	Southampton

Source: Wellcome Trust.

as in Yorkshire ('White Rose' partnership) with Sheffield (2 awards), York (1 award) and Leeds (1) or in Scotland Glasgow (1), Dundee (1) along with Edinburgh, and, in Wales, one centre based in Cardiff, could expect to be candidates for development outside the 'golden triangle' of Cambridge–Oxford–London.

This is predominantly where, given appropriate quality bids, centres of expertise may be expected to become centres of excellence,[5] and the move, first successfully demonstrated with the establishment of numerous centres of research in Stanford University (Gibbons, 2000) of specialist research away from large university teaching departments gets under way. This also applies to another big casualty of the changing research mode, clinical research in hospitals. In the UK, the latter is underlined in Wellcome's policy of funding clinical research facilities (CRFs) because

> while the UK has the inestimable advantage of a National Health Service ... the financial pressures on the NHS and healthcare reforms have created many obstacles to patient-oriented research. Not least of these is the enormous pressure on beds; patients requiring treatment obviously take priority, leaving no spare capacity that could be used by researchers (Wellcome Trust, 2002, p. 13).

The Southampton facility (Table 3, November 2001 entry) is one of five CRFs, the first opened earlier that year in Edinburgh; the others will be at Manchester, Birmingham and Cambridge. Thus regionalisation of special clinical research as well as medical and bioscientific centres of excellence is occurring as a policy initiative being implemented by the Wellcome Trust in response to changes induced in the traditional model by government health policy. Specifically, it is seeking to maximise patient treatment capability on internationally low public expenditure, at the expense of clinical research. Thus the government's Culyer Task Force found ways of segmenting costs of NHS research and paying for it with a levy on healthcare purchasing. These funds are only to be used for recurrent costs, while foundations and research councils cover direct research costs. Absence of funding for fixed capital developments led to Wellcome Trust policy to invest in CRFs. They are modelled on US General Clinical Research Centres (GCRCs) of which there are 78. Their existence was initiated by US health insurance companies' refusal to pay for research in hospital beds, and now even outpatient clinics to which research moved are too busy for research. GCRCs cost $170 million per year to run and are funded through grant applications from centres to the National Institutes of Health. The Wellcome Trust programme is funded at £20.5 million.

Thus, as government funding constraints

have placed the NHS under ever-greater financial pressure, clinical research capability is facing diminished capacity. Providers are thus becoming more entrepreneurial in their response and seeking significant funding from non-public sources, notably foundations like the Wellcome Trust co-funded for CRCs thus far by university hospitals and local health service administrations (the NHS Trusts). Having large patient databases for research is a necessity in the new world of molecular medicine and rational drug design. Thus it is not difficult to see the evolution of regionalised 'knowledge value chains' from basic university or research institute centres of excellence such as the Sanger Institute for genomic and post-genomic research, through to medical and clinical research at centres of excellence in university hospitals or schools of biosciences, to biotechnology institutes or centres, and gene centres or gene parks where exploitation and commercialisation are conducted by academic entrepreneurs interacting with clusters of DBFs.

The model can be observed in Greater Boston where all these facilities are in place, where over $1 billion in research funding alone is spent annually, much of it in collaborative partnerships among universities, special research centres of excellence (such as the Whitehead or Dana–Farber Institutes at MIT and Harvard), large hospitals like Massachusetts General or Brigham and Women's, GCRGs, incubation and successful start-up and more mature biotechnology firms like Ariad (founded by Nobel Laureate David Baltimore, discoverer of the key cell regulator enabling Glivec to be produced by Novartis), AlphaGene, Dyax, Genetics Institute, Genome Therapeutics, Genzyme and Progenitor. Reference to the University of Siena database (Orsenigo *et al.*, 2001) shows that non-regional research partners are Aventis, Bayer, Bristol–Myers Squibb, Chugai Pharma, DuPont, Merck and Pharmacia (now Pfizer). These are increasingly engaged as investors, developers (but see below, section 5), distributors and marketers for the products and services of the regional biopharmaceuticals innovation system in Massachusetts, focused on Greater Boston and with its epicentre at Cambridge (Cooke, 2001a). Massachusetts has for many years had a regional science policy to support with tax-breaks and other incentives high-technology industry, once seen as responsible for the 'Massachusetts miracle' in mid-sized computers and now experiencing a resurgence through the promotion of biotechnology, biomedical and venture capital 'clusters' (Best, 2000).

4. Regional Science Policy: The Basic Model and Some National Variants

It has been argued thus far that bioscience underwent a cognitive, methodological and technological evolution that appears to have been expressed as an empirical punctuation point around 1992, although much of that change had been in the pipeline well before that. Some move into Mode 2 knowledge production was enforced as transdisciplinary research networks among research centres of excellence, academic entrepreneurs and successful start-up DBFs began accessing dynamic externalities in the form of knowledge spillovers from co-location in geographical proximity to exploit opportunities for rational drug design. However, such networks remained hierarchical both because of élite science (the 'star' system; Zucker *et al.* 1998) and the continuing involvement of big pharma companies, less as originators than as developers of therapeutic solutions coming from biotechnology. The argument then evolved to discussion of a 'knowledge value chain' in life sciences spanning the arc from basic post-genomic and proteomic research through clinical research and treatment to innovation and commercial exploitation by clustered DBFs. It was then argued that this exists in a few regions of the world, that centres of excellence are competitive and attract or possess large financial resources and that their regional and technological innovation system governances have explicit or implicit science policies. Other governances will seek to emulate these leaders and, indeed, have begun doing so.

In the US and Europe, the regional 'clusters of excellence' include Southern California, centred on San Diego, in northern California it is Silicon Valley and, in Massachusetts, Boston. In Europe, such clusters are found, on a smaller scale than the US, in the UK at Cambridge in the Eastern England regional development agency (RDA) area, possibly also Oxford (South East RDA) and Scotland (a triangle including Dundee, Edinburgh and Glasgow), in Sweden Stockholm–Uppsala and in Germany, Munich in Bavaria although two other BioRegios also exist (Dohse, 2001; Cooke, 2001a, 2002). In these innovative 'biotech cities', it is vital to recognise the regional innovation systems in which they operate. These supply finance (for example, Bavaria sold its state energy company and established a high-tech fund which invests in biotechnology [and ICT] research and commercialisation activity); in Scotland, as we shall see, 'regional' funds for implementing its science policy for biotechnology are pooled among its RDA, Parliament and university funding body the Scottish Higher Education Funding Council (SHEFC). As part of its modest move to regionalise administratively, the Swedish national government in 2001 established VINNOVA, the Swedish Development Agency for Innovation Systems with responsibilities to invest in *regional*, technological systems in biotechnology and other advanced technology sectors. There is also a unique cross-border policy and R&D body (Øforsk) to exploit the new bridge, by building an Öresund regional biosciences innovation system between Denmark's 'Medicon Valley' near Copenhagen and the Ideon Science Park bioscience cluster at Lund near Malmö, where AstraZeneca has a large R&D facility. We shall explore a further Nordic case in depth below, which is the case of Finland.

Observing these developments, regions with aspirations and some perceived or actual potential to emulate the élite can relatively easily be identified. Two in the US are worth briefly exploring. These are North Carolina and Maryland, both of which have emergent regional biotechnology innovation systems. In the case of North Carolina, there has been an effective science policy since the 1950s when the Governor got approval for the Research Triangle Park (RTP). This had the limited objectives of attracting R&D jobs with no presumption that synergies would flow among the facilities locating there. Subsequently, major support was given to boosting the research capabilities of the three universities, Duke, UNC Chapel Hill and NC State, but especially the last two public ones. Duke's private endowment has ensured that its medical school and bioscientific research profile have prospered, while in the 1990s NC State was the recipient of major state funding to develop it as a technology campus with industrial R&D laboratories co-located with science and technology departments. In between, in 1981, the North Carolina Biotechnology Center (NTBC) was established on RTP (as, at approximately the same time, were the NC Supercomputer and Electronics Centers). NTBC was not a research but a commercialisation facility. In early 2002, NTBC housed some 30 biotechnology businesses, including sites of Aventis, BASF, Bayer, Biogen, Eli Lilly and Glaxo SmithKline among the 90 in the RTP and broader Raleigh–Durham–Chapel Hill area, and 142 in the state.

Duke University Medical Center is prestigious in basic and clinical biomedical research with cancer and urology being leading fields for which the centre is ranked sixth in the US. Basic scientific research is wide-ranging and operates in 38 laboratories including biochemistry, cell biology, genetics, immunology, microbiology neurobiology, pharmacology and cancer biology. The Duke Comprehensive Cancer Center is accredited by the National Cancer Institute (NCI) and conducts clinical research, patient care and teaching in cancer immunobiology, prevention, detection and control, cell regulation and transmembrane signalling, cellular and structural biology, experimental therapeutics, molecular oncology and cancer genetics. UNC School of Medicine is unofficially ranked 22nd in the US and its strongest research field is biomedical engineering in which expertise is found in medical imaging, biomedical computer communication, medical informatics, neuroscience engineering, bioelectronics and sensors, physiological system modelling, biomaterials and real-time computer systems. The Lineberger Comprehensive Cancer Center is one of the NCI national network of Cancer Center Programme facilities specialising in biomedicine.

Maryland is also an important US centre for bioscience, hosting 210 bioscience businesses, half in research services, testing and contract manufacturing, two strong university systems, organisations like FDA and NSF, and a large number of federal research laboratories, notably the NIH system. Much of this activity is clustered along the I-270 'Technology Corridor' (Bethesda–Rockville–Frederick) and around Johns Hopkins University in Baltimore. The Howard Hughes Medical Institute research laboratories are nearby in Chevy Chase. NIH has 25 institutes and centres, including the US National Human Genome Research Institute at Bethesda and the National Cancer Institute at Rockville. The Johns Hopkins University is ranked first among US universities in receipt of federal R&D funds, the School of Medicine is first in receipt of NIH extramural funding and is unofficially ranked second nationally after Harvard Medical School. Its research expertise is focused on AIDS, biomedical engineering, cancer, clinical immunology, genetics, molecular biology, neuroscience, organ transplantation and urology. The University of Maryland, Baltimore, is a rapidly expanding biomedical research centre in partnership with the University of Maryland Medical School System, the Veterans Administration Medical Center and the Medical Biotechnology Center, specialising in molecular genetics and human molecular biology. There is also a UM Biotechnology Institute specialising in basic science applications to health, marine environmental and agricultural biotechnology, protein engineering and structural biotechnology.

4.1 Science and Technology Policies: Market Facilitating

Both states inherited buoyant technology markets from past public investment decisions. On the basis of these strengths and to a high degree influential upon them, both states are among the 13 in the US to have adopted state-wide strategic science and technology policies, from between 1991 and 1995 (American Association for the Advancement of Science, 1999). The main goal of each policy has been enhancement of economic growth and improving standards of living by capitalising on the state's research base. Policies recognised the importance of sustaining and strengthening the R&D capacity of university research and training. In North Carolina, building on the success of Research Triangle Park, strategy focused on further stimulating exploitation of biotechnology and other technologies, and continuing to strengthen R&D capacity. New strategies emphasised stimulating indigenous entrepreneurship and promoting generative rather than redistributive growth. Maryland's strategy included recommendations for exploiting the comercialisation potential of technology from its strong universities and federal research laboratories. Both Maryland's and North Carolina's policies were initiated by their Governors, but others were the result of private initiative. Usually, they began by analysing the strengths and weaknesses of the state economy and research infrastructure. In many cases, they then went on to identify knowledge-based industry clusters, arguing that the state's economic base was passing from an old to a new economy character. Strategic policies were proposed to meet the challenge. In North Carolina's case, this involved seeking input from six task forces and nine focus groups, using the North Carolina Alliance for Competitive Technologies as the governance body for the process. Both Maryland and North Carolina included specific outcome measures, such as quantifiable growth rate of technology businesses, industry support for university R&D and new start-up companies.

However, the AAAS assessment of these policies was that they were insufficiently detailed and mostly failed to address issues of social exclusion.

In 2000, North Carolina published *Vision 2030: Science and Technology Driving the New Economy* based on a new approach emphasising *visioning* based on a state-wide foresight options process. It will be shown later that this is becoming a more widely adopted approach to regional science policy. It also involves cluster identification and regional stakeholding to attempt to commit industry and university administrations to invest in co-funding actual initiatives intended to be implemented. UNC Chapel Hill organised regional conferences, focus groups, cluster analyses and global benchmarking and produced the North Carolina Innovation Index. Recommendations included evolving a knowledge economy through supporting venture capital, public funding and tax incentives, marketing North Carolina globally as a knowledge economy and designing a globally competitive R&D tax credit. Maryland's newest policy statement *The Maryland Technology and Innovation Index* was launched in late 1999 with similar style and content, using comparative benchmarking indicators addressing performance, dynamics and resources using the Maryland Technology Alliance of private-sector, academic, federal and state government organisations as the catalyst.

4.2 Science Strategy: Science-led Growth from Below

Devolution in the UK has opened up a responsibility for democratically elected executives in Scotland, Northern Ireland and Wales to formulate science policies. Wales developed the EU's first Regional Technology Plan in 1994 and relies on an updated version under the Regional Innovation Strategy 2 programme from Brussels. The Welsh strategy has guided the establishment of expenditure patterns on technology and innovation under the Structural Funds Objective 1 action lines. This includes establishing a Knowledge Exploitation Fund, technology counsellors in universities and other infrastructures in support of *innovation* rather than basic science strategy. Northern Ireland was in a better position because of the existence of the Industrial Research and Technology Unit (IRTU) which, through its annual corporate planning process, designed technology and innovation, if not science policy in the region. It has now become a division of Invest Northern Ireland, the new integrated regional development agency. It is noticeable that, despite its peripherality and political troubles, Northern Ireland has developed a discernible science and technology policy not unlike but more piecemeal than Scotland's. Thus biosciences and ICT (especially telecommunications and Internet software) have been supported with research funding from IRTU, contest-successes for UK grants to enhance academic entrepreneurship and the construction of nine incubators for the two target sectors. The necessity for regional science policy in both Wales and Northern Ireland is demonstrated by the evidence that, at £34 and £24 per student respectively, the UK government's low investment in science funding there compares unfavourably with the £44 per head in England and £58 in Scotland. Research performance, measured since 1986 in the UK Research Assessment Exercise, explains the disparities to some extent. In this context, it is noteworthy that Northern Ireland's most significant biomedical research initiative, the University of Ulster's £14.5 million Centre for Molecular Biosciences, was equally co-funded by the Northern Ireland 'Support Programme for University Research' fund and Atlantic Philanthropies, an Irish American foundation which, since 1982, has invested $1.3 billion in higher education world-wide, 28 per cent of which was in Northern Ireland and the Republic of Ireland. The donor, Charles Feeney, also funded the Sinn Fein office in Washington. The university's vice-chancellor, bemoaning a 20 per cent decline in the region's funding for academic research through the UK system said: "With devolution, we have found greater awareness of the importance of the research base than when we had direct rule" (Farrar, 2002).

Scotland was first in the UK to seize the opportunity to develop a regional science policy, its Minister of Science publishing in January 2001 *A Science Strategy for Scotland*. It was preceded by a report in 2000 from the Science Strategy Review Group and informed by Scotland's Science Policy Unit. The report shows that about £800 million is spent on scientific research in Scotland annually and that Scottish universities won £141 million or 11 per cent of the UK Research Councils budget in 2000, a percentage point or so above the country's share of the UK population or GDP. The *Science Strategy* makes it clear that, although Scotland's economy performs at about the UK norm, market forces alone cannot be relied on for economic growth to occur but that Scotland's basic science advantage and government activity more generally have to be directed increasingly at sustaining world scientific leadership in a few feasible areas and raising commercialisation and entrepreneurship opportunities arising from science. The report prioritises bioscience and genomics, medical research and e-science as the three areas of world leadership in basic science that the Executive will support in particular. This means maximising targeted science research expenditure for these areas, including improving relationships between university and biological research institute research facilities in Scotland. To assist this, the Executive commits itself to investment in Scotland's joint Science Research Investment Fund.

Making an important commitment towards science funding in the UK as a whole, it aims to assist in setting in place a more transparent research funding methodology to ensure that underfunding of the kind widely perceived to have bedevilled UK science for decades cannot happen again. Scotland's problems of low industrial R&D and a high proportion of small businesses are to be moderated by connecting to economic growth initiatives such as the Scottish Executive's *The Way Forward: A Framework for Economic Development, The Knowledge Economy Cross-cutting Initiative* and the *Digital Scotland Task Force*. Accordingly, it commits to keeping the 'Proof of Concept Fund' (see below), setting up a National Health Service Technology Transfer Office, revitalising UK-originated small business research and technology awards, assisting academic entrepreneurship, using Foresight to identify future scientific challenges and opportunities, and recruiting investment and scientists from overseas.

This is clearly a more interventionist set of commitments than are discernible in the more 'market-following' policies described previously. As in Northern Ireland and Wales, government has to do more because of market arrest, in a context of greater reliance on market forces where they are strong—which in the UK means, in effect, the aforementioned 'Golden Triangle'. Scotland was advantaged in bringing forward its fairly robust commitments to science support by preceding work done by Scottish Enterprise. The £40 million 'Proof of Concept Fund' established in 1999 is a good illustration. It allows scientists in the prioritised sectors, among which biotechnology and ICT were the first to benefit, to buy-out teaching and administration time to conduct research leading to academic entrepreneurship. The fund was formed from contributions by Scottish Enterprise, the Scottish Executive and SHEFC. Notably, no private co-funding was committed to the fund. Scottish Enterprise estimated in late 2002 that 14 biotechnology projects had been funded and that, since March 1999, 28 new biotechnology companies have been created, equivalent to a growth rate of 30 per cent per annum. This compares favourably with the European average of 17 per cent per annum over the same period. Scotland is home to 20 per cent of the biotech companies in the UK and is recognised as one of the fastest-growing regions for start-ups. Thus far, policy to support commercialisation of bioscience has been successful; also, we have seen that Scotland has received major funding, including both Research Council and Wellcome Trust grants in support of its leading university centres of excellence and their research. Scotland now has a strategic science policy and it remains to be seen if the effectiveness shown without one can be enhanced consequentially.

4.3 Science Strategy: Science-led Strategy from Above

The last case to be explored is that of Finland, a small country that has emphasised the importance of developing centres of expertise in its regions, supporting university-centred basic research, commercial exploitation and cluster-building in biotechnology as it did with global success in relation to ICT and the rise of Nokia to global prominence in mobile telephony. The model is also one in which foresight and envisioning play a role in bringing about a consensus among business, academia and industry to invest in centres of expertise in locations that already show some comparative advantage. Centres of expertise in biotechnology arose from a Ministry of Education national research programme on biotechnology in 1987.

The aim was to develop four regional centres of biotechnological expertise by 1992, planned to be affiliated to those Finnish universities assessed to have the appropriate potential. The selected centres were at Helsinki, Turku, Kuopio and Oulu. The programme was evaluated and continued in 1996, then extended to 2000. Financing came from the Ministries of Trade and Industry, Agriculture and Forestry, and Social Affairs and Health as well as Education. Other centres were added such as Tampere and Seinajoki. The arrangement for enlargement of the network is one whereby, if a municipality is sufficiently committed to serious long-term investment in biotechnology—by funding a number of chairs in universities—for example, then provided it passes exacting tests of expertise, it can become eligible for designation and funding as a centre of expertise. This has led to an excess of demand for centre designation and the programme has been terminated in consequence. Centres specialise within biotechnology: Oulu, Turku, Tampere and Kuopio focus on medical research and co-operation with the pharmaceutical industry; Helsinki and Seinajoki specialise in agro-food biotechnology and some agro-food R&D is also performed in Kuopio, Oulu and Turku (Academy of Finland, 2002).

Tekes, the state technology agency has invested some $90 million in biotechnology, some 27 per cent of its total investment portfolio. The centres of expertise programme receives $4.1 million annually from Tekes and the Academy of Finland. Thus some 40 per cent of these two agencies' budgets is in support of biotechnology. Also, the National Programme for Research on Biotechnology, begun in 1988, invests an annual amount of some $13.5 million in biotechnology. Further expenditure on the genome research programme and the cell biology research programme attract $4.5 million and $1.8 million annually over 6- and 3-year programme periods respectively. In 1993, the Ministry of Education set a new centres of excellence standard, seeking to identify 10 'top units'. By 2000, 26 had actually been established of which 9 are in biosciences and biotechnology. In March 2001, a further $102 million rising to $151 million by 2006 was committed by Tekes, the Academy of Finland, Sitra (Finnish national R&D Fund), Finnish Bioindustries and a substantial group of pharmaceuticals companies to 'Medicine 2000' addressing biomedicine, medicine development and pharmaceutical development research and technology.

Finland's commitment to evolve a strong biosciences and biotechnology capability is remarkable, with proportionately comparable shares of total national R&D budgets (some 40 per cent) as the US. The fact that its agro-food firms are responsible for nutraceuticals innovations like the anti-cholesterol product Benecol (Raisio Ltd), lactobacter drinks and UHT infant food (Valio Ltd), and xylite sweeteners (Danisco–Cultor Ltd) suggests where current strength lies. Orion Pharma and Orion Diagnostica are the two leading biotechnology players, the former having the leading Parkinson's treatment

Comtess newly released, the latter targets the global point-of-care (POC) market for *in vitro* diagnostic products. Orion Pharma collaborates with all the Finnish centres of excellence, but particularly the regional centres at the Universities of Helsinki and Kuopio, the Helsinki Biotechnology Institute and, increasingly, with the Universities of Tampere and Oulu—the latter also being Orion Diagnostica's main research partner. The Finnish national innovation system is highly integrated but state-led with a knowledge value chain involving the Finnish Academy funding basic research, Sitra funding R&D, VTT conducting research and technology transfer, Tekes funding technology development and centres of excellence in universities working directly with large firms, start-ups and spin-offs in clusters on state and locally funded science and medical technology parks like Hermia at Tampere, Oulu Technopolis, and Medipark or DataCity and BioCity at Turku. In the report on Finnish life sciences by Tulkki *et al.* (2001), these regional innovation systems are presented as worked models of the Finnish view of the functioning of Silicon Valley. The key difference is the involvement of large firms and public investment in the commercialisation process, substituting for an arrested market for key innovation support services. In this respect, it has been influential upon the German regional biotechnology clusters commercialisation programme BioRegio that similarly sought a 'corporatist' version of the 'basic research–academic entrepreneurship–venture capital' model that was pioneered in California (Dohse, 2001).

5. Conclusions

This paper started with a question about the existence and observability of a new phenomenon, regional science policy. Its likelihood was implied by a number of important changes in global politics (ending of the Cold War), scientific research funding (major transfers from defence to healthcare), knowledge production (Mode 1 to Mode 2) bioscientific research approach (molecular biology), drug research methodologies (chance discovery to rational drug design), R&D leadership (big pharma laboratories to university centres of excellence) and innovation leadership (big pharma to DBF clusters). So developed are these relationships that inevitable ethical clashes have arisen. Thus, troubled US firm Bristol–Myers Squibb, in trying to forge closer DBF links, invested $2 billion in 20 per cent of ImClone to access Erbitux, a colon cancer drug. One of ImClone's key shareholders was Martha Stewart, doyenne of US home-making who in 2004 was found guilty of insider dealing. The problem arose when FDA approval was withheld due to faulty clinical trialing by ImClone. This case has caused big pharma to question its absorptive capacity; it has become mainly marketer-distributor to DBFs like ImClone, or Celltech with a similar deal with Pharmacia (now Pfizer, which in 2003 halted the deal), and Isis Pharmaceuticals with Eli Lilly. Attempts by Bristol–Myers to take over the development due diligence function with ImClone failed.

Clearly, with some 500 DBFs world-wide researching 1300 compounds for new biotechnology products, it is no surprise that up to 30 per cent of big pharma R&D budgets are now spent on alliances with extramural partners when the top 20 pharmaceuticals firms in 2001 spent $28 billion on intramural R&D for a yield of only 28 new drug approvals. Pfizer, currently the world's largest pharmaceuticals firm, has over 1000 alliances with DBFs and universities in response to the drought. So the knowledge-based clusters and the university or research institute centres of excellence at their hearts continue to be the pacemakers in molecular bioscience research and rational drug design. The paper showed also how changes in funding regimes for healthcare, diminishing the traditional 'free-rider' system of clinical research in hospitals in favour of development of clinical centres of excellence was hastening this process. Moreover, the vast amounts of Research Council and foundation research funding for centres of excellence accelerate it even further. Of course, such regional clusters, drawing on national funding to meet global market demand, are by no means ubiquitous. This is because, abundant though funding is, it is increasingly excellence-driven when it comes to funding allocations. Under such circumstances, alliance and partnership-based cluster governance have been shown to be an asset and the functional presence of regional innovation systems with the full knowledge value chain in place and the lobbying and grantsmanship expertise that comes with a sophisticated science and innovation support system are invaluable.

While regions became familiar with the importance of regional innovation systems and strategies in the 1990s (Cooke, 2001c), the current evolutionary position in Life Sciences research requires learning to apply those skills to creation of the infrastructure of excellence that provides the foundation for regional technological systems—namely, strong and varied basic and applied research capabilities. The logic of this points to the future rise of the formulation and implementation of regional science policy. This is evolving as in the UK where two northern English regions have established Regional Science Councils and one, the North West, published its Regional Science Strategy in 2002. Finland decentralised development through its science strategy; Scotland built from below. Both are peripheral, with relatively weak market mechanisms but a strong science base. Each has developed focused science policies with strong public funding targeted at a few world-class Life Sciences sectors. North Carolina moved to the kind of foresight-led, envisioning, stakeholder plus action leadership process pioneered in Massachusetts, then adopted in Scotland and later in Germany's regional cluster solution to its biotechnology innovation deficit, BioRegio. National funding bodies have to respond by making more transparent the allocation of research funding, as demanded in Scotland's science strategy. They can devolve regional funds or designate annual *tranches* for regional science development. Most scientists react in horror to this citing criteria of excellence and equivalence in regard to research grants, infrastructure funding or investment in the ever-developing centres of excellence located in regional clusters. Safeguards would be needed to prevent the target-inflation and excessive spread of investment revealed in the Finnish programmes. But equally, if regions show enterprise in mobilising strategic capabilities, they merit appropriate reward for so doing. This exists in the US Experimental Program to Stimulate Competitive Research (EPSCoR) that assists 22 less favoured states by accepting marginally lower-scoring science grant applications than elsewhere (www.her.nsf.gov/epscor). Regional science policy is beginning to prove a key precondition for the fulfilment of regional development visions in the knowledge economy.

Notes

1. For the UK, results from the DTI study of clusters reveal extraordinarily high location quotients for biotechnology businesses in Cambridge and Oxford, also the main university and other public R&D locations (DTI, 2001). In the US, the Brookings Institution reports (for example, Cortright and Mayer, 2002) consistently show there to be some nine main Life Sciences clusters in the US. In Canada, the work of Niosi and Bas (2001) shows high concentrations and intra-cluster innovative interaction in Montreal and Toronto. Comparable results exist for Germany (Kettler and Casper, 2000); France (Lemarié *et al.*, 2001), Sweden (VINNOVA, 2001) and Israel (Kaufmann *et al.*, 2003). The sample of one regional biotechnology 'cluster failure' written about by Orsenigo (2001) suggests Italy's *exceptionalism* regarding advanced technology activity among G7 countries—something underlined by Tavoletti (2004) who shows the highest unemployment for all skill/qualification categories (including workers having received *no* formal education at any level) in all Italian regions to be that of post-doctoral students.
2. To magnify the 'multiscalar' dimension here would require a further paper. An anonymous referee suggests that Bunnell and Coe (2001) and Mackinnon *et al.* (2002) "need to be acknowledged and cited" and this is duly done. They say it is wrong to emphasise the regional level and wrong to overlook regional specificity. This is an improvement upon positions such as that of Bathelt (2003)

in the same journal that only nations have specificity and that they are also *closed* systems. The geographical scale debate seems in danger of generating more heat than light. At a general level, Cooke *et al.* (2000) showed how multilevel governance of innovation in Europe has resulted in *innovation* measures often having been taken regionally by stimulus from European Commission policies, while member-states devoted more attention and many more resources to science policies. The regions in question are supralocal, subnational or meso-level entities with policy legitimacy in the relevant field. In the EU, for example, many, but not all, happen to be EU NUTS 2 regions—a statistical artifice in some cases, an isomorph of a legitimate meso-level entity in others. In yet others, NUTS 2 areas aggregate to a legitimate meso-level policy entity. Here, regions are meso-level policy entities with legitimacy for implementing innovation measures emanating supranationally, or from state-level (for example, in England) or from within the region (such as N. Ireland's 'Think, Create, Innovate' strategy). Economists who specialise in science, innovation and technology studies have little or no understanding of the presence of regions and their institutions as, albeit often weak, actors in innovation and, as this paper shows, tentatively in science policy too—hence the need to reiterate it. Geographers, of course, are more fortunate.

3. As a consequence of the disaster that resulted in deformed births after mothers had taken thalidomide to control morning sickness, trialling of candidate treatments goes through three lengthy phases. The first is usually mammalian, the second with small groups of patients with the drug's target disease, and the third a large sample of patients with controls that yield statistically valid results.
4. An anonymous referee indicates the unlikelihood of readers of *Urban Studies* being technically literate in bioscientific terminology. Hence, 'bioinformatics' involves capturing molecular or genetic information on databases that identify chemical compounds that may inhibit particular disease-causing molecular or genetic combinations. 'Biosciences' are those that involve study of medicine, biology, biochemistry, pharmaceutical and other treatments, medical devices and instruments, research and testing relevant to intervention with and improvement of human, animal and plant life. 'Big pharma' is a term used by academe, government and industry as an abbreviation of large pharmaceuticals companies. 'Biotechnology' involves tools like genetic engineering to create treatments that counter diseases (such as the example of Glivec in Table 1). 'Orphan drugs' are used in diseases or circumstances which occur so infrequently that there is no reasonable expectation that the cost of developing and making available a drug for such disease or condition will be recovered from sales of such drugs. Firms gain exclusivity and fee waivers for producing such drugs. 'High throughput screening' occurs when a compound interacts with a target in a productive way so that the compound then passes the first milestone on the way to becoming a drug. Compounds that fail this initial screen go back into the bioinformatics library from whence they came, perhaps to be screened later against other targets.
5. Centres of excellence are generally considered to be more exacting, higher-quality designations than centres of expertise. Although numerous governments, like that of Finland, began with the former, regional politics caused the latter to replace them so that more regions could qualify. This action is criticised further on in this paper.

References

ACADEMY OF FINLAND (2002) *Biotechnology in Finland*. Helsinki: Academy of Finland.

AMERICAN ASSOCIATION FOR THE ADVANCEMENT OF SCIENCE (1999) *US R&D Budget and Policy Programme*. Washington, DC: AAAS.

ASHCROFT, B., DUNLOP, S. and LOVE, J. (1995) UK innovation policy: a critique, *Regional Studies*, 29, pp. 307–311.

BATHELT, H. (2003) Growth regimes in spatial perspective 1: innovation, institutions and social systems, *Progress in Human Geography*, 27, pp. 789–804.

BEST, M. (2000) Silicon Valley and the resurgence of Route 128, in: J. DUNNING (Ed.) *Globalization, Regions and the Knowledge Economy*, pp. 459–484. Oxford: Oxford University Press.

BUNNELL, T. and COE, N. (2001) Spaces and scales of innovation, *Progress In Human Geography*, 25, pp. 569–589.

BURNELL, S. (2002) Biomed researchers applaud NIH budget, *UPI Science News*, 31 January, p. 12.

COOKE, P. (2001a) Biotechnology clusters in the UK, *Small Business Economics*, 17, pp. 43–59.

COOKE, P. (2001b) New economy innovation systems: biotechnology in Europe and the USA, *Industry and Innovation*, 8, pp. 267–289.

COOKE, P. (2001c) Regional innovation systems, clusters and the knowledge economy, *Industrial and Corporate Change*, 10, pp. 945–974.

COOKE, P. (2002) *Knowledge Economies*. London: Routledge.

COOKE, P. (2004) *Globalisation of biosciences: knowledge capabilities and economic geography*. Paper presented at the *Association of American Geographer's Centennial Conference*, Philadelphia, March.

COOKE, P. (forthcoming) Rational drug design, the knowledge value chain and bioscience megacentres, *Cambridge Journal of Economics*.

COOKE, P., BOEKHOLT, P. and TÖDTLING, F. (2000) *The Governance of Innovation in Europe*. London: Pinter.

CORTRIGHT, J. and MAYER, H. (2002) *Signs of Life: The Growth of Biotechnology Centres in the US*. Washington, DC: The Brookings Institution.

DICKEN, P. (2001) *Globalising processes and local economies*. Paper presented to the *Institute for Regional Innovation (Invent) Conference 'Global–Local: Regional Economic Development in an era of Globalisation and Localisation'*, Vienna, December.

DOHSE, D. (2001) Technology policy and the regions: the case of the BioRegio contest, *Research Policy,* 29, pp. 1111–1133.

DTI (DEPARTMENT OF TRADE AND INDUSTRY) (1999) *Biotechnology Clusters*. London, Department of Trade and Industry.

DTI (2001) *Business Clusters in the UK: A First Assessment.* London: DTI.

DUNNING, J. (ed.) (2000) *Globalization, Regions and the Knowledge Economy*. Oxford: Oxford University Press.

FARRAR, S. (2002) Sinn Fein backer pays for centre, *The Times Higher Education Supplement*, 8 February, p. 3.

GEREFFI, G. (1999) International trade and industrial upgrading in the apparel commodity chain, *Journal of International Economics,* 48, pp. 37–70.

GIBBONS, J. (2000) The role of Stanford University, in: C. M. LEE, W. MILLER, M. HANCOCK and H. ROWEN (Eds) *The Silicon Valley Edge: A Habitat for Innovation and Entrepreneurship*, pp. 200–217. Stanford, CA: Stanford University Press.

GIBBONS, M., LIMOGES, H., NOWOTNY, S. *ET AL.* (1994) *The New Production of Knowledge.* London: Sage.

GRIFFITH, V. (2002) Biotech groups stuck on a rollercoaster ride, *Financial Times*, 1 February, p. 29.

HUMPHREY, J. and SCHMITZ, H. (2000) *Governance and upgrading: linking industrial cluster and global value chain research.* Working Paper 120, Institute of Development Studies, Sussex University.

IRVINE, J. and MARTIN, B. (1984) *Foresight in Science: Picking the Winners*. London: Pinter.

KAUFMANN, D., SCHWARTZ, D., FRENKEL, A. and SHEFER, D. (2003) The role of location and regional networks for biotechnology firms in Israel, *European Planning Studies,* 11, pp. 823–840.

KETTLER, H. and CASPER, S. (2000) *The Road to Sustainability in the UK and German Biotechnology Industries*. London: Office of Health Economics.

LAGENDIJK, A. (2001) Scaling knowledge production: how significant is the region?, in: M. FISCHER and J. FRÖHLICH (Eds) *Knowledge, Complexity and Innovation Systems*, pp. 79–100. Berlin: Springer-Verlag.

LATOUR, B. (1998) From the world of science to the world of research?, *Science,* 280, pp. 208–209.

LAWRENCE, S. (2001) *Health Funding Update.* New York: The Foundation Centre.

LEMARIÉ, S., MANGEMATIN, V. and TORRE, A. (2001) Is the creation and development of biotech SMEs localised? Conclusions from the French case, *Small Business Economics,* 17, pp. 61–76.

LUNDVALL, B. and BORRÁS, S. (1997) *The Globalising Learning Economy.* Luxembourg: European Commission.

MACKINNON, D., CUMBERS, A. and CHAPMAN, K. (2002) Learning, innovation and regional development: a critical appraisal of recent debates, *Progress in Human Geography,* 26, pp. 293–311.

NIOSI, J. and BAS, T. (2001) The competencies of regions: Canada's clusters in biotechnology, *Small Business Economics,* 17, pp. 31–42.

NOWOTNY, H., SCOTT, P. and GIBBONS, M. (2001) *Re-thinking Science*. Cambridge: Polity.

ORSENIGO, L. (2001) The (failed) development of a biotechnology cluster: the case of Lombardy, *Small Business Economics,* 17, pp. 77–92.

ORSENIGO, L. PAMMOLLI, F. and RICCABONI, M. (2001) Technological change and network dynamics: lessons from the pharmaceutical industry, *Research Policy,* 30, pp. 485–508.

PAMMOLLI, F., RICCABONI, M. and ORSENIGO, L. (2000) Variety and irreversibility in scientific and technological systems, in: A. NICITA and U. PAGANO (Eds) *The Evolution of Economic Diversity*. London: Routledge.

POWELL, W., KOPUT, K. and SMITH-DOERR, L. (1996) Interorganizational collaboration and the locus of innovation: networks of learning in biotechnology, *Administrative Sciences Quarterly,* 41, pp. 116–145.

SCOTTISH EXECUTIVE (2001) *A Science Strategy for Scotland.* Edinburgh: Scottish Executive.

SENKER, J. and ZWANENBERG, P. VAN (2001) *European biotechnology innovation systems.* Final Report to EU-TSER Programme, Science Policy Research Unit, University of Sussex.

SWANN, P., PREVEZER, M. and STOUT, D. (Eds) (1998) *The Dynamics of Industrial Clustering*. Oxford: Oxford University Press.

Small Business Economics (2001) Special Issue ed. G. Fuchs, *Small Business Economics,* 17, pp. 1–153.

TAVOLETTI, E. (2004) Higher education and high unemployment: does education matter? An interpretation and some critical perspectives, in: P. COOKE and A. PICCALUGA (Eds) *Regional Economies as Knowledge Laboratories,* pp. 75–86. Cheltenham: Edward Elgar.

TEICH, A. (1999) The science of health and the health of science, *Information Impacts*, 12 November.

TULKKI, P., JÄRVENISVU, A. and LYYTINEN, A. (2001) *The Emergence of Finnish Life Sciences Industries*. Helsinki: Sitra.

VINNOVA (2001) *The Swedish Biotechnology Innovation System*, Stockholm, VINNOVA.

WELLCOME TRUST (2002) Clinical research facilities: an introduction, *Wellcome News, 12 February.*

ZUCKER, L., DARBY, M. and ARMSTRONG, J. (1998) Geographically localised knowledge: spillovers or markets?, *Economic Inquiry,* 36, pp. 65–86.

‘Embryonic’ Knowledge-based Clusters and Cities: The Case of Biotechnology in Scotland

Joseph Leibovitz

[Paper first received, March 2003; in final form, July 2003]

1. Introduction

In recent years, the concept of industrial clusters has gained substantial interest from academics, policy-makers and commentators. Cluster-inspired thinking has influenced a wide range of academic research in management, economic geography and urban studies (Brown and Duguid, 2002; Humphrey and Schmitz, 2002; Malecki, 2002; Porter, 2000). The recent study commissioned by the Department of Trade and Industry (DTI, 2000a, 2000b) to examine the extent to which clusters exist and function in the UK economy exemplifies the importance that policy-makers attribute to clusters. It is widely believed that industrial clusters can help to improve the performance of urban and regional economies by strengthening the competitiveness of firms, thereby generating growth, employment and productivity gains (Porter, 2001).

Biotechnology has been particularly attractive to policy-makers because of its association with the ‘knowledge-based economy’. It is often perceived as a growth industry that has the potential to reposition national and regional economies competitively and, as a result, has been the subject of a range of policies and initiatives coming out of European Union institutions, national governments and regional and local economic development agencies (Biotechnology

Scotland, 2003; Cooke, 2002; Niosi and Bas, 2001). In the UK, the biotechnology clusters initiative was announced in the 'Competitiveness' White Paper (DTI, 1998) and was further supported by Lord Sainsbury's report (DTI, 1999) and the initiation of the Biotechnology Exploitation Platform Challenge and the Biotechnology Mentoring and Incubator Challenge which aim to provide expert and financial support to the industry and to facilitate the transfer of knowledge from research institutes to the commercial world. The attraction of knowledge-based industries such as biotechnology has been particularly pronounced in less-favoured regions. It is believed to have the potential to enable such regions to take the 'high road' to economic development, contributing to higher value-added activities, wages and standard of living. As the Scottish Executive's Minister for Enterprise and Lifelong Learning has put it

> In Scotland, we cannot compete through low cost or a low skilled workforce. We must promote our most innovative businesses and encourage people to develop their science and technology ideas. Our biotechnology cluster has a key role to play in doing this. Its innovation can help drive forward the knowledge economy, creating wealth and high-quality jobs (Scottish Executive, 2000).

While the cluster theory and policy approach has been considerably influential, the spatial qualities, characteristics and dynamics of clusters are less clearly understood. In particular, the relationship between clusters and urban economic change requires further theoretical and empirical scrutiny: do cities play an important role in the development of clusters? Are there defining urban characteristics that may support or hinder the development of clusters? What is the potential of clusters to contribute to urban economic development? In this paper, I examine the major locational dynamics affecting the biotechnology industry in Scotland, with a particular focus on Scotland's Central Belt (that is, the region containing the metropolitan areas of Edinburgh and Glasgow).[1] The paper poses several interrelated questions: What are the major processes driving the development of biotechnology firms in their location? What role does the urban environment play in the location and competitiveness of those industries? And what is the balance between local or regional linkages and external connections? In addition, the paper raises issues pertaining to contemporary approaches in local and regional economic development policy.

2. Clusters, Scale and Place: An Analytical Framework

The starting-point for the analysis is Porter's now well-known arguments regarding the economic utility of industrial clusters. Porter defines clusters as

> geographical concentrations of interconnected companies, specialised suppliers, service providers, firms in related industries, and associated institutions ... in a particular field that compete but also co-operate (Porter, 2000, p. 15).

Fundamental to Porter's approach is the argument that cities and regions which exhibit cluster tendencies are likely to be more economically competitive because clusters are said to foster innovation, productivity and sustained rates of employment growth. However, weaknesses in Porter's conceptualisation include its ambiguity or silence regarding the specific processes and factors that encourage spatial agglomeration; the various spatial scales at which clustering processes operate; and the extent and importance of external relationships that urban and regional economies develop (Martin and Sunley, 2003). One danger is that such ambiguities would translate themselves into blinkered policy formulae and to attempts to imitate, emulate and import models of (seemingly) successful cluster development from elsewhere. Yet industrial development trajectories are characterised by considerable diversity and cluster development—to the extent that it is desirable—

needs to be viewed with sensitivity to the spatial and historical particularities of different economic sectors.

Here, the recent contribution by Gordon and McCann (2000) is useful in outlining at least three 'ideal types' of industrial clustering as a way of capturing the 'spatiality of clusters' and the geographical dynamics that may underpin the development of clusters (see also Malmberg and Maskell, 2002; Parr, 2002).[2] The first type, associated with the notion of *agglomeration economies*, emphasises the economic utilities (cost reduction, greater efficiency gains, etc.) that accrue from the geographical concentration of firms (Chinitz, 1961). The benefits include the development of a local pool of specialised labour, the provision of shared inputs and the maximisation of flows of information and ideas (Krugman, 1996b). An important distinction is drawn between 'urbanisation economies' which refer to the advantages emanating from the presence of diverse urban assets (major infrastructure, airports, services, educational institutions and so on) that benefit a wide range of economic activities, and 'localisation economies' which are associated with the geographical agglomeration of specialised services and sets of skills, and which confirm advantage to specific industries or sectors.

The second ideal type, the *industrial complex,* is characterised by relatively stable trading linkages between firms, where localised backward and forward linkages are prevalent (Aydalot and Keeble, 1988). Spatial concentration allows for minimising transaction costs, with emphasis on the quality of transport and communication means. The emergence of just-in-time and other methods of flexible production is said to have reinforced the logic of spatial proximity in supplier–customer relations (although not in all sectors of the economy), even in the face of technological change and developments in communication systems (Gertler, 2001). Particular advantages often attributed to the local industrial complex model include higher degrees of continuity, stability, predictability and planning in the linkages between firms as opposed to the more atomised traditional agglomeration. This might enable firms to establish long-term horizons in their strategic decision-making and invest in labour and machinery, thus strengthening their competitive position in the market (Bellardi, 2002; Garafoli, 2002; Whitford, 2001).

Thirdly, the *social network* model, based on the work of Granovetter (1985) and other economic sociologists (Best, 1990; Piore and Sabel, 1984), places the premium on close collaboration and trust between firms and related institutions, so that market failure can be overcome, risk can be spread and innovation and learning are facilitated through collaboration (Cooke and Morgan, 1998). Geographical concentration is said to be important to the development and reproduction of such relationships as trust is strengthened by local common identity and tradition, and by spatial proximity (Storper, 1997).

While these three 'ideal types' focus on on-going and *local* relationships driving growth, it is also important to consider the role of three particular other possible factors underpinning the dynamics of clusters. First, the previous models focus on a 'snapshot' of economic linkages and as a result can produce analysis that is static in nature. In other words, they neglect the historical and path-dependent nature of urban economic change (Krugman, 1996a; Martin and Sunley, 1996; Massey, 1984). Existing industrial structures reflect forces of inertia, previous investment decisions, the gradual creation of certain sets of skills and the potential development of specialised local labour markets, and 'accidents' of history. Studies of innovation and research on the evolutionary nature of economic change have shown that innovation is often a result of a cumulative process where the competencies of firms and institutions are heavily reliant on past legacies, chance discovery, organisational exploitation and exploration, and 'critical junctures' in the economic history of the firms and regions (Arthur, 1994; Danneels, 2002; Nelson and Winter, 1982; Prevezer, 2001).

Related to this is a second important element which explains the relationship between cities and cluster development—namely, the role of public-sector investment and public policy in supporting (or hindering) the growth of clusters. In particular, public-sector 'anchors' may have a crucial role to play in 'kick-starting' certain cities and regions onto specific trajectories of economic development. For example, government laboratories and research institutes, major hospitals and universities may play a particularly significant role in providing the skill and science base for cluster growth in biotechnology. While this may not be a sufficient condition for the 'success' of clusters, it nonetheless endows cities and regions with important physical and social infrastructure which makes the conditions for clusters more favourable (Lambooy, 2002; Lambooy and Boschma, 2001; Smith *et al.*, 2000).

Thirdly, there may be a tendency in discussion about clusters to overemphasise the role of localised relationship. For relatively small and exposed economies such as Scotland's, the significance and possible implications of *external* linkages need to be considered. For certain clusters and industrial sectors, linkages which are international in scope may be as significant, if not more so, than local ties. If major customers driving the demand for services and products offered by producers, services and research facilities of a cluster are a few multinational enterprises, then the business strategies of local firms will need to be geared towards establishing networks and links which are international in scope. The economies of cities and regions, and of clusters, thus need to be analysed by paying attention to the spatial organisation of production chains and its possible implications (Amin and Thrift, 2002; Oinas and Malecki, 2002; Scott, 2001). On the one hand, external control may limit the capacity of local firms to control their own destiny and may make them vulnerable to take-overs, acquisitions and mergers. By implication, it may increase the vulnerability of cities and regions to mobile capital. On the other hand, integration into international production networks may help firms to penetrate markets, to access resources and to achieve internal economies of scale and scope (Brown and Duguid, 2002; Cooke, 2003; Humphrey and Schmitz, 2002).

These three additional perspectives, in conjuncture with the three 'ideal types' of cluster development, present an analytical framework which captures the locational factors shaping the spatiality of clusters whilst sensitising analyses to the particular path-dependent nature of urban and regional economic change. This analytical framework is particularly appropriate in cases when clusters are in their early growth (or 'embryonic') stage, such as that of biotechnology in Scotland.

3. Biotechnology in Scotland: Defining Characteristics, Geographical Patterns and Path-dependent Evolution

Biotechnology has a long history and its roots are often traced to the fermentation of foods and drinks, activities which remain important in terms of outputs and value (Hacking, 1986; McKelvey, 1996). The modern rise of biotechnology is associated with the advent of antibiotics during and following the Second World War (Saliwanchik, 1988) and, more recently, with a range of "enabling technologies involving the practical application of living organisms in products, processes and services" (Scottish Enterprise, 1999, p. 2). One of its defining properties is that it creates generic 'technology platforms' that can alter the potential of goods and services across a wide range of industries. Biotechnology is still, however, in an early stage in terms of industrial maturity and life cycle (CRIC, 2001). It typically serves five key markets: healthcare, chemicals, veterinary, food/agriculture and environmental industries. In addition, a range of services and related support institutions are closely associated with the biotechnology industry, to include research and innovation-support institutions; potentially sophisticated end-user companies that might form joint venture and strategic alliances to allow for market penetration; support and supply firms; and a range of legal, financial and technical

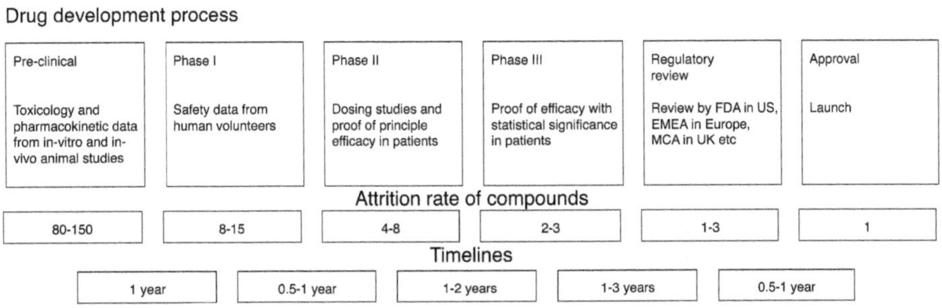

Drug development process

Pre-clinical	Phase I	Phase II	Phase III	Regulatory review	Approval
Toxicology and pharmacokinetic data from in-vitro and in-vivo animal studies	Safety data from human volunteers	Dosing studies and proof of principle efficacy in patients	Proof of efficacy with statistical significance in patients	Review by FDA in US, EMEA in Europe, MCA in UK etc	Launch

Attrition rate of compounds

80-150	8-15	4-8	2-3	1-3	1

Timelines

1 year	0.5-1 year	1-2 years	1-3 years	0.5-1 year

Figure 1. The time-scale of the drug development process. *Source*: *Financial Times* (2000).

support organisations, both public and private.

A number of key trends and a set of corresponding challenges typify the industry and these include

—high barriers to entry borne out of high cost, high risk and lengthy 'idea to market' processes—by some estimates, it is not unusual for 8 years to pass from the pre-clinical stages of the drug development process to the final approval stage (Figure 1);
—consolidations, mergers and acquisition on national and international scales, so that the market in pharmaceuticals, for instance, is dominated by a small number of multinational companies;
—pressures towards product diversification in order to minimise risk;
—political sensitivities and suspicious public attitudes towards the industry;
—changing demographics leading to increasing demand for therapeutic and diagnostic developments;
—technological convergence;
—the importance of new business formation and the increasing demand for 'bio-analysts'; and
—the significance of intellectual property regulations in knowledge transfer and research commercialisation (Newell, 2002; Oakey, 1993; OECD, 1996; Rifkin, 1998; Saliwanchik, 1988).

The biotechnology cluster in the UK contains some 230 core firms employing 17 000 people, with almost 75 per cent of employment concentrated in the South East and Eastern regions of Britain (DTI, 2001a). A recent report, commissioned by the Economic and Social Research Council argues that "the UK is widely perceived to be already relatively successful in the economic performance of its fledgling biotechnology activity" (CRIC, 2001, p. 4), based on the competitiveness of its specialist biotechnology firms and large pharmaceutical companies, and scientific advances such as the Human Genome project (for instance, the revenues of British biotech firms in 2000 were over 2 billion euros, compared with 1.3 billion euros generated by Swiss firms, the second-most successful group; *Financial Times*, 2001).

In Scotland, the Department of Trade and Industry has identified 24 biotechnology firms (11 per cent of the UK total), employing 1300 people (7 per cent of the UK total; DTI, 2001b) and has defined the cluster as 'embryonic' (that is, small in relation to the UK and situated in an early development stage). A more expanding definition employed by Scottish Enterprise includes support and supply services, medical device firms and academic and research institutes, and estimates the number of those employed in the cluster at some 24 000, although this has to be treated carefully because it tends to exaggerate the picture. Cloning technology, stem cell research, clinical trials and neuroscience research are among the cluster's particular strengths. Since the late 1990s, the number of core biotechnology firms in Scotland has grown steadily, from 49 in 1999 to 97 in 2002 (Table 1). The majority of the

Table 1. The Scottish biotechnology cluster: number of organisations and jobs (in brackets)

Year	Core biotech firms	Support and supply	Medical devices	Academic and research
1999	49 (3211)	93 (2976)	90 (4000)	25 (2542)
2000	74 (3677)	163 (4969)	92 (4771)	53 (5013)
2001	86 (3897)	192 (5571)	99 (6714)	51 (8225)
2002	97 (4152)	227 (6403)	101 (6756)	52 (8099)

Source: Biotechnology Scotland (2003).

Scottish biotechnology companies are small, averaging 49 employees (Griffin, 2001). Within Scotland, the biotechnology sector displays an urban orientation, with Edinburgh firms accounting for 27 per cent of the total, while Glasgow's and Dundee's shares stand at 14 and 13 per cent, respectively (Table 2).

There is little doubt that support for clusters in general, and biotechnology in particular, has occupied an important place within economic and industrial policy discourse in Scotland. This has been reflected in an impressive range of initiatives and policy measures designed to create an institutionally 'thick' environment within which biotechnology could flourish. Interventions and strategies have also been formulated at a range of geographical scales, from national economic strategies through to regional and localised initiatives. Scotland's national economic development framework, entitled *Smart, Successful Scotland,* has reflected a qualitative shift in the premises of the nation's economic growth from a strategy emphasising the attraction of inward investment on the basis of low-cost advantages, low value-added and labour-intensive activities, to one which attempts to redress issues of productivity, competitiveness and innovation (Scottish Executive, 2001). Support for biotechnology occupies an important place within this framework and the industry seems to play a symbolic role in policy discourses that stress the importance of clusters as 'entities' through which the Scottish economy can assume "greater entrepreneurial dynamism and creativity", increase the level of e-businesses, foster higher degrees of research commercialisation and innovation, and achieve "global success in key sectors" (Scottish Executive, 2001, pp. 10–11).

More specific support for 'key sectors' such as biotechnology has been entrusted to Scottish Enterprise, the nation's economic development agency. As stated in its Action Plan, the cluster approach is predicated on the premise that clusters can deliver the key drivers of economic competitiveness in a knowledge-based economy

> Scottish Enterprise is assisting a number of key industries to enhance their international competitive position through the application of a cluster based approach. This approach focuses on three primary drivers: competitiveness—productivity/efficiency, innovation, and new firms formation/commercialisation. This approach also helps us develop strong national and international cluster networks and increase recognition for Scotland as a centre of excellence in specific key industries (Scottish Enterprise, 2003a, pp. 23–24).

Furthermore, initiatives take more localised forms supported by Scottish Enterprise's local agencies. For instance, investment to the tune of £1.4 million is expected in order to develop a Centre for Biomedical Research in Edinburgh as 'a life science knowledge hub' accompanied by other elements of business and physical infrastructure. A Science Park is also being planned for the City of Dundee, arguing that "it will encourage millions of pounds in international biotechnology investment and facilitate spin outs from research (Scottish Enterprise, 2003a, p. 27), while Glasgow's West of Scotland Science Park is

Table 2. Geographic distribution of biotechnology firms and related organisations in Scotland, 2002

	Number of core firms	Number of organisations	Total employed
Edinburgh and Lothian	23	119	8702
Glasgow	12	86	3874
Tayside (Dundee)	11	47	3502
Lanarkshire	5	29	1187
Forth Valley	7	28	1262
Ayrshire	8	22	1932
Grampian	6	22	911
Dunbartonshire	3	20	372
Renfrewshire	1	19	855
Fife	2	16	537
Highlands and Island Enterprise	4	11	830
Scottish Borders	3	8	312
Dumfries and Galloway	1	1	130

Note: Area boundaries correspond to those of Scottish Enterprise 'Local Economic Companies' administrative units.
Source: Scottish Enterprise.

promoted as Scotland's leading location for developing bioscience business. As Figure 2 illustrates, localised support for biotechnology has taken the shape of considerable investment in physical and social infrastructure and in promotional efforts.

Despite the excitement surrounding the potential growth and significance of Scotland's biotechnology cluster, it should be noted that its relative contribution to urban economies is still limited, employing (even according to the more expansive definition) some 8700 people in Edinburgh and less than 4000 in Glasgow.

Scotland's science base provides an important foundation for the biotechnology industry. It is overrepresented in the UK in terms of its share of university graduates in the life sciences, 'producing' a yearly average of 18 per cent of the UK's postgraduates in these fields (Griffin, 2001). Public-sector anchors have played an important role in the development trajectory of the biotechnology cluster in Scotland, influencing its scale, scientific expertise, research capacity, reputation, timing and location (Collier, 2000). The example of the Roslin Institute in Edinburgh is illustrative. The Institute ventured into biotechnology in the early 1980s when it shifted its research orientation from farm animal breeding and production to technologies which could be used to create 'transgenic' animals which produce human proteins in their milk. The lack of interest, at the time, from the big pharmaceutical companies in commercialising this area of research led to the Institute's setting of its spin-out company, Pharmaceutical Proteins Ltd (PPL) in 1987 to take the technology to markets. PPL gained world fame in 1997 when it created Dolly the Sheep, the first mammal cloned from an adult cell. By 2000, PPL Therapeutics had established its position as a specialist in the production of therapeutic proteins in the milk of transgenic animals, was engaged in collaboration with the German pharmaceutical firm Bayer in order to carry a lung disease drug through clinical trials and had extended the potential geography of its operation to include the manufacturing and commercialisation of its drugs internationally, research and farming facilities in New Zealand and Virginia, and investing in a new manufacturing plant near Edinburgh.

A further example of a spin-out of the Roslin Institute's cloning technology is Roslin Bio-Med, created in 1998 with addi-

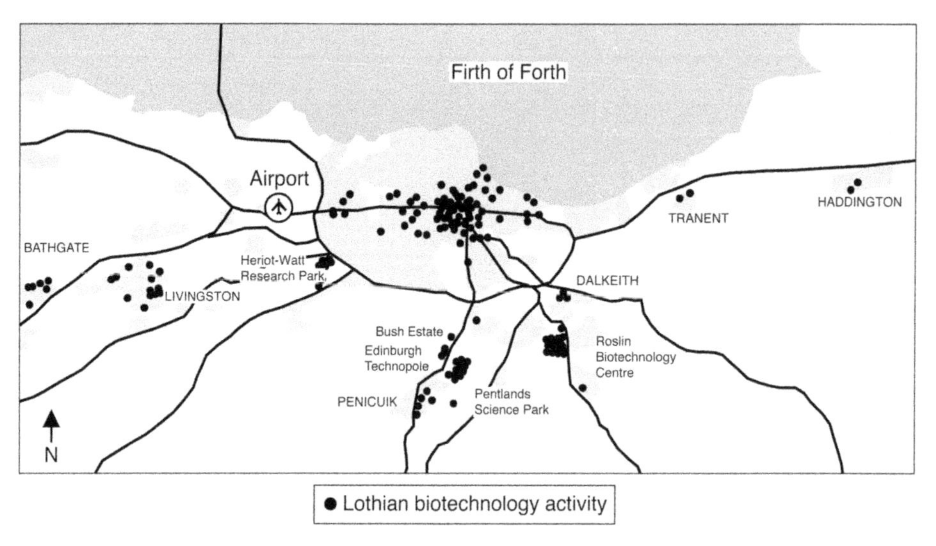

Figure 2. Edinburgh's biotechnology cluster, taken from Scottish Enterprise's promotional literature.

tional investment by the venture capital firm 3i. By 1999 it was acquired by the US biotech company Geron, helping to generate an investment of £12.5 million by the parent company towards research on nuclear transfer within the Roslin Institute, enabling the Institute to increase the size of its research team and focus its research strategy. Other activities by the Institute include its involvement and investment in the development of the Roslin BioCentre to the south of Edinburgh as a site for life science companies, thus adding a property development dimension to its scientific contribution.[3] While public-sector anchors have played an important role in influencing the timing, the scale and the scientific foundations of biotechnology in Scotland, the role of cities and spatial processes requires further elaboration.

4. Clustering and Spatial Dynamics

4.1 Agglomeration Economies

Agglomeration economies work as the most significant clustering factor in the case of biotechnology industries. In particular, Glasgow and Edinburgh are said to provide a pool of educated and high-quality labour force, largely generated by university graduates. However, this assertion deserves further qualification. The role of the universities and public-sector anchors as centres of research excellence and providers of well-educated scientists is clearly significant in providing the essential scientific infrastructure so central to biotechnology firms. Their effects can be divided to 'substantive' and 'reputational'. Substantive effects accrue through the generation of scientific research, the training of high-quality scientific research staff, the spinning-out of technology, ideas, expertise and firms, and the receipt and distribution of research funding. Reputational effects should not be underestimated, especially for clusters in their early growth stages. Here, interview with firms revealed that the 'aura' and prestige associated with the presence of key institutions such as Roslin help to place Scotland's biotechnology on the map of potential business partners, suppliers, customers and investors.

At the same time, the embryonic nature of the industry in Scotland means that agglomeration economies in themselves are not sufficient in providing a critical mass of skilled and experienced labour in a number of occupations. For many firms, 'people fresh out of universities', as they have put it, do not possess the appropriate experience, especially in management capacities. For instance, even on the scientific side, a number of firms have faced difficulties in recruiting experienced scientific staff and have had to expand the geography of their recruitment efforts to outside Scotland.

In addition, the very specialised nature of many of Glasgow's and Edinburgh's biotechnology firms means that general scientific know-how is judged to be insufficient for their needs. While this concern might raise questions as to the capacity and willingness of local firms to invest in the training of labour, it also points to the difficulties associated with the growth of an embryonic cluster: the lack of critical mass means that the supply of experienced labour is limited, whilst the potentially skilled workers might perceive the area as providing insufficient career opportunities in biotechnology. Importantly, however, interviewees have stressed that this somewhat problematic image of the Scottish biotechnology sector is shed once workers do decide to relocate, primarily because the quality of life in Scotland is perceived to be high compared to that in the south of England. Significantly lower levels of congestion, relatively lower property prices and recreation opportunities have all been cited as important advantages in Glasgow's case in particular.

The picture of labour recruitment and retention is one of diverse experiences, based on the maturity, the stage of development and areas of specialisation of firms. For firms that are already 'established' (that is, in a rather more advanced stage in terms of product life cycle, sources of finance and so on), one of the major benefits is the quality of labour already employed within them. They perceive this labour to be highly skilled and significantly valuable to the competitiveness of the firm, thus placing a high premium on staff turnover. In some cases, long-term staff retention has been cited as one of the more fundamental factors in the survival and subsequent competitiveness of firms. This, in turn, has locational implications because such firms perceive their current location in the Glasgow and Edinburgh metropolitan areas as an important asset and a considerable source of competitiveness. As one executive noted

> In the workforce of about 50 people that we have, there are about 15 people who have worked with the company between 8 and 15 years. And that's a lot of experience. We've been fortunate that although we've experienced a lot of turbulent times that there's a backbone of people who have stayed with the company, and were able to use their experience to ensure that the quality of our products are maintained, and to help train new people ... in the correct way ... *So by far and away, our strongest advantage is the group of people that we've got here.*

Furthermore, in some cases when firms have been faced with decisions regarding the possibility of relocation to other areas (for instance, as a result of mergers and acquisition), the corporate decision has been to remain in the existing urban area. The example of the Glasgow-based Rhone Diagnostics is a case in point, which illustrates a wider trend. Although founded by French entrepreneurs in the late 1980s, faced with the option of relocating the firm to France, a decision was taken to remain and later expand the Glasgow facilities. The supply and high-quality of university graduates, lower overhead costs, good accessibility and proximity to Glasgow airport, as well as public-sector support, have all been important factors in the firm's locational decision (interview with a senior executive).

While agglomeration economies play a significant role in providing large pools of qualified labour, they do not tell the entire story of the geographies of recruitment. Skill levels are an important 'sorting' mechanism for the geography of recruitment. For the most part, low-skilled labour force such as cleaners and service providers are recruited locally. For higher-skilled labour force, however, especially in areas of cutting-edge research and expertise in biotechnology, the geography of recruitment extends far beyond local labour markets. In some cases, for firms that have reached certain levels of maturity and expertise, and relatively secured sources of finance, recruitment efforts are international in scope. In those cases, the search for the best quality of labour means an international scale of competition for workers. For several firms in Edinburgh, for instance, non-UK nationals constitute more than half of the workforce.

The recruitment geography of firms thus does not always bear close relationship to the dynamics of local labour markets. In the case of Edinburgh, for instance, recent research has raised concerns as to the possible effects of an overheating local labour market where demand for qualified labour outstrips supply (Turok and Bailey, 2003). Yet in the context of biotechnology, while inhibiting somewhat the ability to recruit locally, and increasing the overall cost of labour, this is not judged to be a major factor for internationally competitive firms. For those firms, competition in biotechnology markets is international in scope, and therefore competition on salary levels is judged against comparable locations world-wide, against which Edinburgh compares favourably (interviews with senior biotechnology executives). In addition, student placement schemes—operating on the basis of agreements between educational institutions and businesses—facilitate the utilisation of scientific expertise at lower costs.

However, tight property and labour markets might affect the long-term competitiveness of early-growth firms in Edinburgh, as many of them cannot afford to make special payment provisions to compensate for higher living costs. At the moment, innovative biotechnology firms have to rely on the quality and nature of their work, particularly in areas of cutting-edge biotechnology, clinical and therapeutic research, in order to attract and retain their skilled staff. In other words, firms in Edinburgh and Glasgow rely on the innovative and intellectually challenging aspects of their activities as sources of staff attraction rather than merely high salaries.

The development of specific science parks in Edinburgh and Glasgow is one particular off-shoot of agglomeration factors. Indeed, this has been one of the key policy instruments in supporting biotechnology in Scotland in recent years, with the West of Scotland Science Park in Glasgow and the Pentlands Science Park near the Roslin Institute in Edinburgh providing the primary examples. The particular advantages of these parks relate to the ability of firms to economise on service provision (with an important premium placed on safety and security issues due to the sensitive nature of biotechnology undertakings), accessibility advantages and the availability of adequate commercial space and tailored facilities (such as storage and laboratories).

There are also elements related to the prestige and image associated with science parks. They are considered attractive and pleasant environments in which to work, where the proximity to businesses in related activities (primarily technology-oriented firms) is said to contribute to a 'feel good' factor associated which such locations: "It does look the part", as one interviewee remarked in relation to the West of Scotland Science Park.

Likewise, in Edinburgh, science parks were said to provide an attractive environment in which to work. However, their location in out-of-town sites is at times seen as a disadvantage, not only because they reinforce dependency on cars, but also because at the moment these are isolated greenfield developments, removed from shopping facilities and other urban services readily available in city locations. The significance of this relative isolation need not be overstated, as firms reported that their employees were content with living in Edinburgh and commuting outwards to the new out-of-town locations. In addition, gradual retail development around those industrial estates increases the attractiveness of their sites, and goes some way to ameliorating their relative isolation.

One question is whether the geographical proximity of firms and institutions in related activities—biotechnology in this instance—gives rise to any tangible benefits and enables closer collaboration on a range of activities (research and development, sharing of experience and best practices, and so on).

Clustering in business parks is perceived to play a more significant advantage to the suppliers to the industry (for example, suppliers of chemicals, biological agents, raw materials) who benefit from the concentration of a number of their customers in a few key sites, thus economising on distribution costs. The benefits from such clustering effects, even in knowledge-intensive industries, seem to be confined to pragmatic issues of services and infrastructure and (perhaps surprisingly) fail to generate a strong social-network effect despite the obvious potential for this to occur (see below). The following comment captures the predominant view regarding the advantages offered by science parks

> The fact that there is a piece of land which is designated 'bio-manufacturing area', which means it's got at least outlining planning consent and decent roads to get goods out, will tend to make people put their bio-manufacturing plant there, because some of the hassle has been taken out of it ... Is there any other advantage to that? I hardly think so (senior manager, biotechnology firm, Edinburgh).

A further issue relates to the relationship between the supply of and demand for appropriate commercial space. Again, the diversity of the biotechnology cluster defies simple generalisations. Companies at an early-growth stage, with no credit record, have experienced particular difficulties finding appropriate premises and lease terms. As a result, they are faced with difficult demands from landlords regarding the timing and cost of leasing a property, such as substantial deposits on leases. The evidence suggests that this problem is particularly severe in Edinburgh due to overheating property markets. As one executive recalls

> When we set up this office everything was done on [our manager's] credit card because we had no credit record, so we couldn't buy telephone, computers, couldn't set up an account with a travel agent ... basically having no track record meant that you couldn't get any credit terms with any supplier ... Unless you've been an operating business for three years, you can't rent mobile phones as a business. So the things that limit us ... are effectively about establishing lines of credit when you have no credit record, and that was a huge problem for us, and continues to be.

In addition, congestion in Edinburgh and perceptions of inefficient public transport are causing difficulties, affecting even the recruitment efforts of firms

> I don't think transport helps as we're trying to help our newest recruit to find somewhere to live. Then it's all about "don't live in Fife because commuting over the bridge is a problem", and if you live out in the west the M8 is a problem, and in the east coming in on the by-pass is a problem, and the train service to Berwick is appalling.

On the other hand, road connections to both Edinburgh and Glasgow airports are perceived to be of satisfactory quality, enabling firms to reach these crucial facilities easily.

4.2 The Geography of Industrial Linkages

The industrial complex dimension seem to be the weakest locational factor of the three 'ideal types' identified in the analytical framework. Indeed, most of the trading and industrial linkages tend to be external and, in many cases, international as the major customers of products, research outputs and technologies tend to be multinational pharmaceutical companies. These multinational players capitalise on the high-quality workforce and research capacity of Scottish firms, and the quality of products, and provide access to capital, markets and economies of scope. At the same time, the integration of Scottish firms into an international production chain raises concerns related to external control and to the prospect for a 'deep' localised Scottish biotechnology cluster to be developed in the long term. As one interviewee summarised, "There is no driving force to direct us to purchase or to supply locally". And as another senior manager commented, it is important not to exaggerate the significance of local linkages

> One of the problems is that when you have a highly specialised area of science it's highly unlikely that people are going to be located on your doorstep. As long as you're *able* to travel, *can* travel and *want* to travel ... then it isn't a problem.

As Table 3 illustrates, the key market, financial, technology and regulatory drivers influencing the dynamics of biotechnology tend to be non-local and include the volatile conditions of financial markets, the growth of strategic alliances and mergers and acquisitions on an international scale, genome research and regulations regarding drug approvals. Thus, although much has been made in the cluster and industrial milieu literature, as well as in policy frameworks, of the advantages of localised linkages in the context of an innovation system (for a critical review, see Gertler, 2001; and Oinas and Malecki, 2002), the evidence from the Scottish biotechnology case shows that such linkages are relatively weak (see also Cooke, 2001) and, in a sense, confirm the assertion that "it could truly be said that the biotechnology sector is the first science-intensive set of industrial activities which has been truly globalised 'from birth'" (CRIC, 2001, p. 4).

A particular aspect of this is university–industry relations. Although many firms *originate* from universities and research institutes, there is not an evident persistence of strong ties between the firm and its 'originator' in terms of research and science collaboration. To be sure, firms use a range of analytical university services such as scanning facilities and laboratory equipment to test samples, but in many instances their research links to universities are weak. While some local firms finance research students, this is not a widespread practice, primarily because these firms do not have the resources to support such activities. In fact, in some cases, local firms ceased to support university researchers in response to financial pressures, holding the view that such activities are peripheral to their operation. However, external linkages and acquisitions (see below) may enable some firms in Glasgow and Edinburgh to fund research collaboration with local universities. Thus, somewhat ironically, external linkages may contribute to strengthen the ties between local research institutes and firms, thus solidifying this aspect of industrial clusters.

The limited significance of localised research links between the 'science base' and its potential user stems from a combination of factors. The limited financial resources which typify firms in an embryonic cluster have been mentioned above. Other factors include the potential incompatibility between the research strategies of universities and the need of local firms. Interviews with university officials and representatives of firms have revealed that major research institutions compete in an international market-place for the production of basic scientific research that is not always immediately applicable to the research needs of fledgling biotechnology firms. Furthermore, universities perceive that their contribution to the local economy is not necessarily through their direct ties to local industries, but through their ability to attract high-quality researchers and to draw research income from a range of sources, including multinational pharmaceutical firms with no direct links to local economies. Issues of intellectual property, licensing and incentives for academic staff to set up start-up firms are also perceived as impediments to stronger university–industry 'knowledge flows'.

While the overall picture is of weak local industrial linkages, some inputs are sourced locally and in Scotland, especially the more transport-cost-sensitive ones. Quality considerations are important, however. Thus, commodities such as chemicals and solvents, as well as stationery and the most basic supplies, tend to be sourced locally as much as possible. Some of the higher-value and specialised raw materials, however, are likely to be sourced internationally, including from the US, Japan and China.

A particular question revolves around the

Table 3. 'Non-local' drivers of the biotechnology industry: key dynamics during 2000

Event	Implications
Market and financial dynamics	
European biotech indices significantly outperformed the London Stock Exchange, NASDAQ and the pharmaceutical sector	Investors differentiated biotechnology from the high-technology stocks
European biotechnology companies raised 6 billion euros in public funding	Growth in the 'wealth' of European biotechnology companies
American biotechnology companies raised 30 billion euros in public funding	Wider 'Atlantic' gap disadvantages European companies
Eight out of top 10 European biotechnology companies have dual stock market listing in 2000 compared with 3 in 1999	Biotechnology companies need a broad, global investor base; expensive to manage
Merger of GlaxoWellcome and SmithKline Beecham	Biotechnology firms have fewer potential partners
Strategic alliances increase by 54 per cent	Big pharmaceutical firms increasingly utilise biotechnology companies to fill product and technology gaps
Rapid growth in 'Biotech-to-biotech' strategic alliances	Big pharmaceuticals no longer necessarily the preferred partner
Shire acquired Biochem Pharma; Celltech acquired Medeva; Ellan aquired The Liposome Company and Dura; Evotec acquired Oxford Asymmetry	Major biotechnology firms seek critical mass through mergers and acquisitions
Products and technology	
Human Genome unveiled	Unprecedented amounts of data become available; Gene 'land grab' gathers pace
Celltech, Protherics and Shire win product approvals; 27 products in Phase III of the European pipeline	European biotechnology industry begins 'to deliver'
Europe remains commercial GMO-crop-free zone	Key technology absent from European farming; European agrifood companies face uncertainty
Regulatory and public policy issues	
UK backed liberal therapeutic cloning law	Potential for UK to become international hub for stem cell research
Europe introduced orphan drug status and designated 14 products as such	Level playing-field created for European biotechnology firms *vis-à-vis* Japan and the US
EU member-states increased spending on genomics	Strengthening of the biotechnology sector within the knowledge-based economy
Some EU member-states challenged biotechnology patents directive implementation	Threat to integrated intellectual property protection in Europe

Source: Ernst & Young (2001, pp. 10–11).

nature of the relationship between local biotechnology firms and their major customers and end-users, usually multinational pharmaceutical firms. These relationships tend to be relatively stable, a function of the very long time-span which characterises the product development process in the industry. However, relationships do collapse from time to time, mostly in cases when products—most typically drugs—fail (either through failing to pass regulatory requirements, or by failing to perform). Big pharmaceutical companies understand that certain proportions of products will fail prior to arrival at the market. The problem for small biotechnology firms is their (typical) dependency on a small number of products—in many cases, a single product—and the need for constant flows of capital to sustain further research, development, trials and investment. Many such firms remain, in fact, loss-making operations for several years.

Success brings its own uncertainties, however: a publicly quoted company could be acquired by a larger multinational pharmaceutical company and this might cast a shadow over the location, size and future function of a Scottish biotechnology firm. Acquisitions mean in practice loss of strategic corporate control and, because of that, a particular difficulty for an embryonic cluster is the virtuous cycle of dependency on global forces. Other development scenarios include the evolution of biotechnology firms into small but significant players in the pharmaceutical supply chain as developers and manufacturers of drugs (Cooke, 2003; Humphrey and Schmitz, 2002).

While attempting to gauge which scenario is more likely to unfold entails a measure of speculation, is it possible to provide a careful assessment of past events. First and foremost, the currently highly specialised nature of the biotechnology cluster has implications for the manner in which firms develop and establish their position within an international production chain. Put in other words, an embryonic cluster such as the Scottish biotechnology industry provides an opportunity to observe closely the path-dependent nature of industrial change and its spatial underpinnings.

A case in point to illustrate the interplay of path dependency, local history, changing market conditions and the geography of the firm is that of Cruachem in Glasgow. It was established in 1979 by five PhD students from the University of Glasgow. Although the firm was first located in Livingston (west of Edinburgh), it subsequently relocated to Glasgow's West of Scotland Science Park in 1987. In the following years, the firm opened a site in Virginia, in the US, in order to supply the American markets with biochemical products. In 1992, it achieved the ISO 9001 accreditation, followed by the Pharmaceutical Code of Practice quality accreditation which affirms the capacity of the firm to supply the pharmaceutical industry with high-quality material. In the mid 1990s, the firm faced difficulties as it failed to generate profit. This, in turn, led to a merger of Cruachem with the American firm Symphony Pharmaceuticals in 1997 to create the ANOVIS Group. The Virginia site was subsequently closed, a process accompanied by redundancies on both sides of the Atlantic. Cruachem itself saw a reduction of its labour force from 60 employees in 1996 to 30 in 1997, as the firm refocused its activities largely around chemistry and chemical processes and withdrew from instrument production such as DNA synthesisers. In 1999, the firm restored its profitability and by 2000 had achieved a 20 per cent increase in growth. As a result, the group as a whole became attractive to acquisition and, in May 2001, was acquired by the American firm TransGenomics.

The example of the Glasgow-based firm Rhone Diagnostics (RD), is also telling in that it demonstrates how local firms can benefit when their acquisition by foreign firms is driven by considerations of added value, access to technologies and expertise, market expansion and product upgrade. Rhone Diagnostics was recently acquired by the German firm R-Biopharm, as part of its own development strategy. The acquisition has been driven on the basis of RD's unique expertise, its utilisation of technology and the quality of its products. The consolidation has in fact increased the likelihood of further growth in the Scottish premises of RD, as the German firm has decided to consolidate many of the operations in Glasgow, thus effectively doubling the Glasgow-based workforce.

Thus, from the point of view of industrial linkages, *local* spatial clustering effects are generally weak. What is driving competition and growth in the sector has little to do with local factors. The number of direct competitors who are located either locally, or even within Scotland, is marginal. In most cases, competitors are spread in North America and Europe, so that there is little to talk about in terms of innovation-stimulating local competition. In some instances, local firms have benefited from trends in the nature of competition and interfirm rivalries that are truly international. For example, to the good fortune of some local Scottish firms, several potential competitors have pulled out of the market-place in recent years, thus leaving them with only a few competitors globally. As one executive readily admitted

> We find ourselves, through no activities of our own, in the fortunate position where customers are looking around for a supplier they can rely on, and they're coming more and more to us.

The significance of external non-local connections has bearings on the process of innovation that takes place within the biotechnology cluster in Scotland. The literature on innovation systems has recently emphasised the importance of the *local* properties of the innovation process, essentially arguing that geographical proximity is important because it facilitates trust and reciprocity between firms and related institutions (Acs and Varga, 2002). Yet, the weak local linkages characterising the Scottish biotechnology sector mean that innovation is less dependent on intensive local interaction or local competition than might be expected.

Rather, the innovation capacity and competitiveness of firms are highly dependent on a complex constellation of internal scientific expertise, management skills, access to capital and favourable market conditions. Thus, while scientific expertise is an important ingredient of the performance of biotechnology firms, it is not a sufficient condition for their ability to retain their competitiveness. The ability to utilise and appropriate science and to reap commercial benefits out of scientific expertise is crucial in a highly competitive market

> The important thing is not just to have the science, or even the initial or original idea, but what you do with the science afterwards. From our customers' point of view—the big pharmaceutical companies—they are only interested in one thing: is there a market for the product and does it work? There're not terribly fussed about the technology behind it ... and they don't really care if you're a pioneer or not (senior director, Edinburgh-based biotechnology firm).

The skills of the research and scientific staff are an important source of competitiveness for biotechnology firms. Innovation in the Scottish biotechnology cluster tends to be internalised by the firm and thus is dependent on internal scientific expertise, management strategies and corporate practices. For some firms, this means 'special treatment' of their research and development staff to enable a certain flexibility, although this has to be carefully balanced with the firm's ability to generate enough revenue to be able to survive and to make 'business sense' out of innovative practices. As one biotechnology manager has put it

> The philosophy is of giving them freedom to try things. There has to be. But there has to be a commercial end-point. As an example, we use embryonic stem cells that we know can be turned into other kinds of tissues or organs ... we don't know how that process is completely controlled ... It's an exciting area, and an offshoot of [confidential] technology, and my research director said "This is an important research area; we need to have a research programme in this area" ... and I said, "That's fine, but I want an end-point: what is it we're trying to do?" So we must

> have a focus. He did some desk work and concluded that the end-point would be to turn these stem cells to … insulin producing cells. So at least we now know that we're starting from A and trying to get to B. There's purpose and focus to his work, and we know there's a huge market if we make it. So there has to be a commercial rationale to do the work.

As a consequence of the highly volatile nature of the biotechnology sector, there is a certain danger that firms with research expertise will eventually become merely contract manufacturers of more or less standardised pharmaceutical goods, thus shedding their research and cutting-edge scientific and innovative capacity. On the other hand, pharmaceutical companies rely quite heavily on small biotechnology firms for their future products. This is partially an outcome of the preference of scientists to work in smaller organisations where their research expertise and ideas constitute a considerable element in the firm's innovative capacity. It is perceived that the exploitation of innovative ideas and expertise is more easily achieved in small companies rather than large organisations.

A further aspect of the importance of external connections relates to the extent and quality of competition and collaboration with the UK's dominant biotechnology cluster in the South East and Eastern regions. The dominance of the South East and Eastern regions does not have straightforward implications. First, firms do not find it particularly disadvantageous to be located away from the main UK cluster in terms of disconnection from 'where things are happening'. Collaboration takes place between Glasgow- and Edinburgh-based firms, and firms and research institutes located in the South East and Eastern regions of the UK. Scottish firms also use specialised equipment and material suppliers from the south without particular reservations and concerns. In addition, Glasgow and Edinburgh firms form part of UK-wide and, in many cases, Europe-wide networks composed of a range of suppliers, customers, research institutes and intermediaries (legal advice, specialised service firms and so on).

Secondly, the dominance of the southern UK biotechnology cluster does not necessarily result in severe recruitment difficulties or 'brain drain'. Most firms have been able to recruit scientific and technical staff without much difficulty because of the availability of what is judged to be a high-quality graduate supply in Scotland. The perception of managers and company directors is important in this regard

> We're finding that Glasgow itself is a good city in attracting young people … it's becoming a well-known city for students, entertainment, socialising … we've even had requests from our people in France to come to work in Glasgow because they've heard that it's a good place and a good environment to work in.

Thus, perhaps somewhat surprisingly, firms do not perceive the dominance of the south of Britain biotechnology cluster as an impeding factor in their ability to recruit highly skilled labour. The marked specialisation of biotechnology firms and their potential capacity to innovate provide strong enough pull factors, even against competition from the Oxford–Cambridge cluster. The following represents a typical comment made by a senior manager

> It's not an issue, and it's not an issue because unless you're excited about the vision and idea of [our company] then you won't get a job here anyway. And there isn't another company like this.

Thus, when firms are able to find market niches and concentrate on health-related areas neglected by large pharmaceutical firms or other biotechnology competitors, they also benefit from an ability to attract highly skilled research staff who are attracted by the innovativeness and the 'unique science' offered by particular firms.

A further issue relates to the availability of venture capital and sources of finance within reasonable geographical proximity. The concentration of the venture capital industry in the south of Britain is often mentioned as a constraining factor with respect to biotechnology in Scotland (Cooke, 2001). Yet, there are ambiguities about whether the relative weakness of local sources of finance is an impeding factor in the growth of firms. Indeed, in a considerable number of cases, it has been judged that access to early-stage finance was not the major constraint on the development of firms. This reveals an interesting gap between perception and reality

> My perception of what the barriers are to small biotech companies operating out of Glasgow and Edinburgh, before I joined this one, were completely wrong in that finance hasn't been an issue. There's plenty of money out there if you have a good enough idea, a good enough science and good enough team.

A substantial concern relates, however, to the gap in perceptions and expectations between investors and start-up firms. Sources of finance tend to be short term and often require return more quickly than is possible in biotechnology. They are also cyclical in nature. For example, the current crisis in financial markets creates severe difficulty in raising capital. The biotechnology industry often needs large, sustained investment for a long time and, as one manager has put it, "the money needs to be patient money". There are 'funding windows', but often these do not correspond to the timing of scientific discovery. As another biotechnology entrepreneur explained, the funding of new premises and facilities can become an acute problem because the investment community tends to view capital investment with suspicion: "financing real assets for a loss-making company is a real difficulty, especially in light of shareholders' perceptions and expectations".

At the same time, Scotland's institutionally 'thick' policy environment provides a dense network of sources of finance for early-growth firms and for entrepreneurs. Yet again, Scotland's somewhat different position in that regard is worth mentioning because public-sector agencies—in this instance, Scottish Enterprise—have provided the foundation for an important institutional infrastructure in support of entrepreneurial development. The establishment of Intermediary Technology Institutes (ITIs) is one example of public-sector proactivism designed to correct market failures. ITIs provide a bridge between a range of actors in order to encourage technology development and uptake, and to provide financial support in areas deemed to offer future market opportunities. It is anticipated that £450 million will be allocated to the emerging ITI network over the next 10 years (Scottish Enterprise, 2003b). Other examples of public-sector-led finance for research commercialisation and business start-ups include the Proof of Concept Fund—with an investment of £5.2 million in early-stage ideas and currently supporting some 29 biotechnology projects throughout Scotland—and the Scottish Health Innovations initiative designed to facilitate the exploitation of intellectual property and to assist National Health Service staff in taking new ideas to the market.

4.3 The Significance of Local Social Networks

Overall, the extent and significance of local social networks in the cluster are judged to be limited. Partially, this is a reflection of the cluster's early growth stage, where many firms are preoccupied with their own survival and are not in a position to allocate significant resources towards informal interaction. The lack of critical mass—that is, the relatively small size of the sector in absolute terms—also limits the opportunity and economic rationale for the development of local social networks.

Thus, while some networking activities take place at the national (that is, Scottish) level, the local level is not seen by most actors in the biotechnology cluster as significant enough to warrant stable and long-lasting informal social networking. While most firms have pointed to membership in industrial associations and periodic and sporadic participation in networking events, joint training initiatives and trading missions led by Scottish Enterprise, these seem to be of marginal significance to the overall performance and activities of the cluster.

The overriding factor limiting the presence and significance of localised social interaction and collaboration relates to the fragmented, highly specialised and embryonic nature of the biotechnology industry in Scotland. In the words of one managing director

> The difficulty is that we're all doing different things. We're a special company in the world, let alone in Scotland. I mean, we're the only company doing what we do in the UK, so there's not a lot of point in talking to somebody in Glasgow because they don't know what we do.

At the same time that it is hard to talk of localised forms of dynamic innovation-inducing collaboration in the cluster, there is a certain recognition that the diversity of biotechnology prevents the sort of cut-throat competition (for technologies, expertise, products and labour) that might plague other regions. In this respect, firms may eventually find it easier to develop relationships of trust and collaborative learning, where social networks can be sustained.

Perceptual issues play an important part in the process of community building that may underpin the development of stronger local social networks in the Scottish biotechnology cluster. Here, the somewhat more privileged position of Dundee and Edinburgh, compared with that of Glasgow, has been noted by respondents. It has been suggested that, for a range of historical, institutional and organisational reasons (the presence of flagship institutions, local policy initiatives, the nature of university research, the initiative of key individuals, leadership and so on), Dundee and increasingly Edinburgh are able to position themselves perceptually as the locus of 'up and coming' biotechnology clusters. Enabled by the presence of the Roslin Institute in Edinburgh and by the University of Dundee and strong public-sector leadership in the case of Dundee (primarily through Scottish Enterprise's efforts), both cities are said to nurture gradually a sense of community identity within the sector, and to project it externally. The development of local industrial associations in the form of Bio-Dundee and the Edinburgh Bio Alliance may reflect the incremental strengthening of social networks.

Glasgow has lagged in building comparable institutional support for the biotechnology sector, although its recent Bio-Science Initiative may be able to address this issue. However, the success, in terms of occupancy rates, of Glasgow's West of Scotland Science Park has worked to provide the city with an important showcase for its own achievements.

The significance of institutionalised social networks in Edinburgh and Dundee is twofold: first, networking events help to address challenges and correct market imperfections in areas where local expertise is not readily available. The provision of 'educational services' on training issues, business management expertise and product development processes has been cited as particularly important. Secondly, institutional efforts of existing networks place particular emphasis on promotional issues, as a way of marketing places as technology hubs. Consequently, existing localised social networks have less to do with encouraging and reinforcing social capital as a potential innovation factor and more to do with economic development efforts. Thus, networking events and supporting institutions have been received with a measure of scepticism by biotechnology firms

> As with most of these kinds of things, they are more heavily populated by people who see them as a source of revenue ... like lawyers, accountants, and so on ... and this isn't really what it ought to be about ... (senior manager, biotechnology firm).

More fundamentally, the spatiality of social networks is driven by the logic and geography of industrial linkages, as discussed above, whereby external linkages are considerably more significant than local ones. A case in point has been the merger between the Scottish Biomedical Industrial Association and its English counterpart. The direct reason for this was the perception by firms and actors in Scotland that "there wasn't enough of a cluster here to be viable as a trade body" and that "meeting with other biotech people across the UK and Europe is just as important, and even more important, as meeting people [from the industry] in Stirling and Edinburgh". Furthermore, some within the industry have argued that efforts towards local networks are misguided because they unnecessarily fragment an already small industry. The importance of business networks and industrial association continues to be conceived in terms of lobbying efforts *vis-à-vis* government regulation and less in terms of supporting trust-based interfirm collaboration. This, in turn, drives the scale of engagement away from and 'above' the local level

> Most of the things we need to influence are European-wide. [The EU] is where most of the regulations come out of; so we need to influence Europe and we need to influence the UK government so that they vote the right way in Europe.

While the evidence of collaborative local social networks is limited, they nonetheless exist in a variety of specific forms. One form of local social relations contributing to the viability of firms materialises in the form of investors and venture capitalists who serve constructively on the managing board of newly formed firms and provide advice and expertise on business strategy. There is some evidence that this is happening, especially in spin-out companies who have reflected positively on the role of venture capitalists in providing valuable advice and experience. Secondly, informal linkages exist between National Health Service and university consultants who provide advice and scientific support for firms involved with clinical trials.

Finally, the development of flagship science parks has been mentioned above within the context of agglomeration economies. It is interesting to note that the development of such parks is also envisioned in terms of providing institutional support for local social networks to emerge. For instance, the development of the Biomedical Research Park in Edinburgh aims to exploit the assets of the city, such as the research hospital and the medical school in the south of the city, and to accommodate physical requirements such as suitable commercial property and space. It is also, importantly, about sending 'softer' signals to the market

> It can help position Edinburgh as being serious in its attempt to work collaboratively between people in industry and academia, in the same way that if you ask people internationally where's [*sic*] important in the biotech industry they would say Boston or San Diego ... If you go to Boston, the Medical School is located next door to the major teaching hospital which is located next door to a major research park, which can attract, I think, not just commercialisation, but also mobile investment from the major 'pharma' and biomedical companies in the world (senior official, economic development agency).

5. Conclusion

An examination of an embryonic cluster provides an opportunity to tease out the path-dependent aspects of economic change and to generate an analysis that is sensitive to the influence of historical legacies and key institutions on the spatiality of cluster development. Historical legacies and public-sector anchors have influenced the initial location, the types of expertise and the innovative properties of biotechnology in Edinburgh and Glasgow.

Of the three 'ideal type' locational factors identified in this paper, agglomeration economies play the most important role (Table 4). Edinburgh and Glasgow contain large and diverse pools of skilled labour and high-quality university graduates, and infrastructure

Table 4. Main locational elements underpinning biotechnology activities in Edinburgh and Glasgow

	Edinburgh	Glasgow
Public-sector anchors	Advantaged by the presence of flagship institutions	Lacks the presence of a 'Roslin-type' institution, but universities play an important role in providing 'science base'
Labour market issues	Adequate supply of science graduates; small critical mass limiting the range of locally available senior managers	Size and diversity of local labour market an advantage; small critical mass limiting the range of locally available senior managers
Infrastructure	Disadvantaged by congestion and escalating property prices; helped by development of science parks	Enjoys good access to the airport; affordable property although requires new investment in specially designed premises
Industrial profile	Key strengths in cloning technologies and stem cell research; manufacturing	Key strengths in contract research and clinical trials; manufacturing-oriented firms
Sources of finance	A mixed geography of local and national sources, including venture capital and Scottish Enterprise support programmes	A mixed geography of local and national sources, including venture capital and Scottish Enterprise support programmes
Localised interfirm linkages	Weak	Weak
External linkages	Strong	Strong
Social ties	Weak locally, although benefits from improving leadership and organisational capacity	Weak local ties; late development of organisational capacity and weak associational structures compared with Edinburgh

plays an important role in enabling firms to economise on the provision of shared services. At the same time, the small critical mass of the cluster means that localisation economies are weak and that there is little to talk about in terms of dynamic and innovation-stimulating local competition.

At the moment, the extent and effectiveness of social networks as a locational factor are relatively marginal. This is partially a result of the embryonic nature of the cluster, partially an outcome of many firms' preoccupation with survival and partially a reflection of the lack of a deep-rooted culture of interfirm collaboration. However, there are some signs of change—albeit limited—with various institutions serving as forums for localised social interaction.

The industrial complex dimension seem to be the weakest locational factor of the three 'ideal types' identified above. Most of the trading and industrial linkages tend to be external and in many cases international as the major customers of products, research outputs and technologies tend to be multinational pharmaceutical companies. These multinational players capitalise on the high-quality workforce and research capacity of Scottish firms and the quality of products, and provide access to capital, markets and economies of scope. At the same time, the integration of Scottish firms into an international production chain raises concerns related to external control and to the prospect for a 'deep' Scottish biotechnology cluster to be developed in the long term. Regulatory issues are significant not only in shaping the trajectories of industrial development, but also in inducing competition and dictating the spatial pattern of interaction along supply chains and between firms in related activities. Furthermore, 'soft' locational factors in the form of Scottish identity and the image of Edinburgh and Glasgow as vibrant places are apparent (as one founder of a biotechnology firm commented "locating a business here is about the other 'parts' of your life and lifestyle factors, and it's not necessarily about whether or not it's the right location for your business"), although it is hard to estimate their precise significance.

The evidence from this research suggests the potential for considerable diversity to exist within an industrial cluster when firms tend to be highly specialised. From a public policy point of view, this calls into question straightjacket approaches to cluster growth and suggests, as Brown and Duguid (2002) have recently argued, that policy-makers should neither attempt to imitate the experience of other success stories, nor concentrate too narrowly on the traditional strengths of the local economy. Rather, existing local competencies, needs, expertise and talent should be complemented by the development of new technologies. Furthermore, the importance of international industrial links suggests that policies should avoid focusing too narrowly on the strengthening of local ties. At the same time, the significance of agglomeration economies means that policies to increase the diversity of local skills and expertise, and to improve the quality of infrastructure and services, will contribute to the development of biotechnology. Public policy thus needs to be carefully attuned to the spatial nuances underpinning biotechnology in Scotland and elsewhere.

Notes

1. The research on the basis of which this paper is written involved 20 interviews with representatives of biotechnology firms in Glasgow and Edinburgh, experts on the industry and its science base, university commercialisation officials and policy officers in economic development agencies. Interviews covered the history of the industry in its location, the sources of innovation, the range of local and external linkages, the profile of the labour force and sources of recruitment, and the effectiveness of policy. Evidence was also gathered from previous reports, media coverage, consultancy and market analysis, industrial directories and data-sets, policy documents, and company statements.
2. Alternative typologies are, of course, possible. For example, Smith *et al.* (2000) have suggested four types of 'localism'.
3. A similar story could be told of the role of other important public-sector anchors—such as the Moredun Research Institute for animal health research in Edinburgh and the local universities in Glasgow and Edinburgh—in providing the science base and opportunities for research commercialisation and spin-outs (see, for instance, Collier, 2000).

References

ACS, S. J. and VARGA, A. (2002) Geography, endogenous growth and innovation, *International Regional Science Review,* 25, pp. 132–148.

AMIN, A. and THRIFT, N. (2002) *Cities: Reimagining the Urban.* Cambridge: Polity Press.

ARTHUR, W. B. (1994) *Increasing Returns and Path Dependence in the Economy.* Ann Arbor, MI: University of Michigan Press.

AYDALOT, P. and KEEBLE, D. (1988) *High Technology Industry and Innovative Environments.* London: Routledge.

BELLANDI, M. (2002) Italian industrial districts: an industrial economics interpretation, *European Planning Studies,* 10, pp. 425–437.

BEST, M. H. (1990) *The New Competition.* Cambridge: Polity Press.

BIOTECHNOLOGY SCOTLAND (2003) *Framework for Action 2002–2003.* Glasgow: Scottish Enterprise.

BROWN, J. S. and DUGUID, P. (2002) Local knowledge—innovation in the networked age, *Management Learning,* 33: 427-437.

CHINITZ, B. (1961) Contrast in agglomeration: New York and Pittsburgh, *American Economic Review,* 51, pp. 279–289.

COLLIER, A. (2000) Cell seekers: the Scottish biotechnology industry, *CA Journal,* December, pp. 18–25.

COOKE, P. (2001) Biotechnology clusters in the UK: lessons from the localisation and commercialisation of science, *Small Business Economics,* 17, pp. 43–59.

COOKE, P. (2002) Biotechnology clusters as regional, sectoral innovation systems, *Inter-*

national Regional Science Review, 25, pp. 8–37.

COOKE, P. (2003) Biotechnology clusters, 'Big Pharma' and the knowledge-driven economy, *International Journal of Technology Management,* 25, pp. 65–80.

COOKE, P. and MORGAN, K. (1998) *The Associational Economy*. Oxford: Oxford University Press.

CRIC (CENTRE FOR RESEARCH ON INNOVATION AND COMPETITION) (2001) *Biotechnology in the UK: Scenario for Success in 2005*. Manchester: University of Manchester and UMIST.

DANEELS, E. (2002) The dynamics of product innovation and firm competences, *Strategic Management Journal,* 23, pp. 1095–1121.

DE PROPRIS, L. (2002) Types of innovation and inter-firm co-operation, *Entrepreneurship and Regional Development,* 14, pp. 337–353.

DTI (DEPARTMENT OF TRADE AND INDUSTRY) (1998) *Competitive Future: Building the Knowledge Driven Economy*. London: The Stationery Office.

DTI (1999) *Biotechnology Clusters*. London: Biotechnology Directorate.

DTI (2000a) *Business Clusters in the UK: A First Assessment, Volume 1*. London: DTI.

DTI (2000b) *Business Clusters in the UK: A First Assessment, Volume 2*. London: DTI.

ERNST & YOUNG, (2001) *Integration: European Life Sciences Report 2001*. London: Ernst & Young International.

FELDMAN, M. P. and RONZIO, C. R. (2001) Closing the innovative loop: moving from the laboratory to the shop floor in biotechnology manufacturing, *Entrepreneurship and Regional Development,* 13, pp. 1–16.

Financial Times, (2000) Mammoth task to untangle the Genome, 17 November.

Financial Times, (2001) Special Reports: Eurobiotechs, 25 April.

GAROFOLI, G. (2002) Local development in Europe: theoretical models and international comparisons, *European and Regional Studies,* 9, pp. 225–239.

GERTLER, M. (2001) Best practice? Geography, learning and institutional limits to strong convergence, *Journal of Economic Geography,* 1, pp. 5–26.

GORDON, I. R., and MCCANN, P. (2000) Industrial clusters: complexes, agglomeration and/or social networks?, *Urban Studies,* 37, pp. 513–532.

GRANOVETTER, M. (1985) Economic action and social structure: the problem of embeddedness, *American Journal of Sociology,* 91, pp. 481–510.

GRIFFIN, H. (2001) Biotechnology: brave new world, *Focus: Scotland 2001*. Edinburgh: The Winchester Group.

HACKING, A. J. (1986) *Economic Aspects of Biotechnology*. Cambridge: Cambridge University Press.

HOLLINGSWORTH, J. R. (2000) Doing institutional analysis: implications for the study of innovations, *Review of International Political Economy,* 7, pp. 595–644.

HUMPHREY, J. and SCHMITZ, H. (2002) How does insertion in global value chains affect upgrading in industrial clusters, *Regional Studies,* 36, pp. 1017–1027.

KRUGMAN, P. (1996a) Making sense of the competitiveness debate, *Oxford Review of Economic Policy,* 12, pp. 17–25.

KRUGMAN, P. (1996b) Urban concentration: the role of increasing returns and transport costs, *International Regional Science Review,* 19, pp. 5–30.

LAMBOOY, J. G. (2002) Knowledge and urban economic development: an evolutionary perspective, *Urban Studies,* 39, pp. 1019–1035.

LAMBOOY, J. G. and BOSCHMA, R. A. (2001) Evolutionary economics and regional policy, *Annals of Regional Science,* 35, pp. 113–131.

MALECKI, E. J. (2002) Review of: van der Berg, L., BRAUN, E., and VAN WINDEN, W. 'Growth clusters in European metropolitan cities: A comparative analysis of cluster dynamics in the cities of Amsterdam, Eindhoven, Helsinki, Leipzig, Lyons, Manchester, Munich, Rotterdam and Vienna', *Urban Geography,* 23, pp. 499–500.

MALMBERG, A. and MASKELL, P. (2002) The elusive concept of localisation economies: towards a knowledge-based theory of spatial clustering, *Environment and Planning A,* 34, pp. 429–449.

MARSHALL, A. (1920) *Principals of Economics*. London: Macmillan.

MARTIN, R. and SUNLEY, P. (1996) Paul Krugman's geographical economics and its implications for regional development theory: a critical assessment, *Economic Geography,* 72, pp. pp.259–292.

MARTIN, R. and SUNLEY, P. (2003) Deconstructing industrial clusters, *Journal of Economic Geography,* 3, pp. 5–35.

MASSEY, D. (1984) *Spatial Divisions of Labour: Social Structures and the Geography of Production*. London: Macmillan.

MCKELVEY, M. D. (1996) *Evolutionary Innovations: The Business of Biotechnology*. Oxford: Oxford University Press.

NELSON, R. and WINTER, S. (1982) *An Evolutionary Theory of Economic Change*. Cambridge, MA: Harvard University Press.

NEWELL, P. (2002) *Biotechnology and the politics of regulation*. Working Paper No. 146, Institute of Development Studies, University of Sussex.

NIOSI, J. and BAS, T. G. (2001) The competencies

of regions: Canada's clusters in biotechnology, *Small Business Economics,* 17, pp. 31–42.

OAKEY, R. P. (1993) Predatory networking: the role of small firms in the development of the British biotechnology industry, *International Small Business Journal,* 11, pp. 9–22.

OECD (ORGANISATION FOR ECONOMIC CO-OPERATION AND DEVELOPMENT) (1996) *Intellectual Property: Technology Transfer and Genetic Resources.* Paris: OECD Publications.

OINAS, P. and MALECKI, E. J. (2002) The evolution of technologies in time and space: from national and regional to spatial innovation systems, *International Regional Science Review,* 25, pp. 102–131.

PARR, J. B. (2002) Missing elements in the analysis of agglomeration economies, *International Regional Science Review,* 25(2), pp. 151–168.

PIORE, M. J. and SABEL, C. F. (1984) *The Second Industrial Divide.* New York: Basic Books.

PORTER, M. (2000) Location, competition and economic development: local clusters in a global economy, *Economic Development Quarterly,* 14, pp. 15–34.

PORTER, M. (2001) Regions and the new economics of competition, in: A. J. SCOTT (Ed.) *Global City Regions: Trends, Theory, Policy,* pp. 139–157. Oxford: Oxford University Press.

PREVEZER, M. (2001) Ingredients in the early development of the US biotechnology industry, *Small Business Economics,* 17, pp. 17–29.

RIFKIN, J. (1998) *The Biotech Century.* London: Victor Gollancz.

SALIWANCHIK, R. (1988) *Protecting Biotechnology Inventions.* Madison, WI: Science Tech Publishers.

SCOTT, A. J. (2001) Globalization and the rise of city-regions, *European Planning Studies,* 9, pp. 813–826.

SCOTTISH ENTERPRISE (1999) *Biotechnology Report.* Edinburgh: Scottish Enterprise.

SCOTTISH ENTERPRISE (2003a) 2003/04 *Operating Plan: Growing Businesses* (www.Scottish-enterprise.com).

SCOTTISH ENTERPRISE (2003b) *Biotechnology Framework for Action.* Edinburgh: Scottish Enterprise.

SCOTTISH EXECUTIVE (2000) Biotechnology to drive knowledge economy, Press release, 14 November (www.scotland.gov.ac.uk).

SCOTTISH EXECUTIVE (2001) *A Smart, Successful Scotland: Ambitions for the Enterprise Networks.* Edinburgh: The Stationery Office.

SMITH, H. L., MIHELL, D., and KINGHAM, D. (2000) Knowledge-complexes and the locus of technological change: the biotechnology sector in Oxfordshire, *Area,* 32, pp. 179–188.

STORPER, M. (1997) *The Regional World.* New York: Guildford Press.

TUROK, I. and BAILEY, N. (2003) *Twin Track Cities? Linking Prosperity and Cohesion in Glasgow and Edinburgh.* Glasgow: ESRC Cities Project Report.

WHITFORD, J. (2001) The decline of a model? Challenge and response in the Italian industrial districts, *Economy and Society,* 30, pp. 38–65.

Knowledge-based Clusters and Urban Location: The Clustering of Software Consultancy in Oslo

Arne Isaksen

[Paper first received, February 2003; in final form, November 2003]

1. Regional Clusters: From Diffused Development to Central Location

The period since the mid 1970s represents a transition to a new phase of capitalist development that is seen to be distinctly different in its organisation and in its geography of production from the preceding Fordist period (Amin, 1994). The changes in the geography of production consist, in particular, of a resurgence of regional economies, as a number of local areas experienced new growth based in agglomerations of small and medium-sized enterprises (SMEs) (Piore and Sabel, 1984). The growth of new, and the revitalisation of old, industrial agglomerations—or regional clusters as they have come to be called (Porter, 1998)—has led researchers from several academic disciplines to focus on the territorial foundations of the economy. A strong causal relationship is drawn between regional resources and firms' innovation ability and competitiveness (Storper, 1997). To some extent, places, rather than the actors within them, have become the focus of analysis in economic geography (Amin and Thrift, 2002).

The past few years have witnessed an increased interest in regional clustering in the so-called knowledge-intensive industries (Cooke, 2002). These are industries with comparatively high R&D-intensity and services that are large users of embodied technology and have comparatively many workers with higher education (OECD, 2001a)—in short, high-tech manufacturing and knowledge-intensive services. The 'new' clusters in knowledge-intensive industries seem to have a somewhat different location pattern from the 'old' clusters. Until the 1990s, the growth of 'post-Fordist' regional clusters was seen to result in a reorganisation of production away from the former centres of capitalist accumulation—i.e. the large city-regions. The growth of industrial districts in Italy in the 1970s—for example, led to a break with the territorial concentration of manufacturing production in the two preceding decades. A process of diffused development took place—i.e. a more decentralised production based on local entrepreneurship, flexible small-firm networks and varying local know-how and resources—which particularly occurred in the Third Italy (Garofoli, 1992). Scott (1988) generalised about the development in the Third Italy and other areas and saw a series of new industrial spaces coming into existence during the 1970s and 1980s as a result of a new and more flexible regime of accumulation. According to Scott, the new industrial spaces formed alternative centres of capitalist accumulation and led to a marked reagglomeration of production. Flexible production sectors, including design-intensive craft industries as in the Third Italy, high-technology industries and knowledge-intensive business services, started to spring up in areas that were either socially or geographically insulated from the old 'Fordist' industrial regions in the centre of the nations.

The postulation of a large-scale geographical decentralisation of industry and jobs away from central areas does not agree with the reality in most Western countries. From the 1980s—for example, major cities in western Europe have gained population and have seen significant economic growth in leading sectors (Sassen, 2000, p. 42). This may reflect the fact that the acclaimed success stories of local networks of SMEs in less central areas are too few and that the flexible production sectors that form new industrial spaces have too few jobs to influence substantially the overall geography of production. However, the recent growth of major cities also to some extent rests upon the fact that clusters in knowledge-intensive industries (or knowledge-based clusters, in short) are seen to be more centrally located than, according to Scott (1988), is the case with new industrial spaces.

> The new economy sectors are highly skewed in geographical terms. They are overwhelmingly found in or near large, well-diversified services and knowledge-based cities or specialist research university campus cities or towns (Cooke, 2002, pp. 130–131).

This paper examines the cluster-building mechanisms in the software industry in Oslo, the capital region of Norway. The industry is narrowly defined as NACE 722, 'Software consultancy', which is one of the knowledge-intensive industries as defined by OECD (2001a). The Oslo area contained 60 per cent of the 24 500 jobs in this industry in Norway in 2001, while the area had 26 per cent of all jobs. The concentration of software consultancy jobs in Oslo reflects the industrial structure of the area. Like most large cities, Oslo has relatively few manufacturing jobs (compared with the national average), but comparatively many jobs in knowledge-intensive services. Additionally, the Oslo region demonstrated much faster job growth than the rest of Norway during the 1990s. Oslo also has a relatively innovative industry. Thus, studies demonstrate that firms in the Oslo region introduce more innovations new to the market; the region has more innovations in some high-tech sectors and 'creates' more new and innovative firms than the Norwegian average (Isaksen and Aslesen, 2001). The paper explores three main questions related to the concentration of the software industry in Oslo

(1) Which activities constitute the software industry in the Oslo area?
(2) What are the mechanisms triggering the development of a software cluster in Oslo?
(3) Which theoretical approaches are most relevant in explaining the development of the software cluster in Oslo?

The paper builds on personal interviews with leaders (mostly managing directors and sales managers) in 14 software firms in the Oslo area, including a good number of the largest firms.[1] The interviews aimed above all at 'unpacking' the sector—i.e. determining what firms actually do 'on the ground'. Thus, the interviews asked about what types of activities the firms perform, how innovation and production are conducted, what the relations between the software firms and their clients are like, etc. The interviews were followed by a telephone survey to a randomly sampled population of software firms and firms in other knowledge-intensive industries, which obtained answers from 800 firms. The paper draws on a few questions in the telephone survey, first of all on answers from 123 firms in NACE 72 'Computer-related services' in Oslo.[2]

The rest of the paper is divided into four parts. Section 2 discusses theoretical approaches that may be relevant guides in exploring cluster-building forces in the Oslo software industry. More specifically, the section contrasts two traditions of territorial innovation models (Moulaert and Sekia, 2003), which are "models of regional innovation in which *local institutional dynamics* play a significant role" (Moulaert and Sekia, 2003, p. 219; authors' emphasis). The two traditions are the industrial district model and research focusing on the growth and dynamics of regional clusters in new, knowledge-intensive industries, respectively.

Section 3 examines how production, innovation and customer relations actually take place in the Oslo software industry. Section 4 departs from the analysis of the characteristics of the Oslo software industry and investigates the clustering mechanisms in this industry in Oslo. Finally, section 5 discusses wider empirical and theoretical lessons from the case study.

2. Cluster Mechanisms in 'Old' and 'New' Regional Clusters

The notion of a regional cluster refers to geographically bounded concentrations of interdependent firms (Rosenfeld, 1997; OECD, 2001b) and may be used as a catchword for older concepts like industrial districts, innovative milieux and local production systems. During the 1990s, regional clusters were widely recognised as environments able to stimulate the productivity and innovativeness of firms and the formation of new businesses. The concept refers to some economic, sociocultural and institutional conditions and processes that are seen to promote innovation activity and competitiveness of cluster firms (see Camagni and Capello, 2002).

There is no space here to do justice to all the schools of thought and all the nuances of the comprehensive literature on regional clustering or of the different 'territorial innovation models'.[3] The paper focuses on some key contributions from the 'classical' industrial district tradition and a more mixed tradition focusing on regional clustering of high-technology manufacturing and knowledge-intensive business services (KIBS). The second strand of research includes work on new industrial spaces (Scott, 1988) and work drawing on the innovation system and the learning economy literature to analyse 'new economy clusters' (Cooke, 2002) and 'consultancy clusters' (Keeble and Nachum, 2002).

The first tradition focuses on research on old regional clusters in craft-based manufacturing industries, which have a much more decentralised location pattern than knowledge-intensive clusters. The two strands of research are contrasted in order to highlight important factors stimulating the clustering of knowledge-intensive industries in urban areas. Thus, the theoretical part of the paper aims to identify important forces underlying the growth and dynamics of the mainly centrally located clusters in knowledge-intensive industries and to compare these with those of industrial districts having a more diffused location pattern. In that endeavour, the paper focuses on three factors seen as important in explaining why firms cluster in space and why firms in certain places manage to retain a high level of international competitiveness. These factors are

(1) external economies associated with the geographical agglomeration of numerous firms in the same or related industries.
(2) institutional infrastructure—i.e. the col-

lective institutional, social and cultural factors underpinning regional competitive advantage; and

(3) local learning and adaptation—i.e. the processes of knowledge creation, knowledge spillovers and innovation.

2.1 Local Linkages and Local Buzz

Regional clustering has traditionally been associated with the role of local linkages between specialised firms in generating competitive advantages. Local networks of specialised firms create external economies—i.e. the achievement of effective production through extensive, external division of labour between firms specialising within various phases of a production chain. Marshall maintains that

> very important external economies can often be secured by the concentration of many small businesses of a similar character in particular localities; or ... by the localisation of industry (Marshall, 1890, p. 266).

Through co-operation, fairly small firms can achieve the same type of economies of scale as achieved internally in large companies, at the same time as business networks stimulate mutual learning and innovation co-operation, and make possible more flexible production.

However, researchers in different ways challenge the primary importance attached to local linkages in stimulating the competitiveness of cluster firms. Thus, several authors claim that work on regional clusters has tended to underemphasise the importance of wider extra-local connections (Alderman, forthcoming; Amin and Thrift, 2002; Bathelt *et al.*, 2002). External connections to the global economy are seen to play an important role in bringing in ideas and knowledge in sustaining competitive advantage, which may be of particular importance in clusters of knowledge-intensive industries. Amin and Thrift (2002, p. 60; emphasis added) find it "hard to accept that local personal contacts are of primary importance for business in the *knowledge economy*". Wired workers and knowledge entrepreneurs are seen to be among the most widely connected or mobile people, always on the move and dependent on distanciated connections. Clusters will rarely break down into localised networks according to Amin and Thrift (2002). Rather, clusters are increasingly structured around flows of goods, people, information and technology across cluster boundaries and to some extent also across national borders. The flows occur within transnational corporations, between firms and project teams scattered around the world, linked up virtually and through placements, and through electronic space.

However, local and extra-local linkages may be different—for example, including the exchange of distinct types of information. Thus, Leamer and Storper (2001) suggest a double-edged geography of the Internet age. They see a tendency towards agglomeration of the production of specialised, customised and innovative products, and of higher-order activities of invention, innovation and management, on the one hand, and spreading out of mass-produced, standardised products and service activities on the other. Agglomerations are still important as the economy is increasingly "dependent on the transmission of complex uncodifiable messages, which require understanding and trust that historically have come from face-to-face contact" (Leamer and Storper, 2001, p. 641). The idea seems to be that it is not local input–output networks between—for example, clients and sub-contractors that create local agglomerations in a more knowledge-based economy, but networks between people occupied with invention, innovation and management, who need to exchange uncodifiable knowledge.

Malmberg and Maskell (2002) put forward a basic critique of territorial innovation models as regards the role attached to local collaboration in triggering learning and innovation. The authors maintain that the collaborative element is emphasised at the expense of rivalry. A strong competitive environment is seen as the main stimulus for innovation and product differentiation in companies (Porter, 1998). By co-locating, rivals are well aware of each other's products, production processes, prices and innovations. Thus, firms in clusters "may dislike each other and refuse to talk but can still, indirectly, contribute to each other's competitive success in global market" (Malmberg and Maskell, 2002, p. 444).

Cluster firms contribute indirectly to each other's learning by spontaneously observing and monitoring the activities and improvements of nearby firms. Firms rely in this sense on the *local buzz* (Bathelt *et al.* 2002); they receive lots of specific information and inspiration by being in a milieu containing many actors with related, but complementary and heterogeneous, skills and knowledge. A cluster firm may copy an adjacent competitor, add a few new elements in the competitor's product or process, which are then copied and bettered by still other companies, summing up to a more or less continuous process of copying, learning and incremental innovation. Importantly, Malmberg and Maskell (2002, p. 442) suggest that rivalry, observation and comparison are key learning features in the early development of clusters—i.e. in embryonic clusters (Trend Business Research, 2001, p. 17). Dense local collaboration may be more important in clusters that have become more established and mature.

The suggestion put forward by Malmberg and Maskell (2002) regarding local buzz and rivalry as key features in the early development of clusters may be further elaborated by use of the distinction between localisation and urbanisation economies (Moseley, 1974). Localisation economies refer to externalities associated with the presence in a place of a mass of other producers in the same or closely related industries. Over time, the co-location of firms in an area dominated by one or a few related industries (as is the case with most industrial districts) *may* lead to the development of social institutions that facilitate dense collaboration. Urbanisation economies, on the other hand, are found where there is a diverse industrial base, extensive infrastructure and services supporting it and a concentration of institutions that generate new knowledge—that is, in large urban centres. Firms in large urban areas may be more able to rely on the local buzz in picking up new ideas and information than firms in smaller centres. The propensity for knowledge-spillovers and for observing, monitoring and imitating firms is seen to be higher in urban regions and innovative firms are expected to be located mostly in these regions (Isard, 1956; Armstrong and Taylor, 1993). The idea is that big urban agglomerations attract a large and differentiated variety of activities and thus become particularly suitable as breeding-places for innovations (Glaeser *et al.*, 1992; Brouwer *et al.*, 1999).

The argument of urban regions as centres of growth and innovation in new industries agrees with arguments in the geographical variant of the product cycle theory (Norton and Rees, 1979; Lundquist, 1996). Firms in new industries, which are in the earlier stages of a product's life cycle, are seen to need, in particular, inputs from R&D institutions, demanding customers and a range of firms able to offer specialised supplies and services. Products and production technology are rapidly changing, which further underlines the need for the mainly new and small firms in new industries to grasp information, know-how and technology from external players and to take part in intended and unanticipated learning processes. Thus, location in an urban region containing many rivals, clients, specialised suppliers, R&D institutes and multifaceted industry is seen to stimulate learning processes.

In brief, while knowledge-intensive industries have both local and global connections, clustering in space is seen to be important as these new sectors rely quite considerably on the exchange of complex, uncodifiable knowledge and on the local buzz that includes observing, monitoring and imitating rivals. Relevant information and knowledge are often most easily acquired in large cities.

2.2 Territorial and Professional Community

During the past decade, the literature on industrial districts, regional clusters, etc. has focused less on 'economic' reasons, such as external economies stemming from vertical disintegration, and agglomeration economies. Instead, the literature focuses on 'social' and 'cultural' reasons,

> such as intense level of interfirm collaboration; a strong sense of common industrial purpose; social consensus; extensive institutional support for local business; and structures encouraging innovation, skill formation, and the circulation of ideas (Amin and Thrift, 1994, p. 12).

According to Amin and Thrift (1994) these factors are best summed up by the phrase local 'institutional thickness', which consists of two main components. The first component includes support organisations, such as universities, training centres, government agencies and trade associations. Support organisations offer business services and know-how seen as important in keeping the economic machinery moving (Amin and Thrift, 2002).

The second component of institutional thickness is social institutions that foster close, informal collaboration between personnel in firms and organisations. The local players are aware that they are involved in a common enterprise and they have a common industrial purpose and shared cultural norms and values. Social institutions resemble the notion of locally based untraded interdependencies (Storper 1997, p. 5), which include habits, informal rules and conventions that help to co-ordinate firms' activity under conditions of uncertainty. Shared norms and mutual understanding reduce uncertainty that is present particularly in innovation processes and may thus stimulate innovation collaboration.

The type of social institutions may vary between 'old' clusters of craft-based manufacturing and clusters in new, knowledge-intensive industries. In industrial districts, social institutions are embedded in the local community. Thus, Becattini sees

> the industrial district as a socio-territorial entity which is characterised by the active presence of both a community of people and a population of firms ... and ... community and firms tend to merge (Becattini, 1990, p. 38).

The basis for trust and co-operation is individuals being members of a territorial community. Managers and workers typically have a strong local identity and a sense of solidarity. In 'new' clusters, the co-operation between firms typically rests upon corporate mangers and office staff being members of a professional community. The office staff have often been educated at the same business schools or technical universities, or they may have been colleagues in some pioneer firms in the cluster. Silicon Valley is—for example, characterised by exceptionally high mobility of engineers and managers among firms and "similar shared professional experiences continued to reinforce the sense of community in the region even after individuals had moved on to different, often competing, firms" (Saxenian, 1996, p. 31).

The pattern of co-operation may vary between territorial and professional communities. In old clusters, the social institutions may be quite place-specific, supporting the choice of local collaborators under conditions of uncertainty. Social institutions in new, knowledge-based clusters are more group-specific, backing close co-operation with persons having similar educational and professional background also across cluster boundaries.

2.3 Consultant–Client Interaction

The third main factor seen as important in explaining regional clustering and the competitive advantage of clusters is knowledge accumulation and transfer; these stimulate innovation activity in 'cluster firms'. Innovations in traditional industrial districts occur through trustful co-operation between several specialised firms and through informal contact between entrepreneurs, firm leaders and workers. Innovations are mainly seen to be limited to incremental innovations through 'learning-by-doing' and 'learning-by-using', and through trial and error to solve concrete problems on the shop floor (Asheim 1992). Informal, non-scientific and interactive knowledge is seen to be important (Amin, 2000, p. 164). Important parts of the knowledge are seen to remain tacit, bounded by their human and social context, and, thus, localised and 'sticky' (Maskell *et al.*, 1998).

In some knowledge-intensive industries, and particularly in biotechnology, firms are seen to rely more on scientific knowledge and advances in R&D (Cooke, 2002). Scientific knowledge is included in the 'know-why' type of knowledge by Lundvall and Johnson (1994), which is seen as mainly a codified form of knowledge. Maskell *et al.* (1998) regard codified knowledge as globally available. Consequently, scientific knowledge does not seem to be locally embedded and may hardly form the basis for regional clustering. It is pointed out that exchange of information and collective learning among individuals can be organised across many geographical scales (MacKinnon *et al.*, 2002, p. 304). On the other hand, it is increasingly recognised that parts of *new* scientific and technological knowledge are complex and uncodifiable, and thus have important elements of tacitness and firm specificity (Acs *et al.* 2002, p. 4). The knowledge is sticky because humans are the containers for storing and transferring uncodifiable knowledge. Much of the information needed to innovate in sectors building on scientific knowledge is, thus, "available only through access to the right persons, often few in numbers, who are working in a given problem area" (Leamer and Storper, 2001, p. 655). The specific knowledge of importance for innovation is seen to be spatially specific, leading some firms to cluster to gain scarce information. "Capital is mobile, new knowledge is comparatively immobile" (Cooke, 2002, p. 9).

As demonstrated in the above discussion, territorial innovation models focus mainly on supply-side factors—i.e. specific factors in an area that stimulate learning, innovation activity and competitiveness of cluster firms. However, demand-side explanations are seen to be of key importance for understanding the clustering of, in particular, professional consultancy firms in large cities (Keeble and Nachum, 2002, p. 85). The rationale of consultancy firms lies in the provision of customised and often novel information, expertise and knowledge to other firms and organisations. Consultancy firms can be co-producers of innovation (den Hertog, 2002) as they may work together with client firms in their innovation process. Proximity to client firms stands out as important, both to those in the same region and to those accessible through the communication nodality of large urban areas. Innovation pressure develops, in particular, when an area has demanding clients and when several local suppliers compete in serving these clients (Porter, 1998).

Clustering processes may be fuelled by consultancy—client interaction. Professional consultancy firms are seen to be developing increasingly into an informal 'knowledge transfer structure' (Strambach, 2001, p. 66), supplying other firms with vital information and knowledge. The knowledge transfer from consultancy firms is often non-routine flows with ambiguous information content that is notably adverse to extension over long distances (Scott and Storper, 2003, p. 582). Then, firms in regions with a large range, availability and high quality of suppliers of knowledge-intensive services may benefit in their innovation activity compared with firms in regions with fewer such suppliers, which may lead to demand for even more professional services in the initially advantaged regions. A self-reinforcing concentration of consultancy and firms demanding good access to external advice and services may occur in some regions. The concentration can be further strengthened by the fact that location in a specific area recognised as a dynamic consultancy cluster may itself signal "quality and credibility to potential clients seeking reassurance in a very uncertain and imperfect service marketplace" (Keeble and Nachum, 2002, pp. 83–84).

Another cluster-building force of potential importance in consultancy is increased importance of project organisations. Time-de-

Table 1. Characteristics of 'old' and 'new' regional clusters

	'Old' regional clusters	'New' regional clusters
Archetypical examples	'Classic' industrial districts	Knowledge-based clusters
Typical industries	Craft-based manufacturing	KIBS
Typical location	Diffused location pattern	Large cities
Important location factors of 'high-order' activities	Areas with a long tradition of developing and producing diversified quality products	Where new and scarce scientific knowledge develops, in transport nodes and near large markets
Important external economies	Local collaboration between specialised firms	Local buzz and rivalry
Type of social institutions	Place-specific, territorial community	Group-specific, professional community
Main process of knowledge spillovers	'Industrial atmosphere'	Innovative interaction with clients and with knowledge organisations
Main cluster-building mechanisms	Supply-side factors; specific, local factors stimulating learning, innovation and the building of competitiveness	Also demand-side factors; close collaboration with demanding clients

limited projects in temporary organisations are seen as an emerging, entirely new type of organisation in industry in general (Ekstedt *et al.*, 1999) and as a dominating practice in—for example, consulting services and the media sector (Grabher, 2002a, 2002b). The project-based organising does seem to contradict important ideas in work on innovation systems, which underlines the importance of long-term relationships within and between firms for successful learning and innovation (see, for example, Cooke 1998). Time-delimited projects in temporary organisations seem to leave too little time for the development of personal relations of mutual trust seen as vital in interactive learning (Grabher, 2002a). Also, organisational learning seems to require a continuity of personnel from project to project in long-term partnering relationships with other firms in the building of tacit knowledge, trust-based relations and shared coding systems that promote the transfer of knowledge between projects and firms (Barlow 2000). Nevertheless, project organisation is also seen as one answer to a growing demand for innovation (Gann and Salter, 2000).

According to Grabher (2002a, p. 209) "repeated project collaboration quite often, though by no means necessarily, takes place in densely-knit clusters".[4] Grabher explains the logic of co-location of project partners in that: spatial proximity allows for significant savings of diverse transaction costs; it provides favourable preconditions for rapid face-to-face interaction that stimulates localised learning processes; and, it facilitates the continuous 'monitoring' of the relevant pool of resources and potential collaborators. However, these 'traditional' arguments only partly capture the logic of co-location. Additionally

> co-location facilitates the emergence of 'interpretive communities' ... [in which] processes of 'negotiating meaning' tie project clusters together (Grabher, 2002a, p. 209).

The theoretical overview above (and summarised in Table 1) provides the point of departure from which to analyse potential clustering effects in the software industry in the Oslo area. In short, knowledge-intensive industries are seen to cluster in large cities because

(1) these new sectors rely quite considerably on the exchange of complex, uncodifiable knowledge, often through time-limited projects;
(2) the knowledge has important elements of tacitness and firm specificity;
(3) new firms and clusters in their early development may rely relatively heavily on local buzz—i.e. on spontaneous observing, monitoring and imitating of nearby rivals and other firms—and the flow of information is often richest in large cities;
(4) some sectors, such as consultancy, deliver non-routine services that benefit from close collaboration with demanding clients; and
(5) interaction between knowledge creators, clients and firms often takes place within short geographical distances, as this kind of knowledge diffusion generally requires face-to-face contact.

3. 'Unpacking' Software Consultancy in the Oslo Region

The theoretical discussion above departs from the need to take context more into consideration—for example, the fact that different processes occur in different industries. Taking this standpoint seriously, in examining the comparatively large number of jobs in the software industry in Oslo, the paper first 'dissects' the sector. That is, the paper analyses the main activities that characterise the sector and how innovation, production and customer relations actually take place 'on the ground'. The paper distinguishes four main activities in the Oslo software industry (Table 2). Individual firms normally perform several of these activities. The activities follow a rough value chain of software production and distribution.

3.1 Platform Suppliers

Platform suppliers deliver generic technology and tools that are the basis for developing software solutions (applications) by other firms. The platform suppliers are mainly large, global and US-based corporations with subsidiary companies or branch offices in the Oslo area. Important customers are other parts of the software industry that use the platform technology to develop their own products and solutions. However, the platform suppliers may themselves also sell applications developed by use of the companies' own technology. Thus, Oracle Norge puts two main solutions on the market, one economy system (ERP, Enterprise Resource Planning) and one sales system (CRM, Customer Relation Management).

The platform suppliers build their competitiveness mainly through large research and development efforts, which primarily take place in the 'home base' of the companies. The technology development and sale of licences (which dominates in the Norwegian branch offices of the US companies) 'drive' other activities. The sale of licences forms the basis of training, support and consulting services. Several of the dominating platform suppliers attained a first-mover advantage by developing standards early on that became a basis for the software industry from the 1980s. The firms attained a high market share, which makes it very hard for new market entrants to challenge the first movers (Wibe and Narula, 2002).

3.2 Software Production

Software production consists of constructing standard solutions, within economy and account, customer relation management, logistics, case handling, portals, web applications, etc., for a large number of customers, who are companies or public organisations. The standard products are put on sale through a net of branch offices and authorised distributors who service local markets in Norway and in other countries. Some companies distribute standard software with little or no modification for individual customers. The software products of

Table 2. Main activities in the software consulting industry

Firms/activities	Products/services	Important clients	Main factors in building competitiveness
Platform suppliers	Basic technology and tools	Software producers and consultants	High R&D effort; first-mover advantage
Software production	Standard software solutions	Organisations that need 'simple' ICT solutions	Continuous upgrading of solutions based on signals from clients
Consultancy	Tailor-made ICT solutions, advice	Advanced ICT users	Re-use of solutions and know-how from project to project
After-sales services	Training, support, running ICT systems	All types of organisations	Dependent on the first three activities

other companies require adaptation for each customer. Even though there are no or only minor modifications, consulting services, such as installation, integration of new solutions in the customers' organisation, training of employees, converting existing data to a new software system, are usually included in the deal. The new solutions normally require changes in ways of working and they require organisational development. The clients' costs on installation, modification, training, etc. amount generally to between one and two times the cost of the software products themselves.

The producers build and maintain their competitiveness by redistributing some of their turnover to product development, which most often consists of a continuous upgrading of existing products. In several firms, about 10 per cent of the turnover is used on development work. The development takes place by using internal competence and often in dedicated R&D departments. In Webmaster Unique, for example, 40 of 160 persons work in product development and product improvement, Agresso has a central R&D division in Oslo employing 200 persons;[5] while in Visma Software, about 100 of 600 persons are developing and improving products. Although product development takes place by using internal competence, signals from clients are important. In some cases, firms have close contacts with demanding Norwegian customers, often found in industries where Norwegian firms or organisations are large and/or early users of software solutions. Development of new products is also based on interpretations of technological and market trends. Some firms obtain information about trends from large, international consulting companies and they co-operate with platform suppliers among other things to obtain knowledge about new technology early on.

3.3 Consulting Projects

The third main group of firms in Table 2, the consultants, has customers that generally are more advanced users of ICT than those who are content with standard solutions. Consulting projects require tailor-made solutions, such as the development and implementation of a new software system for a client. Typically, consulting projects may also include advice on the purchasing and implementation of software products and the analyses of labour processes, organisation of activities, reorganisation and competence needs, and preparing IT strategies.

New tailor-made software solutions are often based on generic tools and/or familiar components and knowledge, which have been developed in previous projects. Based on successful consulting projects, firms may develop standard programmes or solutions, or important knowledge for new tailor-made solutions. Development of standard methods and solutions means converting experience-based (often tacit) knowledge into explicit knowledge (Nonaka and Reinmöller 1998). This knowledge is then used in new consulting projects, which may lead to still more tacit knowledge.

Consulting companies and clients co-operate closely in projects, partly to tailor-make software solutions to the needs of clients, partly so that skills in running the new system remain with the clients after implementation. Some firms work increasingly at the clients' offices and together with their expertise. Some consulting companies also conduct long-lasting projects for clients and act more or less as the IT division of the clients. The consulting firms often specialise in undertaking projects in specific industries. The largest software consulting company in the Nordic countries, TietoEnator AS, aims to service industries in which Nordic firms are world leaders in making use of software solutions.

Consulting companies gain competitive strength through their continuing building of competence and methods, through the effort of individual workers to keep their knowledge up to date and through good routines in diffusing information and knowledge inside the companies. Contact with clients in large projects is important for learning and for the building of competence. To some extent, firms put together project teams consisting of experienced and less experienced employees. Some firms seem to have a sophisticated system for cross-project learning, in which developing and maintaining the firms' intellectual capital are important. Firms have—for example, internal groups responsible for developing dedicated subjects and for diffusing knowledge inside the firms through meetings and websites. Knowledge is also obtained from external specialists. However, firms seek to develop internally what they see as their core competence and competence that can be used in several projects.

3.4 After-sales Services

Most firms perform the last main activity in Table 2, the distribution of products and the running of software systems. Firms train clients in the use of both tailored solutions and many standard software programmes. Companies also offer their clients day and night support in the event of questions and problems with their software solutions. Lastly, some firms operate and manage hardware and software systems for large clients, while others run—for example, clients' accounts and wage systems. A firm like EDB Business Partner operates the old software systems of clients for a definite time-period, while the clients may build competence in running new systems.

The after-sales services are mostly standard services requiring specific routines, such as when firms run clients' systems. However, these services also give some important information on clients' needs to be fed back into product refinement activities. Some firms also see the arrangement of external courses as important in the building of internal competence and in diffusing competence internally and externally. Courses in new technology can most easily find participants in Oslo, which has much of the Norwegian ICT industry, has several potential users and has good transport communications to the rest of Norway.

4. The Clustering of Software Firms in the Oslo Area

The software industry in the Oslo area constitutes a regional cluster due to a concentration of Norwegian software firms in Oslo, and, considerable interaction between the firms. Important inputs in explaining the clustering of software firms to Oslo are found in the above description of the four types of firm. The consulting activity, in particular, fuels the cluster-building process, as this activity is project-based and relation-intensive.

Thus, the clustering process rests first of all on the need for dense interaction between software consulting companies as regards

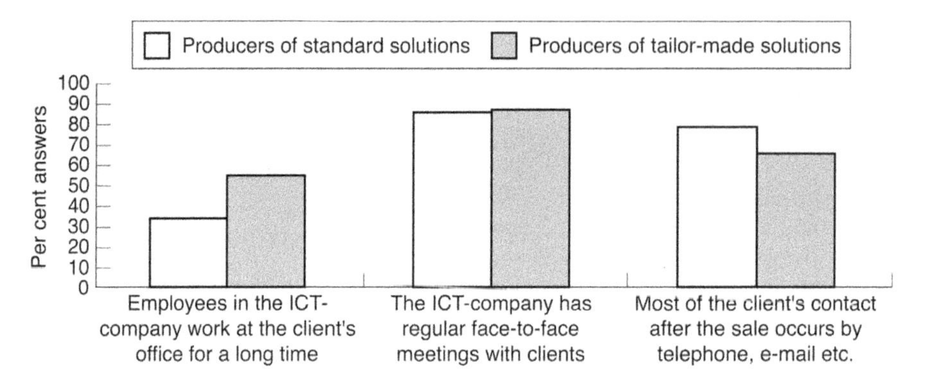

Figure 1. Contact with clients during a project carried out by software consulting companies in the Oslo area ($n = 112$).

large tenders and projects. Companies often have short deadlines and small budgets for preparing tenders. They actively explore possibilities of making alliances. Face-to-face meetings with potential suppliers of components and with other consulting companies are often required. Hence, being located in the same area as several potential collaborators and suppliers is advantageous in preparing tenders. The same applies to the completion of projects. To be capable of carrying out large projects, consulting companies often join together in projects and may help each other out if deadlines are short. In addition, companies may obtain special competence from one-man firms or other consulting companies.

4.1 The Importance of Close Customer Contacts

Another important factor in the clustering process is the need for consulting companies to visit large and important customers so as to influence the decision-makers in order to gain projects. Many of the advanced ICT users, who are actual or potential clients of the consulting companies, are located in the Oslo area. The clients are particularly found in public services; banking, finance and insurance; telecommunications; the energy sector (oil and gas and power plants); and among large manufacturing firms. The Oslo area contained the headquarters of 64 of the 100 largest companies in Norway in 2000 (Jakobsen and Onsager, 2002). Consulting companies often have to visit and 'influence' clients to get orders. Companies often have lasting projects for some clients, in which several employees work on the clients' systems and machines and at their premises. To some extent, consultants 'slide' in at their clients by arranging courses and performing small projects. Through such work, consultants may get a good reputation with the client, which may lead to larger projects. It is then advantageous to be located in an area containing many large, potential clients. Oslo-based consulting companies carry out projects in all parts of Norway and in other countries, and most of the large companies have branch offices in one or several places in Norway to cover different geographical markets. However, as long as the company's main market is in the Oslo area, most of the project work takes place there.

Figure 1, which is based on the telephone survey, shows important aspects of the interaction between software consultants and clients in a project. As regards the production of tailor-made solutions, employees work with the clients' IT systems and in their offices for a long time in more than half of the companies, and more than 80 per cent of these companies have regular face-to-face meetings with clients. This favours a location close to the most important clients.

In 65–80 per cent of the companies, most

of the contact with clients after a contract is signed takes place by telephone, e-mail, etc. Still, the possibilities for the large-scale decentralisation of consulting companies seem highly limited. Producers of tailor-made solutions continually look for new contracts, organise proposals and negotiate with customers about new consulting projects, while producers of standard solutions repeatedly upgrade existing solutions assisted by information from pilot customers, platform producers and the rest of the software industry.

Consulting companies also co-operate with platform suppliers in some projects. The US-based platform suppliers are located in Oslo, which is by far the main economic, political and communication centre of Norway. Yet another cluster-building factor is the large, specialised labour market in the Oslo area. Consulting companies recruit from other software firms, from the universities and university colleges, and from other industries, in particular from industries in which they already have, or are trying to attract, important clients. The large labour market makes it easier to hire and fire employees. Several firm leaders also assert that they get a lot of information 'for free' by being located in a cluster. The concentration of software companies in Oslo results in an information-rich area as regards this industry. Companies obtain a lot of competence in technological and market trends by talking with large suppliers in Oslo and by participating in branch forums, meetings and seminars in this area—i.e. they benefit from the local buzz (Bathelt *et al.* 2002). To some extent, information can be obtained through social interaction. Employees in different companies meet privately and exchange experience. Some of the interviewed firm leaders maintain that a network of managers in software companies exists in Oslo. This kind of social milieu is seen to arise, and to be more fertile, in areas containing a concentration of software firms. It is advantageous to locate where 'the action takes place', where a lot of large projects are carried out, where information and knowledge are flowing, where experienced employees live, etc.

The producers of standard solutions sell their products through a network of distributors. Unlike the producers of tailor-made products, they do not need to have close contact with individual clients. The producers of standard solutions, however, continually upgrade their products. The development goes through several phases, including collecting demands and wishes from clients, prioritising between these demands and wishes, specifying the software solution, designing the software, programming, testing functionality and system, and then releasing and marketing. The mere programming can in principle take place anywhere. It is nevertheless often carried out in the Oslo area where the companies were established and the developers live.

Most of the other phases require face-to-face contact with salesmen, consultants, distributors and pilot customers. These phases are easier to carry out in Oslo where—for example, numerous large clients are found. The producers collect and interpret signals from clients and test prototypes of new software with pilot customers. Distribution of standard products also leads to training, consulting and support activities, which amount to a large share of the clients' costs. Thus, also, producers of standard solutions can serve customers more economically if they are close to them.

4.2 An Embryonic Cluster with Comparatively Small Contact with Knowledge Organisations

The assertions above are confirmed by the results of the telephone survey. Firm leaders assessed the significance of different factors in developing and maintaining the firms' competitiveness (Table 3). The firm leaders value internal know-how and resources highest. However, contact with clients also gains a very high score, pointing to the fact that demand-side explanations may be very important in explaining the concentration of the software industry in Oslo. Although software firms only get about one-third of their turnover from the local market (according to

Table 3. The assessment by software firms in Oslo of the significance of different factors in developing and maintaining firms' competitiveness: evaluation level where 1 means no importance and 6 very important ($n = 123$)

Factors important for firms' competitiveness	Average score
Know-how and resources within the firm or the enterprise	5.6
Contact with clients	5.4
Informal contact with other persons in the software industry	4.1
Contact with suppliers	3.4
Participation in branch meetings, fairs, seminars	3.3
Contact with competitors	3.0
Contact with research, universities, technical colleges	2.5
Contact with finance institutions	2.4
Contact with the public support system	2.1

the telephone survey), the volume of the local market and the presence of demanding clients seem to be important for the clustering of software firms in Oslo. The importance of co-location with important clients is related to the kind of products and services involved in large parts of the software industry. Sixty per cent of the software firms in the survey tailor-make products for clients; much work includes installation and training, in which personnel in the software firms work at clients' offices, and firms may then save time and money in being located very close to the largest market.

The industrial milieu is seen to be of some importance. The milieu includes the four next factors in Table 3: informal contact with other persons in the software industry, contact with suppliers, contact with rivals, and taking part in branch meetings, etc. In fact, contacts with persons in the industry score third-highest when firm leaders assess factors of importance for firms' competitiveness. Thus, information flow in formal and informal settings is seen to be of some importance for firms' competitiveness.

Contact with research institutions, higher education, finance institutions and the support system get the lowest scores. Thus, software firms in general regard contact with R&D institutes to be of minor importance for developing firms' competitiveness. However, the personal interviews revealed that some consulting companies co-operate with universities and research institutes in completing projects, in developing core competence and in recruiting candidates. The interviews also showed that some entrepreneurs in the software industry in Oslo established a firm while working at the university or in an R&D institution. For example, 18 young men from the Department of Informatics at the University of Oslo and from 2 nearby research institutes started the first Internet company in Norway, Oslonett, in 1991 (Steinsli and Spilling, 2002). Half of the co-founders of Oslonett have been involved in other start-ups since 1991. The Department of Informatics was also important as an early user of the Internet technology and for the early training of students in the technology. The entrepreneurs coming from the University of Oslo maintain links with former colleagues at the knowledge organisations—for example, so that they are able to recruit the 'best' students. Case studies thus demonstrate the significant indirect importance of the university and research institutes in Oslo to the development of some types of ICT firms. However, the present successor of Oslonett, which is the result of several mergers and take-overs, will probably not regard the University as being of any vital importance for the company (Steinsli and Spilling, 2002), even though the start in 1991 depended crucially upon university employees.

The low importance ascribed to contact with the knowledge infrastructure is affected by the small size of most firms in the survey. Thus, the importance increases with firm size: the average scores are 2.4, 2.9 and 3.4 for firms with 5–19, 20–99 and more than 100 employees respectively. The importance of contact with research is also higher in firms with an international market than in locally based firms; the scores are 2.2, 2.8 and 3.3 for firms with 50 per cent and more of their turnover in the local, national and international market, in that order. Thus, contact with R&D is clearly seen to be more important for the large and internationally oriented firms.

The telephone survey also demonstrates that software firms in Norway evaluate contact with the knowledge infrastructure as being of somewhat less importance than do firms in some other knowledge-intensive sectors, such as organisational and technical consultancy (parts of NACE 741 and 742), production of instruments (parts of NACE 33) and R&D-activities (NACE 73). The scores are 2.5 for the software firms, and 3.2, 3.2 and 4.7 respectively for firms in the other sectors. Firm managers in the four industrial sectors consider contact with clients of nearly identical importance. However, firm managers in the software industry assess contact with several kinds of players, such as competitors, finance institutions and the public support system to be of less importance than managers in the other sectors. These results *may* point to the fact that the software industry is a comparatively new and fast-growing industry in Norway, in which firms may not have developed contact with other players to the same degree as in more mature sectors. The software firms in the survey are on average 11 years old, while the organisational and technical consultancy firms are on average 17 years, firms in instrument production 19 years and firms in the R&D sector 21 years. These results seem to coincide with the hypothesis that rivalry, observation and comparison are key features in the early development of clusters (Malmberg and Maskell, 2002, p. 442) and thus important in the Oslo software cluster. Thus, several firm leaders stressed the importance of being located in an area where numerous software projects are carried out in order to pick up ideas and information from other software firms, from platform suppliers and in seminars and informal settings. Local collaboration, alongside institutional adjustment, may become more important as times goes by (Malmberg and Maskell, 2002).

5. Conclusion

This paper examines the reasons for the clustering of the Norwegian software industry in the Oslo area. The clustering of software companies in Oslo rests first of all on the need for close interaction between consulting companies and decision-makers and IT personnel of important customers, among software consulting companies themselves and between consulting companies and platform suppliers. The fact is that consulting activity is project-based and involves much coalition-building and face-to-face contact, which is facilitated when players co-locate. Much of the interaction between clients and software companies and between software companies, seems to involve uncodifiable and complex knowledge. Persons have to meet face-to-face when preparing tenders, collaborating in projects, discussing new software solutions, training clients in new software and so on. Knowledge spillovers in informal settings also seem an important advantage for software firms in Oslo and firms base their activities upon pre-existing ensembles of software firms and knowledge. The Oslo software industry follows the same development as many clusters that are in the early stages of their 'life cycle'; new industries create localised externalities, which causes new activity to cluster around them (Storper and Walker, 1989). Concentration is also stimulated by the large, specialised labour market in Oslo.

While knowledge spillovers and localised externalities are important, the most straightforward reason for the clustering of software firms in Oslo is the need to be where the largest market is. Most activities in the software industry need to be close to decision-makers in large public organisations and in private companies, and software consultants often need to work on the clients' IT systems. Software solutions are typical products that cannot be dropped at the doorstep, but are services that have to be delivered by one human to another (Leamer and Storper, 2001, pp. 649–650). Thus, the clustering of software firms in Oslo is strongly influenced by the benefits of proximity and accessibility to clients in Oslo. Demand-side factors then seem to be very important in understanding the concentration of this kind of professional service firms.

The cluster-building factors in the Oslo software industry agree very much with the characteristics of 'new' regional clusters in Table 1. Demand-side factors are of key importance for explaining the clustering of software firms in Oslo: firms benefit by being in the largest market as much collaboration with clients demands face-to-face interaction and working at clients' IT systems. Software firms also pick up a lot of information in formal and informal interaction with other local software firms and suppliers, by participating at branch meetings in Oslo and by observing and monitoring nearby rivals. However, firm managers in the software industry evaluate contact with knowledge organisations, suppliers and competitors to be of less importance for the firms' competitiveness compared with firm managers in some other knowledge-intensive sectors. This result points to the fact that the software industry in Oslo for the time being may be a rather weak and embryonic cluster.

The Oslo software industry may, however, develop into a stronger regional cluster later on. The software industry has grown very fast during the past decade and contains many young firms. The development of external economies and local collaboration that characterises a strong regional cluster takes some time. Extensive local collaboration beyond market transactions requires the development of informal rules and conventions shared by local players that help to co-ordinate firms' activity under conditions of uncertainty (Storper, 1997). The 'light' institutions that stimulate 'cluster effects' include organised activity, such as meeting-places, common services, associations and state support, as well as an informal contact network (Amin and Thrift, 2002). Such informal, 'untraded' relationships may over time grow in importance in the Oslo software industry to supplement the client interaction and the local buzz that now initially trigger the cluster-building process.

Notes

1. The 14 firms have nearly 4000 employees in the Oslo area.
2. The telephone survey included NACE 72, and not only NACE 722, in order to obtain a larger sample of firms.
3. At least five distinct strands of research can be identified: research on industrial districts, focusing especially on the Third Italy (Piore and Sable, 1984; Becattini, 1990); the rather close analyses of the GREMI group of economists on innovative milieux (Camagni, 1991); the Californian school of new industrial spaces (Scott, 1988; Storper and Scott, 1989); the Nordic school of the learning economy (Lundvall and Johnson, 1994; Asheim, 1996); and, Porter's influential work on industrial and regional clusters (Porter, 1990, 1998).
4. In the London advertising industry, Grabher (2002b) sees the local cluster as but one element in project-based production contexts besides firms, (non-local) personal networks and global corporations. Thus, project work may involve a range of different scales, but in general project work is quite often organised inside clusters according to Grabher (2002a).
5. Agresso employs about 300 persons in Norway and 1350 persons in total.

References

ACS, Z. J., GROOT H. L. F. DE and NIJKAMP, P. (2002) Knowledge, Innovation and Regional Development, in: Z. J. ACS, H. L. F. DE GROOT and P. NIJKAMP (Eds) *The Emergence of the Knowledge Economy*, pp. 1–14. Berlin: Springer.

ALDERMAN, N. (forthcoming) Mobility versus embeddedness: the role of proximity in low volume capital projects, in: A. LAGENDIJK and P. OINAS (Eds) *Proximity, Distance and Diversity: Issues on Economic Interaction and Local Development.* Aldershot: Ashgate.

AMIN, A. (1994) Post-Fordism: models, fantasies and phantoms of transition, in: A. AMIN (Ed.) *Post-Fordism*, pp. 1–39. Oxford: Blackwell.

AMIN, A. (2000) Industrial districts, in: E. SHEPPARD and T. J. BARNES (Eds) *A Companion to Economic Geography*, pp. 149–168. Oxford: Blackwell.

AMIN, A. and THRIFT, N. (1994) Living in the global, in: A. AMIN and N. TRIFT (Eds) *Globalization, Institutions, and Regional Development in Europe*, pp. 1–22. Oxford: Oxford University Press.

AMIN, A. and THRIFT, N. (2002) *Cities. Reimagining the Urban*. Cambridge: Polity.

ARMSTRONG, H. and TAYLOR, J. (1993) *Regional Economics and Policy*. New York: Harvester Wheatsheaf.

ASHEIM, B. T. (1992) Flexible specialisation, industrial districts and small firms: a critical appraisal, in: H. ERNSTE and V. MEIER (Eds) *Regional Development and Contemporary Industrial Response: Extending Flexible Specialisation*, pp. 45–63. London: Belhaven.

ASHEIM, B. T. (1996) Industrial districts as 'learning regions': a condition for prosperity?, *European Planning Studies,* 4, pp. 370–400.

BARLOW, J. (2000) Innovation and learning in complex offshore construction projects, *Research Policy,* 29, pp. 973–989.

BATHELT, H., MALMBERG, A. and MASKELL, P. (2002) *Clusters and knowledge: local buzz, global pipelines and the process of knowledge creation*. DRUID Working Paper No. 02-12.

BECATTINI, G. (1990) The Marshallian industrial districts as a socio-economic notion, in: F. PYKE, G. BECATTINI and W. SENGENBERGER (Eds) *Industrial Districts and Inter-Firm Cooperation in Italy*, pp. 37–51. Geneva: International Institute for Labour Studies.

BROUWER, E., BUDIL-NADVORNIKOVA, H. and KLEINKNECHT, A. (1999) Are urban agglomerations a better breeding place for product announcements?, *Regional Studies,* 33, pp. 541–549.

CAMAGNI, R. (1991) (Ed.) *Innovation Networks: Spatial Perspectives*. London: Belhaven.

CAMAGNI, R. and CAPELLO, R. (2002) Milieux innovateurs and collective learning: from concepts to measurements, in: Z. J. ACS, H. L. F. DE GROOT and P. NIJKAMP (Eds) *The Emergence of the Knowledge Economy*, pp. 15–45. Berlin: Springer.

COOKE, P. (1998) Introduction: origins of the concept, in: H.-J. BRACZYK, P. COOKE and M. HEIDENREICH (Eds) *Regional Innovation Systems*, pp. 2–25. London: UCL Press.

COOKE, P. (2002) *Knowledge Economies: Clusters, Learning and Cooperative Advantage*. London: Routledge.

EKSTEDT, E., LUNDIN, R. A., SÖDERHOLM, A. and WIRDENIUS, H. (1999) *Neo-industrial Organising: Renewal by Action and Knowledge Formation in a Project-intensive Economy*. London: Routledge.

GANN, D. M. and SALTER, A. J. (2000) Innovation in project-based, service-enhanced firms: the construction of complex products and systems, *Research Policy,* 29, pp. 955–972.

GAROFOLI, G. (1992) Diffuse industrialisation and small firms: the Italian patterns in the 1970s, in: G. CAROFOLI (Ed.) *Endogenous Development and Southern Europe*, pp. 83–102. Aldershot: Avebury.

GLAESER, E., KALLAL, H., SCHEINKMAN, J. and SHLEIFER, A. (1992) Growth of cities. *Journal of Political Economy,* 100, pp. 1126–1152.

GRABHER, G. (2002a) Cool projects, boring institutions: temporary collaboration in social context, *Regional Studies,* 36, pp. 205–214.

GRABHER, G. (2002b) The project ecology of advertising: tasks, talents and teams, *Regional Studies,* 36, pp. 245–262.

HERTOG, P. DEN (2002) Co-producers of innovation: on the role of knowledge-intensive business services in innovation, in: J. GADREY and F. GALLOUJ (Eds) *Productivity, Innovation and Knowledge in Services*, pp. 223–255. Cheltenham: Edward Elgar.

ISAKSEN, A. and ASLESEN, H. W. (2001) Oslo: in what way an innovative city?, *European Planning Studies,* 9(7), pp. 871–887.

ISARD, W. (1956) *Location and Space-Economy. A General Theory Relating to Industrial Location, Market Areas, Land Use, Trade and Urban Structure*. Cambridge, MA: MIT Press.

JAKOBSEN, S. E. and ONSAGER, K. (2002) *Geografiske konsentrasjoner av hovedkontorer—funksjoner, behov og eksterne effekter*. Working Paper A 53/2002, Institute for Research in Economics and Business Administration, Bergen.

KEEBLE, D. and NACHUM, L. (2002) Why do business service firms cluster? Small consultancies, clustering and decentralisation in London and southern England, *Transactions of the Institute of British Geographers,* 27, pp. 67–90.

LEAMER, E. E. and STORPER, M. (2001) The economic geography of the internet age, *Journal of International Business Studies,* 32, pp. 641–665.

LUNDQUIST, K.-J. (1996) *Företag, regioner och internationell konkurrens*. Meddelanden från Lunds Universitets Geografiska Institutioner avhandlinger Nr 129. Lund: Lund University Press.

LUNDVALL, B.-Å. and JOHNSON, B. (1994) The learning economy, *Journal of Industry Studies,* 1, pp. 23–42.

MACKINNON, D., CUMBERS, A. and CHAPMAN, K. (2002) Learning, innovation and regional development: a critical appraisal of recent debates, *Progress in Human Geography,* 26, pp. 293–311.

MALMBERG, A. and MASKELL, P. (2002) The elusive concept of localization economies: towards a knowledge-based theory of spatial clustering, *Environment and Planning A,* 34, pp. 429–449.

MARSHALL, A. (1890) *Principles of Economics.* London: Macmillan.

MASKELL, P., ESKELINEN, H. and HANNIBALSSON, I. (1998) *Competitiveness, Localised Learning and Regional Development: Specialisation and Prosperity in Small Open Economies.* London: Routledge.

MOSELEY, M. J. (1974) *Growth Centres in Spatial Planning.* Oxford: Pergamon Press.

MOULAERT, F. and SEKIA, F. (2003) Territorial innovation models: a critical survey, *Regional Studies,* 37, pp. 289–302.

NONAKA, I. and REINMÖLLER, P. (1998) *The legacy of learning, Berlin.* WJB Jahrbuch 1998. Berlin: Wichenshaft Zentrum.

NORTON, R. and REES, J. (1979) The product cycle and the spatial decentralisation of American manufacturing, *Regional Studies,* 13, pp. 141–151.

OECD (ORGANISATION FOR ECONOMIC CO-OPERATION AND DEVELOPMENT) (2001a) *OECD science, technology and industry scoreboard 2001—towards a knowledge based economy.* Paris: OECD.

OECD (2001b) *Issues paper.* Paper presented at the *World Congress on Local Clusters: Local Networks of Enterprises in the World Economy.* Paris, January.

PIORE, M. J. and SABEL, C. F. (1984) *The Second Industrial Divide: Possibilities for Prosperity.* New York: Basic Books.

PORTER, M. (1990) *The Competitive Advantage of Nations.* London: Macmillan.

PORTER, M. (1998) Clusters and the new economics of competition, *Harvard Business Review,* November/December, pp. 77–90.

ROSENFELD, S. (1997) Bringing business clusters into the mainstream of economic development, *European Planning Studies,* 5, pp. 3–23.

SASSEN, S. (2000) *Cities in a World Economy,* 2nd Edn. Thousand Oaks, CA: Pine Forge Press.

SAXENIAN, A. (1996) *Regional Advantage: Culture and Competition in Silicon Valley and Route 128.* Cambridge, MA: Havard University Press.

SCOTT, A. J. (1988) *New Industrial Spaces.* London: Pion.

SCOTT, A. J. and STORPER, M. (2003) Regions, globalization, development, *Regional Studies,* 37, pp. 579–593.

STEINSLI, J. and SPILLING, O. R. (2002) *On the role of small firms in cluster evolution: the case of internet development in Norway during the 1990s.* Paper, Norwegian School of Management, BI, Oslo.

STORPER, M. (1997) *The Regional World: Territorial Development in a Global Economy.* New York: Guilford Press.

STORPER, M. and SCOTT, A. J. (1989) The geographical foundations and social regulation of flexible production complexes, in: J. WOLCH and M. DEAR (Eds) *The Power of Geography: How Territory Shapes Social Life,* pp. 21–40. Winchester, MA: Unwin Hyman.

STORPER, M. and WALKER, R. (1989) *The Capitalist Imperative. Territory, Technology, and Industrial Growth.* New York: Basil Blackwell.

STRAMBACH, S. (2001) Innovation processes and the role of knowledge-intensive business services (KIBS), in: K. KOSCHATZKY, M. KULICKE and A. ZENKER (Eds) *Innovation Networks: Concepts and Challenges in the European Perspective,* pp. 53–68. Heidelberg: Springer Physica.

TREND BUSINESS RESEARCH (2001) *Business Clusters in the UK: A First Assessment.* London: Department of Trade and Industry.

WIBE, M. and NARULA, R. (2002) Interactive learning and non-globalisation: knowledge creation by Norwegian software firms, *International Journal of Entrepreneurship and Innovation Management,* 2, pp. 224–245.

Like Phoenix from the Ashes? The Renewal of Clusters in Old Industrial Areas

Franz Tödtling and Michaela Trippl

[Paper first received, February 2003; in final form, November 2003]

1. Introduction

Industrial clusters and their relation to regional growth and competitiveness have become a prominent theme in regional studies and policy in recent years (Porter, 1990, 1998; Baptista and Swann, 1998; Steiner, 1998; OECD, 1999; Cooke, 2002; Benneworth *et al.*, 2003; Enright, 2003). Often, such approaches are focused on dynamic regions or high-tech industries representing the growth phase of cluster development (Saxenian, 1994; Swann and Prevezer, 1996; Keeble and Wilkinson, 2000). Less work has been done so far on the transformation of clusters in mature regions and industries. For these, the assumption often is that they are at the 'end' of their life cycle with little prospects for growth and development (Tichy, 2001). However, a closer look at such regions reveals that under certain conditions interesting processes of renewal may occur, opening up new paths for development and innovation (Cooke, 1995; Braczyk *et al.*, 1998; Cooke *et al.*, 2000; Steiner and Hartmann, 2002). In the present paper, we will investigate such renewal processes of clusters in old industrial regions, looking more closely at the transformation of two clusters (automotive and metal) in the region of Styria in Austria. We are interested in particular in the following questions

—To what extent can we observe processes of renewal (i.e. new activities and innovation) in these clusters and, if so, who are the driving actors in this respect?
—To what extent can we identify the formation of innovation networks and what is the role of the regional innovation system?
—How important are institutional factors and policy approaches for cluster renewal?

The Styrian automotive and metal clusters have been selected in this context because they represent two important clusters in the region and because they show interesting differences regarding the renewal process and underlying institutions. We are going to compare their pre-conditions and restructuring performance and relate the findings to the theoretical concepts presented below.

In the following, we will first present some literature background to these questions. Then, the Styrian economy and its innovation system will be briefly characterised before we deal with the two clusters in more detail. For each cluster, we will present the most important actors, the challenges for adjustment, the formation of innovation networks, as well as socio-institutional factors and the role of policy. In the concluding section, we will summarise the most important results and point out the key factors responsible for the processes observed.

2. Conceptual Background

In the literature, we find contrasting views regarding the questions raised. Tichy's (2001) cluster cycle theory, for example, argues that many features of old industrial regions result from the fact that they are 'overspecialised' in mature industries and clusters. There is stagnating demand for their products, fierce competition from low-cost locations, a high degree of concentration in the industry and a low potential for networking and innovation. The underlying problems are a too strong specialisation in formerly successful industries, a too-strong role of dominant companies and too much reliance on old economic and political networks (Grabher, 1991). Policy suggestions derived from this view are a plea for industrial diversification at an early stage, the stimulation of entrepreneurship and competition and the avoidance of an explicit cluster policy. Although Tichy's approach captures many relevant aspects of cluster development, it seems to be, like other variants of product and industry cycles, too deterministic regarding suggested outcomes and the stylised development path. Institutional differences between regions and clusters as well as the role of policy approaches and actors are not taken sufficiently into account.

The literature on industrial districts, innovative milieux and regional innovation systems points to some important factors in this respect which may have relevance also for the renewal of clusters in old industrial regions. The work on industrial districts (Garofoli, 1991; Pyke and Sengenberger, 1992; Asheim, 1996) demonstrates that local industrial specialisation in mature industries need not necessarily lead to a loss of entrepreneurship and innovation. Although we have observed some firm concentration in industrial districts and a penetration by international firms in recent years, this literature shows that SMEs, by forming co-operative networks at the regional level, can achieve scale economies, collective learning and a good innovation performance (Maskell *et al.*, 1998; Maskell and Malmberg, 1999; Asheim, 2000). This allows them to defend niche markets in high-quality products and to open up new markets. An important precondition for networks and collective learning is a certain level of trust among actors as well as support through regional development agencies and the provision of 'real services' in the fields of research, development, training and marketing. Trust as a form of social capital (Putnam, 1993) is not just a precondition but is also a result of networking—i.e. it accumulates through interaction (Sabel, 1992; Wolfe, 2000). For old industrial regions which are usually characterised by a low level of trust, this implies that trust can be 'built'—for example, through deliberate efforts to enhance contacts and create new networks among actors in the region.

Additional insights can be gained from the literature on innovative milieux and regional innovation systems (RIS). Both argue that the innovation process is highly interactive and that innovation networks, both regional and beyond, are of key importance (Camagni, 1991; Edquist, 1997; Lundvall and Borràs, 1998; Braczyk *et al.*, 1998; Cooke *et al.*, 2000; Thomi and Werner, 2001; Bathelt and Depner, 2003). The RIS approach furthermore stresses institutional factors for innovation by pointing out two sub-systems: one for knowledge generation and diffusion and the other for knowledge application and exploitation (Autio, 1998). The first sub-system consists of research and development organisations, education and technology transfer. The second sub-system is made up by the industrial companies and the vertically and horizontally related firms (i.e. the clusters of the region). For a dynamic innovation process, both well developed clusters and organisations of knowledge generation and diffusion are required. Of particular importance according to the RIS approach are the linkages between these elements and sub-systems both within the region and with higher spatial levels (national innovation system, European and global levels.). The RIS concept has been applied to a variety of regions in Europe (Braczyk *et al.*, 1998; Cooke *et al.*, 2000; Kaufmann and Tödtling, 2001; Koschatzky and Sternberg, 2000; Asheim *et al.*, 2003) and North America (de la Mothe and Paquet, 1998). Included were high-performing regions, industrial districts and old industrial areas. It has been demonstrated that the nature of the innovation process, the actors and networks differ quite strongly between these types of region, depending on the respective institutional background. Old industrial areas often have a high density of relevant institutions and organisations, but they face the problem that many of these institutions are too much oriented towards the old industries and clusters. The challenge is to bring in new technological orientations as well as new and more interactive forms of innovation. Still, in the studies cited, it was found that there is considerable renewal and innovation in old industrial areas. In some of them, innovation was triggered by the attraction and the integration of transnational companies (for example, in Wales). In others, such as Styria and Tampere, there was more reliance on a strong sub-system of knowledge generation and diffusion which was combined with an explicit focus on specific clusters as regards knowledge exploitation. In a third type (Basque country, Valencia), regionalised technology and innovation policy (such as technology and innovation centres) has played a key role in the renewal process.

The literature on innovation systems, however, does not provide a sophisticated definition of the concept of 'system' (see also Kaufmann and Tödtling, 2001). Systems are conceptualised in a rather simple and traditional way as being constituted by elements and their relationships. Additionally, little is said about how a system emerges and develops over time. Modern social system theory helps us to gain a deeper understanding of how system-building takes place (see also Bratl and Trippl, 2001). Work in this field (Teubner, 1987; Willke, 1991) has highlighted that there are several stages to differentiate, covering the following essential operations: selective communication, boundary-drawing; building-up of structures; co-ordination of processes, structures and resources; self-thematisation and reflexivity; and the developing of strategic and generative abilities. It is these operations and processes that distinguish systems from other social phenomena. They enable us to deal in a more profound way with the systemic aspects of clusters and innovation systems and to assess their degree of 'systemness'—i.e. the level of integration of their actors by networks, institutions for co-ordination and a collective identity and reflexivity.

Finally, the new theoretical thinking on policy-making emphasises the importance of 'associative governance' (Cooke and Morgan, 1998), more interactive modes of state intervention (Mayntz, 1997; Messner, 1998) and, consequently, a new role for public authorities as a facilitator of systemness, when it comes to the shaping of cluster development and transformation. For old industrial regions, which are often characterised by outdated routines of direct intervention, this implies the need for substantial policy changes and new forms of governance.

Summarising the results from the literature regarding the renewal of industrial clusters in old industrial regions, we can point out the following:

—Clusters in such regions often face the problems of mature industries such as stagnating demand, high competition and a

Table 1. Key indicators for Styria

Area[a]	16 391.93 square km	(19.5 per cent of Austria)
Population (2001)[a]	1 183 303	(14.7 per cent of Austria)
Number of firms (2000)[b]	39 701	(13.5 per cent of Austria)
Rate of unemployment (2001)[b]	6.5 per cent	(Austria: 6.1 per cent)
Employment structure (2001)[b]		
Agriculture	4 236	(1 per cent)
Industry	140 968	(33 per cent)
Services	278 078	(66 per cent)
GDP per capita (index) (1999)[a]		
Austria	100	(Euro 24 300)
Styria	85	(Euro 20 600)

[a]Statistik Austria (2003).
[b]Amt der Steiermärkischen Landesregierung (2002).

'lock in' into old technology paths. Networks often do not exist within the region or they are oriented towards the old development path. Despite these unfavourable features, a renewal of clusters seems to be possible under certain conditions. The following steps and factors might contribute to such a renewal process.

—The renewal of clusters can be supported by a well-developed RIS. Strong institutions of knowledge generation and diffusion might help companies to build bridges to new technology paths (Kaufmann and Tödtling, 2000). This requires that the respective knowledge institutions are not exclusively focused on the old clusters and trajectories but have also a more generic character.

—Clusters in old industrial regions are often characterised by either fragmentation (few links within the region) or by networks oriented towards the old trajectory. Thus, the renewal of clusters implies also a renewal of networks. This includes new links with knowledge suppliers as well as innovation networks among companies. Although the region is an important spatial level regarding these networks, links to the national innovation system and to international partners are also necessary for innovation (Cooke *et al.*, 2000).

—The attraction of leading transnational companies may have a positive effect on cluster renewal, if they bring in complementary knowledge to the cluster and if they can be integrated into regional supplier and innovation networks.

—An active policy is needed to overcome the situation of 'lock in'; market forces alone will not be sufficient to improve the situation. However, a different kind of policy is required from the one applied in the past. Direct intervention and the provision of subsidies and infrastructure are generally not enough. Important are soft measures such as the stimulation of networks and the enhancement of 'systemness' within clusters and an upgrading of the regional innovation system.

In the following, we shall explore the questions and issues raised empirically for the automotive and metal clusters in the old industrial region of Styria.

3. The Styrian Economy in Transition

Styria, situated in the south east of Austria, is the second-largest of Austria's nine provinces and one of the industrial heartlands of the national economy. Some key indicators for the region are shown in Table 1.

The region of Styria is a typical example of an old industrialised area in which a dominating industrial cluster had promoted

Table 2. Growth performance of Styria in comparison with Austria

	Austria	Styria
GDP 1995–99 (1995 = 100)	114.0	114.8
GDP per capita 1995–99 (1995 = 100)	113.5	115.1
Number of companies 1996–2000 (1996 = 100)	104.0	110.4
Rate of unemployment		
1996	7.0	8.4
2001	6.1	6.5

Source: Authors' calculations based on Statistik Austria (2003), Amt der Steiermärkischen Landesregierung (2002).

growth and prosperity for a long time but has led to decline when external conditions changed. In the post-war Fordist era, Styria had been a major growth pole of the Austrian economy and its prosperity had been based on industries such as iron, steel and metal products. However, in the 1970s and 1980s these traditional branches experienced a severe crisis and—due to both an overspecialisation of the region in these industries and a weakly developed adjustment capability—Styria turned into an old industrialised area. Although the 1980s witnessed an intensive restructuring process, it has been only since the mid 1990s that a recovery of the Styrian economy has become clearly visible (see Table 2). Several recent studies (Geldner, 1998; Tödtling *et al.*, 1998; Adametz and Novakovic, 2000; Janger *et al.*, 2000; Amt der Steiermärkischen Landesregierung, 2002) show that a process of catching-up with respect to growth, employment, new firm formation and innovation has set in and that a structural shift from basic to more technology-intensive branches has occurred.

The successful restructuring of the economy seems to be linked to a well-developed regional innovation system. Applying Autio's (1998) categories, we find that Styria has a strong knowledge generation and diffusion system (Table 3). The Technical University of Graz has strengths in machinery, electrical engineering and construction engineering, whereas the University of Leoben has an international reputation for competencies in mining and materials technologies. Joanneum Research is a large contract research institution (owned by the Land), with competencies in environment/energy, information technology, electronics and sensor technologies, materials and medical technology. All these research institutions actively develop interfaces with industry: the universities have liaison offices as well as co-operative R&D institutions attached. There are 15 'Christian Doppler laboratories' engaged in co-operation with firms with the aim of developing new products and processes. Of more recent origin are nine 'competence centres' supported by the federal state. They are co-operative institutions between university institutes and firms with the task of engaging in joint basic and applied research. A new element (since 1995) in the Austrian and Styrian innovation systems are the technical colleges (Fachhochschulen) providing more applied education and training than the universities. In Styria currently 22 degree programmes are offered, covering fields like design, automotive technologies, information management, industrial electronics and business administration. Also present are several institutions providing vocational training as well as a range of technology centres and incubators for start-up firms. Another important actor in the region is the Styrian Development Agency (SFG). It was founded in 1991 as an independent semi-public agency, controlled and financed by the Land government. The agency supports new and growing firms, innovation projects of firms and the

Table 3. The regional innovation system in Styria

Contract research *Joanneum Research* Sustainability and environment Information technology Electronics and sensor technology Materials and processing Economy and technology Medical technology	*Science/university education* *Technical University Graz, Karl Franzens University Graz, University Leoben* Architecture, construction engineering, machinery, electrical engineering, natural sciences Law, social sciences/economics, medicine, humanities, natural sciences Mining, materials	*Technical colleges* *Technikum Joanneum, WIFI Steiermark: 22 degree programmes, including* Design Automotive technologies Information management Industrial electronics Business administration
Co-operative R&D institutions at universities 15 Christian Doppler Laboratories *Competence Centres* Materials Centre Leoben Polymer Competence Centre Knowledge Management Centre Centre of Competence in Applied Biocatalysis Austrian Bioenergy Centre Virtual Vehicle Competence Centre for Timber Construction Acoustic Competence Centre Competence Centre for Interactive e-business	*Companies* *Cluster structures* Automobile Metal and machine building Wood/paper Chemistry/pharmacology Information technology	*Vocational training* *WIFI, BfI, HTL* Very broad range of training programmes in business administration and technology
Technology transfer/consultancy Liaison offices of the universities TTZ Leoben Agiplan, Trigon and other consulting firms	*Technology centres/incubation centres* Three technology centres Several incubation centres (two of them for academic spin-offs) 'Wirtschaftspark Obersteiermark' as network	*Public support/finance* *SFG, Innofinanz* Subsidies Finance Regional development policy Technology policy Cluster policy

setting-up of incubators and technology centres. More recently, it has become involved in the development of clusters such as the automotive cluster described further below. Regarding the knowledge application and exploitation sub-system, we find in Styria cluster structures with different degrees of development. Adametz *et al.* (2000), by making use of a regional econometric input–output model, identify five important clusters (i.e. 'cores' of interlinked economic activity) within the Styrian economy: metals and machine building, automobile, chemistry/pharmacology, wood/paper and information technology.

In the following, two of them—namely, the automotive cluster and the metal cluster—will be examined in detail. As Table 4 shows there are huge differences between them. The metal cluster is rather large in terms of numbers of firms and employees. It includes about 450 firms with a workforce of 20 500. The automobile cluster, on the contrary, only comprises about 60 firms with 7900 employees.[1] More interesting, however, is the fact that they differ significantly with respect to their actual development. While the automobile cluster has been one of the fastest-growing in the Styrian economy in recent years, the metal cluster has shown a far less positive performance. Between 1995 and 1998, the automobile cluster's growth both in terms of employment (+ 23 per cent) and real output (+ 92 per cent) has clearly been above the respective national average. In the same period, the metal cluster has experienced a higher loss of employment (− 1.6 per cent) than the national average and a growth of output (+ 14.5 per cent) below the national average. In the past, both clusters were confronted with major external threats and neither of the two seemed to be equipped with sufficient capacity to meet the new challenges. Nevertheless, both the automobile cluster and the metal cluster have—as we will show in the following—succeeded in adapting to the new circumstances by undergoing a far-reaching transformation.

The empirical base for the following analysis was established in a series of 58 interviews, 30 of which were held with stakeholders from the metal cluster and 28 with actors from the automotive cluster in the years 1998, 1999 and 2001. In the case of the automotive cluster, 15 interviews were taken with representatives of firms. The sample included all leading companies, some other firms of larger size and a selection of smaller companies. Furthermore, 5 actors from universities and non-university research organisations were chosen as interview partners. Finally, 8 interviews were carried out with other selected organisations, including consultancy firms, technology transfer institutes, the regional development agency, the chamber of commerce and the Styrian association of industrialists. Regarding the metal cluster, 17 interviews were carried out with top executives of firms, mainly of the leading companies. Additionally, 6 interviews were conducted with university and other research institutes and 7 with other key regional players amongst which were technology transfer organisations, technology parks and organisations at the local level established to attract inward investment.

4. The Automotive Cluster

In recent years, the automotive cluster has proved to be an 'engine of growth' for Styria. The core of the cluster consists of some larger, internationally known firms. There is Steyr Fahrzeugtechnik—SFT (now belonging to Magna and therefore renamed in Magna Steyr), Austria's oldest motor vehicle company. It is developing and manufacturing complete cars and automotive components, systems and modules, and has a special competence in four-wheel-drive technology. Another home-grown key player is AVL List, a firm engaged in research and development of combustion engines, and in control engineering and acoustics. In the past two decades—with the help of Bund and Land subsidies—the cluster has been successful in attracting foreign direct investment. Firms like DaimlerChrysler, Magna, Johnson Controls and Lear have entered the region. Among the foreign-owned companies,

Table 4. The automotive cluster and the metal cluster in comparison

	Automobile cluster	Metal cluster
Number of firms	60	450
Number of employees	7 900	20 500
Percentage growth in employment (1995–98)	+ 23.3 (national average: + 6.0)	− 1.6 (national average: − 0.8)
Real output (billion Euros)	2.45	3.26
Growth in output (1995–98)	+ 92 (national average: + 25.6)	+ 14.5 (national average: + 20.2)
Output per employee (Euros)	320 000	160 000
Percentage growth in output per employee (1995–98)	+ 50	+ 13.9

Source: Adametz *et al.* (2000)

Magna is the most important one at present. It is a global supplier to the automobile industry and runs several factories in Styria. DaimlerChrysler arrived at the beginning of the 1990s and established Eurostar, an assembly plant which is now owned by Magna. Other firms present include Remus (exhaust pipes), Pankl (precision components for racing cars) and Steirisches Druckgußwerk (aluminum castings), to name but a few. There are numerous SMEs, mainly suppliers of components and parts, as well as some engineering firms.

A well-developed knowledge generation and diffusion system at the regional level with strong competencies in the automotive field (including key actors like the Technical University of Graz, several institutes of Joanneum Research and the technical college for automotive engineering) back up the cluster firms in their innovation processes.

4.1 New Challenges and the 'Fragmentation Trap'

In recent years, the world automotive industry has undergone a substantial restructuring process (Graves, 1994; Dicken, 1998; Jurgens, 1999; Wolters, 1999). Market growth slowed down; segmentation and fragmentation of demand intensified; product and innovation cycles became drastically shortened: all leading to a fierce competition in the industry. Consequently, firms are confronted with the double imperative to lower production costs and to develop strong innovation competencies. As a result of these challenges, the value chains in the automotive sector have been rearranged. Global sourcing for standardised parts and a trend towards system suppliers for more sophisticated components (implying strong localising forces in spatial terms) have gained momentum.

At the beginning of the 1990s, the capabilities of the Styrian automobile industry to respond to this new, demanding environment seemed to be rather limited. One of the main reasons constraining an adequate adjustment lay in the cluster's poorly developed relational dimension prevalent at that time. There was a marked lack of interaction between the regional firms in the industry. Neither input–output linkages nor non-market relationships in significant proportions could be identified. The Styrian automotive sector showed all the features of an agglomeration of disconnected firms rather than of a cluster working through a variety of networks allowing information exchange and collaboration. In analysing the low level of interaction, the specific socio-institutional endowment emerges as a critical explanatory factor. Indeed, until 10 years or so ago, the cluster was characterised by a hostile climate for co-operation as the norms of reciprocity and sufficient levels of trust necessary to encourage professional interaction and collaborative behaviour were absent. Firms lacked a shared vision and identity at the cluster level necessary for taking full advantage of its competencies and collective strength. The cluster seemed to be caught in a 'fragmentation trap', unable to provide the urgently required collective response to the threats of foreign competition and new market conditions.

4.2 Innovation Networks

Thus far, regional trade linkages in the cluster are not very strong. A recent analysis by Adametz *et al.* (2000) shows that its firms receive not more than 16.7 per cent of their inputs from regional suppliers and that only 2.4 per cent of the output is going to Styrian firms. The cluster's export rate is extremely high; about 95 per cent of its output goes to foreign markets. However, there is some indication of rising innovation networks within the region. In the above-mentioned analysis, strong linkages between industry and knowledge suppliers were found. More than 58 per cent of the firms in the Styrian automobile cluster stated that they collaborated regularly with universities and other R&D institutions.

Our own empirical findings support this result. By the end of the 1990s, a number of R&D partnerships can be identified among knowledge suppliers and companies in various research fields. Table 5 covers primarily

Table 5. Selected R&D partnerships between knowledge suppliers and companies in the automobile cluster

	Regional knowledge suppliers	Cluster firms	Knowledge suppliers outside the region—firms outside the cluster/region
Competence centres			
Automotive Acoustics	Institute of Internal Combustion Engines and Thermodynamics (TU Graz)	AVL List Magna Steyr Fahrzeugtechnik	
The Virtual Vehicle	Several institutes of the TU Graz and the University of Leoben Technikum Joanneum One university institute from outside the region	AVL List Magna Steyr Fahrzeugtechnik Pankl Racing Systems Several other firms	
Christian Doppler laboratories			
Engine- and Vehicle Acoustics	Institute of Internal Combustion Engines and Thermodynamics (TU Graz)	AVL List	
Thermodynamics of the Internal Combustion Engine	Institute of Internal Combustion Engines and Thermodynamics (TU Graz)	AVL List	
Fuell Cell Systems with Liquid Electrolytes	Institute of Chemical Technology of Inorganic Materials (TU Graz)	AVL List	OMV
Automotive Measurement Research	Department of Electrical Measurement and Measurement Signal Processing (TU Graz)	AVL List	Geodata ZT TeamAxess Ticketing Austria Mikrosysteme International
Other R&D projects (selection)			
Digital image processing	Joanneum Research	Several companies	
Development of laser-welded gearpox parts	Laser Centre (Joanneum Research)	Magna Steyr	
Development and production of a new oil pump		AVL List Several local SMEs	

complex interorganisational R&D projects with a life-span of several years as well as some less spectacular cases of innovation networks. There are two competence centres (co-operative institutions supported by the federal state with the aim of bringing together several university institutes and a larger number of firms for carrying out joint research) and four Christian Doppler laboratories (designed to foster R&D partnerships between a more restricted number of actors, often in a bilateral form between one firm and one university institute). We find interactions of outstanding intensity between the cluster's leading firms—above all, AVL List and the Technical University of Graz as the main regional knowledge supplier. Of particular interest are the two competence centres mentioned in Table 5, as they reveal an intensive innovation partnering between the cluster's leading companies AVL List and Magna Steyr. Some years ago, such a co-operative endeavour would hardly have been imaginable. Although these firms have been located in close geographical proximity for many years, they always refused to work together fearing a loss of strategic know-how. Their actual co-operation signals the emergence of a network culture and trust within the cluster. But also other companies have—as the qualitative interviews have shown—started to search more actively for synergy potentials with regional innovation partners, indicating the validity of this finding.

4.3 The Socio-institutional Structure of the Cluster: New Institutions and Trust-building

As we have demonstrated above, the Styrian automotive cluster has overcome its fragmented situation and has become a more integrated system in recent years which is able to bundle competencies and resources. This is reflected in the growing number of co-operations among cluster firms and university–industry partnerships. The intensification of collaboration signals that a substantial change in the socio-cultural structure of the automobile cluster has taken place. Indeed, one of the strong findings of our research is that a stock of social capital—the missing ingredient in cluster development in the past—has been activated. This change has partly been brought about by policy interventions at the Land level. In the second half of the 1990s, a comprehensive cluster policy (see also further below) was started with the building-up of relational assets as a key element. A cluster management unit, established in 1996 and run by a key person of the SFG, has performed a critical function in this process. Amongst other things, it has encouraged firms that had so far little or no contact with each other to communicate and to search actively for synergy potentials. It has brought companies together in informal meetings and it has organised conferences and workshops to inform firms about new technological and market developments. The outcome of these careful attempts to manipulate the social context has been rather positive. Firms have met regularly and trust-building at least among a key group of them has set in, giving rise to the exchange of information, ideas and experiences. The crucial point is that a collective identity and shared understandings among enterprises have slowly emerged, revealing a growing systemness in the Styrian automotive cluster. Dealing with processes of social system building in the cluster leads us to also examine the impact of the advisory committee ('Clusterbeirat'), which was formed at the beginning of the cluster development initiative. Members were delegates from manufacturers, larger local companies and SMEs as well as representatives of the political, scientific and educational systems. In regular meetings, these key actors have become engaged in discourses on the cluster's competitive situation and have worked out strategies in a co-ordinated manner to overcome the identified bottlenecks. It was essentially through this institutionalisation of continuous and recursive communication and co-ordination that self-thematisation and reflexivity at the cluster level have set in and that its strategic abilities have become enhanced. Important stakeholders from the business, scientific and political communities now think of themselves as a 'system'; they share goals and have a common vision for the future.

4.4 The Role of Policy: Policy Actors as Facilitators of Networking

The transformation of the Styrian car industry into a more networked cluster and towards a higher degree of systemness also requires an analysis of the supportive role of policy in this process. Styria has been the first Austrian province adopting an explicit cluster policy, dating from 1996 when the regional development agency SFG launched a comprehensive development initiative (termed 'AC Styria') in the automotive sector. The AC Styria story highlights several interesting aspects. The cluster approach pursued to strengthen the Styrian car industry and to promote its adjustment to new market conditions and technological changes has included a rather creative combination of various instruments and tools. In addition to the already mentioned efforts to foster interfirm networking and research–industry co-operation in the region, actions in the following fields have been taken:

—marketing activities to promote the location in international markets and to attract cluster-specific direct investment;
—international institutional networking in the form of benchmarking projects with other European automobile clusters; and
—qualification (establishment of the technical college for automotive engineering, formation of an Internet-based training network).

In our view, the key aspect of the AC Styria initiative is that an entirely new form of policy can be observed, one that has shifted the focus of attention from individual firms to the system level—i.e. to the whole cluster. This move of public policy towards becoming more system-centred has been accompanied by a redefinition of the role of the state and a new mode of intervention. Indeed, traditional functions like attracting business investment have been complemented by modern tasks like information provision, awareness-raising concerning the importance of networking and innovation, brokerage activities and other forms of social co-ordination. Less direct intervention and more indirect animation have been at the heart of the whole process. This does not only hold true for the regional policy level discussed so far, but also for the national one which played and is still playing a critical role as 'animateur' of university–industry partnerships. The lesson to be drawn is that supporting clusters in old industrial regions to develop new prospects requires substantial changes in policy attitudes and new capabilities of public authorities. To be sure, it was not the public sector alone which has steered the adjustment process of the Styrian automotive industry. As we have seen, a negotiation system (Clusterbeirat), consisting of the main private, intermediate and public agents, has turned out to be the key institution governing its renewal. Hence, cluster transformation has been to a large extent a collective and consensual endeavour.

5. The Metal Cluster

The metal cluster is the largest and oldest cluster in the Styrian economy. It is characterised by a rather high degree of heterogeneity as it is part of several value chains including railway, automotive, aviation and gas and oil. Key elements are formerly nationalised firms. There are several companies of the Böhler–Uddeholm Group. Among these we find Böhler Edelstahl (high-speed steels, tool steels and special materials), Böhler Bleche (special sheets and plates) and Böhler Schmiedetechnik (forgings for aviation and space industry, forgings for turbine blades). Also present is Böhler Schweißtechnik (coated electrodes, cored wires, hardfacing alloys) belonging to Böhler Thyssen Schweißtechnik, a joint venture between Böhler Uddeholm and Thyssen Stahl. Other leading firms include several companies of the Voest Alpine group. Present are Voest-Alpine Stahl Donawitz (semifinished material for the group's companies), VA Schienen (ultra-long weld-free rails), VAE Eisenbahnsysteme (switches), Voest-Alpine Austria Draht (wires for the construction industry, the aeronautics and automotive supporting industry and fastening technology) and Voest-Alpine Tubulars (seamless pipes for the oil and gas industries). Other actors located in the cluster, amongst which we also find firms that have spun off from the former nationalised companies, include Böhlerit (hard materials and tools), Stahl Judenburg (high-grade steel, tool steel), Knorr Technik (testing equipment for the steel industry), Obersteirische Feinguß (castings for the automotive, aviation and mechanical engineering industries) and mec.com (components for the car industry, machine tools).

There are also specialised knowledge suppliers providing cutting-edge basic and applied research. The cluster's R&D capacity includes the University of Leoben, the Technical University of Graz, the Austrian Foundry Institute, the Erich Schmid Institute for Material Science (Austrian Academy of Science) and the Laser Centre Leoben (Joanneum Research).

5.1 New Challenges and the 'Integration Trap'

The Styrian metal cluster has had a turbulent past. After the Second World War, large enterprises were nationalised and promoted by public policy. Since there were also favourable demand conditions in this period of national reconstruction, the cluster exhibited a strong growth until the 1960s. In the 1970s, structural changes marked the end of the 'golden age' and they have made the cluster's economic environment more unstable. The iron and steel industry was hit by a severe crisis. Two trends have coincided in an unfavourable manner. On the one hand, new low-cost competitors from newly industrialising as well as eastern European countries have entered the scene. On the other hand, total demand in the steel industry did not increase to the same extent. The outcome has been a considerable oversupply in the world market and, consequently, a deterioration in prices.

How did the cluster react to these major challenges? In fact, the adjustment performance of the metal cluster was extremely weak in the 1970s and early 1980s. There was a recognition lag regarding the new market constellations, followed by unsuccessful attempts to cope with the crisis by reinforcing outdated strategies. Trying to explain this poor reaction, we have to take into account the internal structuring of the local industry at that time. The cluster was dominated by a small number of large, vertically integrated companies under public control and with strong trade unions. A peculiar pattern of interaction could be observed among firms and policy actors. On the one hand, there was a keen rivalry among the large firms resulting in an absence of market links and co-operation between the big companies and SMEs. On the other hand, intense networking took place between the state, the firms and the trade unions. This policy network focused on subsidising large companies and the jobs of a well-protected workforce and it thus can be characterised as a 'distribution coalition'. This led to a serious political and economic lock-in and had a paralysing impact on the whole cluster. Thus, the cluster suffered from a too strongly developed policy network—a situation which we might call an 'integration trap'. This refers to a situation where the cluster relations (integration) prohibit a successful adjustment to changes in its environment. Fearing a loss of their influence, the network's participants preserved existing structures instead of supporting the cluster to shift its competencies into new areas. Indeed, using state-owned enterprises to protect employment and subsidising the declining industries were inappropriate interventions to transform the cluster.

In the 1970s and 1980s, the cluster was also confronted with centralisation processes, supported by policy, and failed attempts to leapfrog into high-technology sectors. The merger and concentration processes had the negative consequence that management left the region. The plants came under 'external control', lost strategic functions such as planning, R&D and marketing/distribution (Tödtling 1990) and were subject to strategic decisions made at distance from the core of the cluster. It was only in the second half of the 1980s and above all in the 1990s that a shift away from the 'big is beautiful' philosophy and the trend to mergers could be observed. Important organisational innovations were introduced by decentralising and privatising companies. The large conglomerates were broken up and management functions returned to the cluster. Today, the privatised and more autonomous companies focus on their core competencies. Most of them have regained their competitiveness and some are even dominant in specific market niches. Thus, we might conclude that the cluster has transformed itself by organisational change and by recycling know-how residing within the firms.

5.2 Innovation Networks

There are both traded interdependencies and innovation links within the region. According to Adametz *et al.* (2000), 22 per cent of the inputs to firms are provided by suppliers located in the region and 10 per cent of the output is delivered to downstream firms in Styria; about 70 per cent is exported to international markets. Regarding innovation networks, the empirical basis shows no clear picture: Steiner and Hartmann (1997), in qualitative interviews with 25 key actors, traced out several learning networks in this cluster. The majority was of vertical nature (i.e. with customers and suppliers), but there were also some horizontal linkages and connections with knowledge suppliers. Regional and larger-scale networks were found to have about the same frequency. In the study by Adametz *et al.* (2000) the metal cluster showed an excellent performance in terms of knowledge-intensive collaboration. The authors reported that more than 70 per cent of the firms co-operate regularly with regional universities and other R&D institutions and that more than 30 per cent of the firms have intensified these collaborative activities within the past 5 years.

Our own empirical work allows to specify the role of the larger, formerly nationalised firms and of the University of Leoben in the cluster and to assess the extent of their insertion into regional innovation networks. Our empirical evidence shows that these key actors from the industrial and scientific spheres are collaborating intensively at the regional level. Neither the University of Leoben nor the larger firms, however, have strong innovation linkages with other firms in the cluster (see also Table 6). Between the University of Leoben and leading companies of the Voest-Alpine and Böhler Uddeholm groups there are long-standing trust-based relations of 'an old boy network' character where new actors seem to have no access. Currently, we find one competence centre and several Christian Doppler laboratories, in which these actors bundle their resources and competencies.

5.3 The Socio-institutional Structure of the Cluster: Barriers to More Systemic Interaction

To what extent can we observe a change of the socio-cultural structures moving the metal cluster towards a well co-ordinated, reflexive system with some degree of common identity? This question is intriguing given the cluster's former petrified socio-institutional endowment and its subsequent break-up. Our empirical finding is that, despite the R&D partnerships identified, the overall systemness of the metal cluster is of a rather low degree. The cluster has succeeded in getting rid of its old socio-cultural tissue, but it has been less capable of doing the next step—i.e. ensuring that a process of new social system building takes place. There are almost no permanent and recursive discourses among key firms and other main stakeholders about the cluster as a whole, addressing its strengths and weaknesses. There is no indication that public and private partners bundle their knowledge and resources and co-operate in defining strategies for the whole system. There is no such insti-

Table 6. Selected R&D partnerships between knowledge suppliers and companies in the metal cluster

	Regional knowledge suppliers	Cluster firms	Knowledge suppliers outside the region—firms outside the cluster/region
Competence centres			
Materials Centre Leoben	Four institutes of the University of Leoben Erich Schmid Institute for Material Science (Austrian Academy of Science)	Böhler Edelstahl Böhler Schweißtechnik Voest-Alpine Austria Draht VAE Eisenbahnsysteme Voest-Alpine Schienen	Three other university institutes from outside the region Eleven firms, including: Miba Sintermetall Plansee Rübig
	Laser Centre Leoben (Joanneum Research)		
Christian Doppler laboratories			
Fundamentals of metallurgy in continuous casting processes	Department of Ferrous Metallurgy (University of Leoben)	RHI Voest Alpine Stahl Donawitz	Voest-Alpine Industrieanlagenbau Voest-Alpine Stahl Linz
Applied computational thermofluiddynamics	Institute of Petroleum Engineering (University of Leoben)	RHI	Windhager Zentralheizung AVL List Lenzing Voest-Alpine Industrieanlagenbau Voest-Alpine Stahl Linz

Table 6.—*Continued*

	Regional knowledge suppliers	Cluster firms	Knowledge suppliers outside the region—firms outside the cluster/region
Fatigue analysis	Institute of Mechanical Engineering (University of Leoben)	Böhler Uddeholm	BMW München BMW Motoren Steyr Voest-Alpine Industrieanlagenbau
Functionally oriented materials design	Institute of Mechanics (University of Leoben)	Voest-Alpine Stahl Donawitz Böhler Uddeholm	Institute of Lightweight Structures and Aerospace Engineering (Vienna University of Technology) Voest-Alpine Industrieanlagenbau
Analysis of deformation and fracture	Institute of Metal Physics (University of Leoben) Erich Schmid Institute for Material Science (Austrian Academy of Science)	Voest-Alpine Stahl Donawitz Böhler Uddeholm	
Sensory metrology	Institute for Automation (University of Leoben)	Voest-Alpine Stahl Donawitz Böhler Uddeholm	VA Mechatronics-VATRON AVL List Voest-Alpine Stahl Linz

tutional innovation as the advisory committee in the automotive cluster acting as a development coalition with the aim of pushing the cluster forward. Hence, selective communication, self-thematisation, a collective reflexivity, the shaping of a common identity as essential operations of system building are only weakly developed at the cluster level.

In our opinion, there are three main factors which impede progress towards a virtuous cycle of growing systemness in the metal cluster. First, as we have already noted above, the cluster is very heterogeneous regarding the markets to which its products are sold. Secondly, we must bear in mind that the cluster hosts a rather large number of firms. Both the presence of many actors and the local industry's insertion into various value chains imply an enormous complexity where such an ambitious task as enhancing systemness is concerned. But there is still another, third, factor hampering system building. The leading firms and important institutes of the University of Leoben as key agents have little interest in becoming engaged in such a process. Indeed, their orientations and strategies are fundamentally at odds with the concept of social networking at the regional level. The strategy of the University of Leoben is to create and maintain co-operative relationships with big enterprises and it sees little reason to reflect its role in the local cluster. The larger firms' focus is primarily on national and international partners. To shape simultaneously the regional environment by forming a coalition with other local private and public actors has no priority for them. One interview partner stated that a development initiative for the Styrian metal cluster would be a 'stillbirth'. As our empirical research revealed, this view is widely shared among the larger companies. Their unwillingness to get involved in comprehensive projects at the regional level may also at least partly be explained by the legacy of the past. As we sketched out, in former times these companies were nationalised and used as instruments of regional policy, forced to secure full employment and thus heavily constrained from acting according to the requirements of the market. To be put into the position once again, of having to do 'something for the local cluster' seems a nightmare for many of them.

5.4 The Role of Policy: From State Firms towards Innovation Support

The revitalisation of the metal cluster can hardly be understood without discussing the role of the state and stressing the fundamental changes that have taken place in the nature of policy-making. Public actors—above all, at the national level—have always exerted a strong influence over the development of the metal cluster. In the past, the state's key role was that of a public entrepreneur and 'guarantor' of employment security and high wages. Permanent political interference and high levels of industrial subsidy as 'standard repertoire' of policy-makers had—as it was shown above—devastating effects on the cluster's adaptability. The central government has been slow to unlearn old policy routines and to adapt its approach. For a long time, overstrong ties between public authorities, enterprises and trade unions hampered necessary policy changes and retarded the transformation of the state. Indeed, it was only in the second half of the 1980s and above all in the 1990s that old-style policy measures were given up and that significant institutional learning processes in the public sector could be observed. On the one hand, privatisation of nationalised enterprises has been an important step in regenerating the metal cluster. On the other hand, attention has been redirected to strengthening the regional innovation system by providing infrastructure and encouraging research–industry interfaces. This new approach has produced positive results and marks a clear transition from a strategy that focused on protecting ailing enterprises through direct intervention to a policy placing emphasis on promoting innovation by animating the formation of networks. To summarise, the state's withdrawal from the metal cluster—brought about by privatising firms and breaking up tight policy networks—and its re-emergence as provider of innovation-related infrastructure and recently as facilitator of university–industry partnerships support the assumption that, in old industrial areas, cluster renewal and policy innovations are closely interwoven.

6. Conclusions

In the literature, there is a strong emphasis on the early stages of cluster development in growth regions, but the renewal of clusters in old industrial regions remains a neglected topic. In the present article, an effort has been made to shed some light on this issue. Drawing on the insights of cluster-related concepts like industrial districts, regional innovation system (RIS) and social system theory has allowed us to identify some critical factors encouraging such a cluster transformation. The literature review highlights the importance of a well-developed RIS, the building-up of innovation networks, changes in the socio-institutional context and a new role for public policy in this respect.

We have examined the process of cluster renewal and respective factors empirically for the automotive and metal clusters in the old industrial region of Styria. Even if their actual performance differs, both clusters have undergone successful restructuring in recent years. The automotive cluster has made the transition from a fragmented towards a more integrated system, whereas the metal cluster has renewed itself by breaking up institutional inertia and through organisational change such as the privatisation of state-owned companies. Taking a closer look at the transformation processes of the automobile and metal clusters reveals both similarities and differences. The following analogies stand out

—A strong RIS, respectively a well developed sub-system of knowledge generation and diffusion with strengths in cluster-related research, has supported the transformation of both clusters to a significant degree.
—The formation of innovation networking is an important mechanism of cluster renewal and relevant for trust-building and collective learning. In both clusters, however, the most spectacular cases of networking seem to be restricted to the larger firms and to knowledge suppliers.
—New and innovation-oriented policy approaches such as the active support of university–industry links (Christian Doppler laboratories, competence centres) as well as the stimulation of networks have played a critical role in inducing and encouraging cluster transformation. This holds true both for the national and regional policy levels.

Besides these similarities, we also find marked differences with respect to the adjustment pattern of the automotive and metal clusters

—The automotive cluster has had a much better performance and a more dynamic development in recent years.
—The restructuring process in the automotive cluster was clearly faster than in the metal cluster, revealing a picture of fast versus slow learners.
—The two clusters showed a different capability to trigger changes in their socio-institutional structures—i.e. to enhance their systemness.

The following factors might explain the different pattern of cluster transformation. First, general industry characteristics have to be taken into account. The automotive industry in general has seen a more dynamic growth of demand in recent decades, compared with the rather stagnating markets of the metals and steel sector. To a considerable extent, the employment losses in the metal cluster are due to these more difficult industry conditions. Secondly, the two clusters are subject to different degrees of 'lock-in'. The Styrian metal cluster has clearly a greater historical burden in this respect. It has played a stronger role within the region in the past and has become a victim of its past success. Firms and other organisations have become too dominant and it has reached a higher degree of 'institutional sclerosis'. The building-up of new innovation networks is much more difficult under such conditions. Finally, the size and the diversity of the cluster are relevant. In the metal cluster, there are many more firms with quite heterogeneous products and any attempt to improve communication and co-ordination among them and to develop joint visions and goals faces considerably more difficulties than in the smaller and more specialised automotive cluster.

However, more theoretical and empirical work is necessary for a dynamic perspective of cluster development, enabling us to gain a deeper and more complete understanding not just of cluster formation in growth regions but also of cluster renewal and its relevant factors in old industrial regions. The following themes and issues need further research:

—The first concerns the issue of radical versus incremental technological change. Should clusters and regions revitalise themselves mainly by finding new uses for their traditional competencies (incremental change) or can they also leapfrog into wholly new sectors and technologies (radical change)? What are the critical conditions and factors in this respect?
—It is widely acknowledged that both regional firms (endogenous restructuring strategy) as well as foreign companies (exogenous strategy) may act as important engines of cluster transformation in old industrial regions. What remains to be explored further is the question of which strategy is suitable for which type of cluster and region. Also, the issue of how endogenous and exogenous elements can be combined for a successful and sustainable restructuring process requires further clarification.
—This paper has demonstrated that the emergence of consensus and intensive collaboration in clusters is of crucial importance for their regeneration. However, competition and conflict also seem to be required to some extent. Further research is needed to investigate their role in stimulating the revival of old clusters and to identify the 'optimal mix' of co-operation and competition in such regions.
—Then, the key concept of 'system building' must be further specified and operationalised. The challenge is to develop an adequate set of qualitative and quantitative indicators allowing us to measure issues like the level of communication, co-ordination and collective reflexivity in clusters as well as their strategic and generative abilities. Furthermore, the potential trade-off between intensive 'system building' in clusters and the danger of regional lock-in remains relatively little understood so far and needs further research.
—A related question concerns the spatial extension of innovation networks. How important are external connections for regional renewal processes to succeed? What 'mix' of local and extra-local networks is most appropriate and why?
—At the same time, we do not know enough about the relationship between institutional dynamics and the transformation of traditional clusters. How do institutions (especially 'soft' factors like cultural and political routines) change and how do these processes lead to new competitive advantages?
—Regarding the methodological approaches, we contend that qualitative case studies of regions and clusters should be combined with more quantitative investigations (covering both longitudinal and cross-sectional analysis). In addition, the systematic comparison of regions with different development trajectories at European and global levels should provide fruitful results and insights.

Note

1. According to ACStyria—the official cluster organisation—the number of companies and employees is higher (150 firms and a workforce of 13 000). The difference results from different cluster definitions. Adametz *et al.* (2000) apply economic linkages as defining criteria, whereas ACStyria use membership of the cluster organisation as the criterion.

References

ADAMETZ, C. and NOVAKOVIC, M. (2000) *Innovation in steirischen Unternehmen*. Bericht im Auftrag der steiermärkischen Landesregierung im Rahmen des Projektes WIBIS Phase 2. Graz: Joanneum Research–Institut für Technologie- und Regionalpolitik.

ADAMETZ, C., FRITZ, O. and HARTMANN, C. (2000) *Cluster in der Steiermark: Liefer verflechtungen, Kooperationsbeziehungen und Entwicklungsdynamik*. Bericht im Auftrag des Amtes der steiermärkischen Landesregierung im Rahmen des Projektes WIBIS Phase 1. Graz: Joanneum Research–Institut für Technologie- und Regionalpolitik.

AMT DER STEIERMÄRKISCHEN LANDESREGIERUNG (2002) *Wirtschaftsbericht Steiermark 2001. Endbericht Version 1.1*. Graz: Joanneum Research–Institut für Technologie- und Regionalpolitik.

ASHEIM, A. (2000) Industrial districts: the contributions of marshall and beyond, in: G. L. CLARK, M. P. FELDMAN and M. S. GERTLER (Eds) *The Oxford Handbook of Economic Geography*, pp. 413–431. Oxford: Oxford University Press.

ASHEIM, B. (1996) Industrial districts as 'learning regions', *European Planning Studies*, 4, pp. 379–400.

ASHEIM, B., ISAKSEN, A., NAUWELAERS, C. and TÖDTLING, F. (Eds) (2003) *Regional Innovation Policy for Small-Medium Enterprises*. Cheltenham: Edward Elgar.

AUTIO, E. (1998) Evaluation of RTD in regional systems of innovation, *European Planning Studies*, 6, pp. 131–140.

BAPTISTA, R. and SWANN, P. (1998) Do firms in clusters innovate more?, *Research Policy*, 27, pp. 525–540.

BATHELT, H. and DEPNER, H. (2003) Innovation, Institution und Region: Zur Diskussion über nationale und regionale Innovationssysteme, *Erdkunde*, 57, pp. 126–143.

BENNEWORTH, P., DANSON, M., RAINES, P. and WHITTAM, G. (2003) Guest editorial. confusing clusters? making sense of the cluster approach in theory and practice, *European Planning Studies*, 11, pp. 511–520.

BRACZYK, H., COOKE, P. and HEIDENREICH, M. (Eds) (1998) *Regional Innovation Systems*. London: UCL Press.

BRATL, H. and TRIPPL, M. (2001) *Systemische Entwicklung regionaler Wirtschaften*. Studie im Auftrag des Bundeskanzleramtes, Abteilung IV/4. Wien: invent.

CAMAGNI, R. (1991) Local 'milieu', uncertainty and innovation networks: towards a new dynamic theory of economic space, in: R. CAMAGNI (Ed.) *Innovation Networks*, pp. 121–144. London: Belhaven Press.

COOKE, P. (Ed.) (1995) *The Rise of the Rustbelt*. London: UCL Press.

COOKE, P. (2002) *Knowledge Economies: Clusters, Learning and Cooperative Advantage*. London: Routledge.

COOKE, P. and MORGAN, K. (1998) *The Associational Economy: Firms, Regions, and Innovation*. Oxford: Oxford University Press.

COOKE, P., BOEKHOLT, P. and TÖDTLING, F. (2000) *The Governance of Innovation in Europe: Regional Perspectives on Global Competitiveness*. London: Pinter.

DICKEN, P. (1998) *Global Shift*. London: Paul Chapman.

EDQUIST, C. (Ed.) (1997) *Systems of Innovation: Technologies, Institutions and Organisations*. London: Pinter.

ENRIGHT, M. (2003) Regional clusters: what we know and what we should know, in: J. BRÖCKER, D. DOHSE and R. SOLTWEDEL (Eds) *Innovation Clusters and Interregional Competition*, pp. 99–129. Berlin: Springer.

GAROFOLI, G. (1991) Local networks, innovation and policy in Italian industrial districts, in: E. BERGMAN, G. MAIER and F. TÖDTLING (Eds) *Regions Reconsidered*, pp. 119–140. London: Mansell.

GELDNER, N. (1998) Erfolgreicher Strukturwandel in der Steiermark, *WIFO-Monatsberichte*, 3, pp. 167–172.

GRABHER, G. (1991) Rebuilding cathedrals in the desert: new patterns of cooperation between large and small firms in the coal, iron, and steel complex of Germany, in: E. BERGMAN, G. MAIER and F. TÖDTLING (Eds) *Regions Reconsidered*, pp. 59–78. London: Mansell.

GRAVES, A. (1994) Innovation in a globalizing industry: the case of automobiles, in: M. DODGSON and R. ROTHWELL (Eds) *The Handbook of Industrial Innovation*, pp. 213–231. Aldershot: Edward Elgar.

JANGER, J., NAGY, M., MARKOWITSCH, J. ET AL. (2000) *Innovation in der Steiermark: Steigerung der Wettbewerbskraft der bestehenden Produktionswirtschaft und Öffnung in Richtung 'new economy'*, IWI-Arbeitsheft 55. Wien: Industriewissenschaftliches Institut.

JÜRGENS, U. (1999) Neue Systeme der Produktentstehung im Spannungsfeld von Regionalisierung und Internationalisierung, in: G. FUCHS, G. KRAUSS and H.-G. WOLF (Eds) *Die Bindungen der Globalisierung—Interorganisationssysteme im regionalen und globalem Wirtschaftsraum*, pp. 162–191. Marburg: Metropolis Verlag.

KAUFMANN, A. and TÖDTLING, F. (2000) Systems of innovation in traditional industrial regions: the case of styria in a comparative perspective, *Regional Studies*, 34, pp. 29–40.

KAUFMANN, A. and TÖDTLING, F. (2001) Science-industry interaction in the process of innovation: the importance of boundary-crossing between systems, *Research Policy,* 30, pp. 791–804.

KEEBLE, D. and WILKINSON, F. (Eds) (2000) *High-technology Clusters, Networking and Collective Learning in Europe*. Aldershot: Ashgate.

KOSCHATZKY, K. and STERNBERG, R. (2000) Cooperation in innovation systems: some lessons from the European Regional Innovation Survey (ERIS), *European Planning Studies,* 8, pp. 487–501.

LUNDVALL, B. A. and BORRÀS, S. (1998) *The globalising learning economy: implications for innovation policy*. Report to the DGXII. Brussels: TSER.

MASKELL, P. and MALMBERG, A. (1999) The competitiveness of firms and regions: 'ubiquitification' and the importance of localized learning, *European Urban and Regional Studies,* 6, pp. 9–26.

MASKELL, P., ESKELINEN, H., HANNIBALSSON, I. *ET AL.* (1998) *Competitiveness, Localised Learning and Regional Development: Specialisation and Prosperity in Small Open Economies*. London: Routledge.

MAYNTZ, R. (1997) *Soziale Dynamik und politische Steuerung*. Frankfurt: Campus Verlag.

MESSNER, D. (1998) *Die Netzwerkgesellschaft*, 2nd edn. Schriftenreihe des Deutschen Instituts für Entwicklungspolitik, Band 108. Köln: Weltforum Verlag.

MOTHE, J. DE LA and PAQUET, G. (Eds) (1998) *Local and Regional Systems of Innovation.* Norwell: Kluwer Academic Publishers.

OECD (ORGANISATION FOR ECONOMIC CO-OPERATION AND DEVELOPMENT) (1999) *Boosting Innovation: The Cluster Approach.* Paris: OECD.

PORTER, M. (1990) *The Competitive Advantage of Nations*. New York: Free Press.

PORTER, M. (1998) *On Competition*. Boston, MA: Harvard Business School Press.

PUTNAM, R. D. (1993) *Making Democracy Work: Civic Traditions in Modern Italy*. Princeton, NJ: Princeton University Press.

PYKE, F. and SENGENBERGER, W. (Eds.) (1992) *Industrial Districts and Local Economic Regeneration*. Geneva: International Institute for Labour Studies.

SABEL, C. (1992) Studied trust: building new forms of co-operation in a volatile economy, in: F. PYKE and W. SENGENBERGER (Eds) *Industrial Districts and Local Economic Regeneration*, pp. 215–250. Geneva: International Institute for Labour Studies.

SAXENIAN, A. (1994) *Regional Advantage: Culture and Competition in Silicon Valley and Route 128*. Cambridge, MA: Harvard University Press.

STATISTIK AUSTRIA (2003) *Statistisches Jahrbuch Österreich 2003*. Wien: Statistik Austria.

STEINER, M. (Ed.) (1998) *Clusters and Regional Specialisation*. London: Pion.

STEINER, M. and HARTMANN, C. (1997) *Knowledge spill-overs and network externalities in clusters: a case study with a 'learning organizations approach'*. Paper presented at the *37th RSA European Congress*, Rome.

STEINER, M. and HARTMANN, C. (2002) *Material and immaterial dimensions of clusters: cooperation and learning as infrastructure for innovation.* Paper presented at the *42nd RSA European Congress*, Dortmund.

SWANN, P. and PREVEZER, M. (1996) A comparision of the dynamics of Industrial clustering in computing and biotechnology, *Research Policy,* 25, pp. 1139–1157.

TEUBNER, G. (1987) Hyperzyklus in Recht und Organisation, in: H. HAFERKAMP and S. SCHMID (Eds) *Sinn, Kommunikation und soziale Differenzierung*, pp. 89–128. Frankfurt am Main: Suhrkamp.

THOMI, W. and WERNER, R. (2001) Regionale Innovationssysteme, *Zeitschrift für Wirtschaftsgeographie,* 45, pp. 202–218.

TICHY, G. (2001) Regionale Kompetenzzyklen—Zur Bedeutung von Produktlebenszyklus- und Clusteransätzen im regionalen Kontext, *Zeitschrift für Wirtschaftsgeographie,* 45, pp. 181–201.

TÖDTLING, F. (1990) *Räumliche Differenzierung betrieblicher Innovation. Erklärungsansätze und empirische Befunde für österreichische Regionen.* Berlin: Sigma.

TÖDTLING, F., KAUFMANN, A. and SEDLACEK, S. (1998) *The state of a regional innovation system in Styria: conclusions and policy proposals*. REGIS Working Paper No. 5, Institute for Urban and Regional Studies. University of Economics and Business Administration, Vienna.

WILLKE, H. (1991) *Systemtheorie*. Stuttgart: Gustav Fischer Verlag.

WOLFE, D. (2000) *Social capital and cluster development in learning regions*. Paper presented to the *XVIII World Congress of the International Political Science Association*, August, Québec City.

WOLTERS, H. (1999) Systeme—Eine Revolution in der Beschaffung, in: H. WOLTERS, R. LANDMANN, W. BERNHARD *ET AL.* (Eds) *Die Zukunft der Automobilindustrie – Herausforderungen und Lösungsansätze für das 21. Jahrhundert*, pp. 61–74. Wiesbaden: Gabler.

INDEX

Page numbers in *italics* represent Tables. Page numbers in **bold** represent Figures

www.ingramcontent.com/pod-product-compliance
Ingram Content Group UK Ltd.
Pitfield, Milton Keynes, MK11 3LW, UK
UKHW010019280225
455677UK00023B/683